SPACE SHUTTLE STORIES

United States
Discovery
USA

SPACE SHUTTLE STORIES

Firsthand Astronaut Accounts from All 135 Missions

Tom Jones

Smithsonian Books
Washington, DC

Contents

AGE ONE: **A departing *Endeavour* hotographed from the ISS at unset.**

REVIOUS PAGES: ***Discovery* leaves he Vehicle Assembly Building top its mobile launcher and rawler-transporter.**

PPOSITE: ***Columbia*, STS-50, ockets skyward on its twelfth nission on June 25, 1992.**

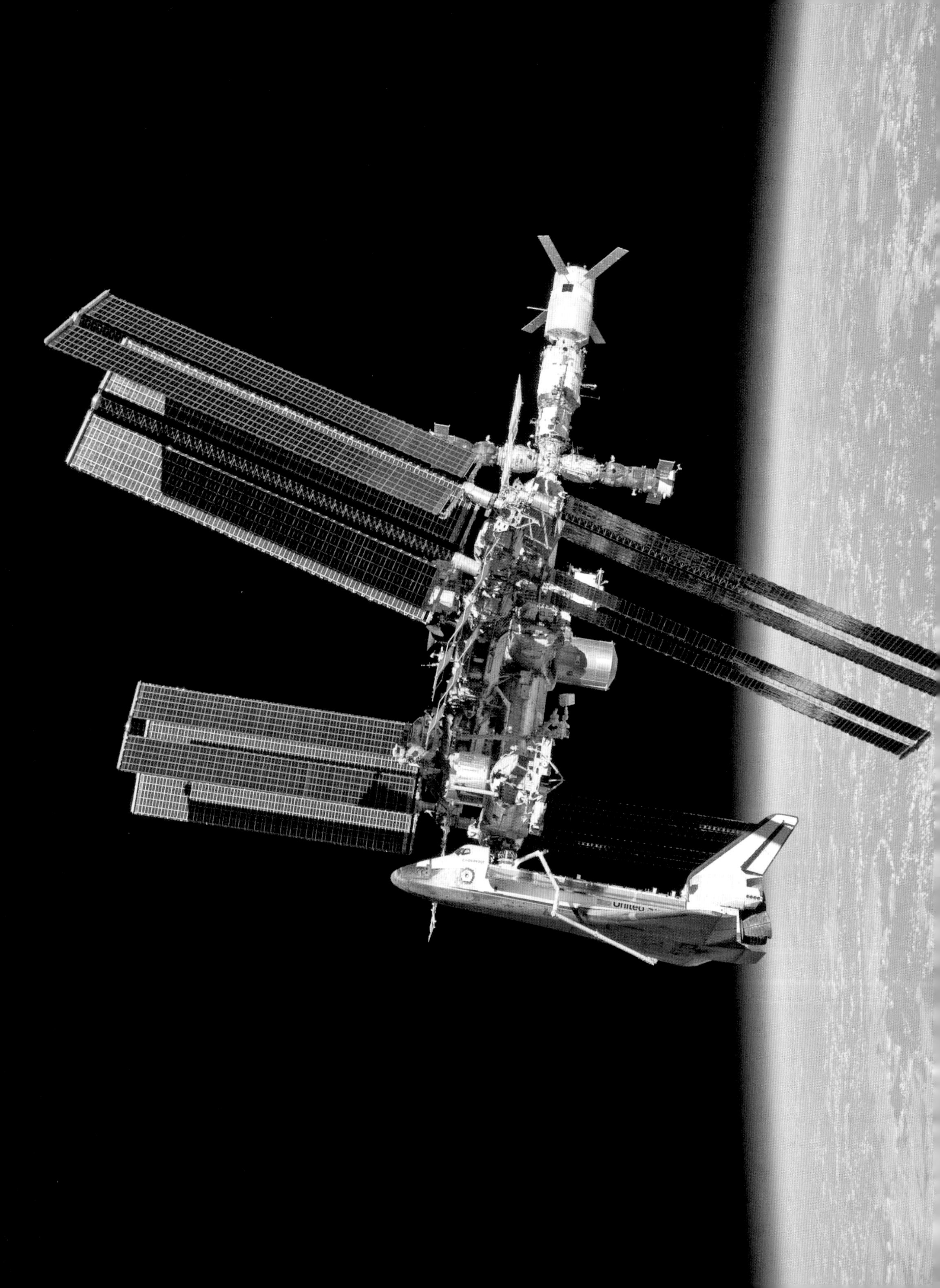

Foreword

For thirty years the space shuttle represented America's prowess in space exploration. It was the brilliant creation of some of the most dedicated engineers, scientists, flight controllers, and operators the nation could muster. At any one time, twenty thousand people worked together to prepare for and execute space shuttle missions. Because people joined and left the program continually over the course of three decades, shuttle veterans today number in the hundreds of thousands.

As any member of the shuttle team can attest, there are thousands of stories surrounding each of the shuttle's 135 orbiter missions. Tom Jones has done a masterful job of selecting a distinctive astronaut story from each of those flights. Many volumes could be written without capturing all the tales from those heady days, but *Space Shuttle Stories* gives us a superb collection of insights from those who flew America's orbiter fleet.

Why is such a book important? The answer lies in the pivotal role that the space shuttle program played in humanity's expansion into space. The program, which ended in 2011, was just one episode in the continuing story of our exploration of the cosmos, but we can better understand the shuttle's importance by putting its thirty-year career in context.

After the pioneering steps into space of Mercury, Gemini, and Apollo, the United States focused its space program toward building a foothold in near-Earth space to exploit the resources there: microgravity, vacuum, and an unfiltered, unparalleled observation point above Earth's atmosphere. Our first attempt at expanding the amount of time that humans could spend in microgravity was Skylab, an experimental, crewed research outpost occupied in 1973 and 1974. We followed Skylab with the Space Transportation System (STS), which everybody knows as the space shuttle. It provided regular transportation to near-Earth orbit and a laboratory for experiments in micro-g. It also carried to space nearly all the elements of a permanently crewed laboratory complex—the International Space Station (ISS)—as well as the astronaut construction crews who put those pieces together. To fulfill the mission of station assembly, the space shuttle had to fly regularly, safely, and much more economically than the expendable, single-use systems that had come before. These requirements presented a challenge that the space shuttle never fully met. But it did provide an iconic and reasonably reliable ride to low Earth orbit for hundreds of crewmembers and an amazing variety of commercial and scientific payloads.

A departing Soyuz crew photographed the ISS and the docked STS-134 *Endeavour* on May 23, 2011.

Much like the Douglas DC-3, the workhorse passenger airplane that ushered in the modern era of air transportation, the space shuttle was to offer a similar transition from the experimental and dangerous to a proven, safe passage to and from orbit. The shuttle proved not to be a space-age DC-3—we will require at least another iteration of spacecraft design to achieve those goals. From a historical and political perspective, the space shuttle was just good enough through its three decades of service to inhibit development of any new US space transportation system. The innovative commercial crew capsules that have taken over the human transportation role from the shuttle function well as orbital taxis but fall well short of the shuttle's broad functionality and flexibility.

In addition to simply ferrying a crew, the shuttle possessed a huge cargo capability that has yet to be matched by any launch vehicle in the decade since it retired. Along with its crew, trained in assembly, maintenance, repair, and even spacewalks, the shuttle carried an integral airlock and an immensely capable robot arm. The shuttle also provided a crucial communications relay between many payloads and their ground controllers. With the Spacelab and SPACEHAB pressurized modules, the shuttle provided laboratories in space to exploit the microgravity and vacuum environments that have no terrestrial counterpart. Laboratory operations were limited only by the length of time the shuttle could stay in orbit and provided necessary proving grounds for technologies and processes that later made the ISS such a productive research platform.

The space shuttle also provided the launch platform for some of the most revolutionary space science and astronomy instruments of the age. Not only did the Galileo spacecraft reveal new secrets of the Jovian system, but the Magellan radar probe of Venus revealed a landscape previously unknown. The Ulysses mission to the poles of the Sun gathered detailed information about the inner workings of our nearest star. The Great Observatories have transformed cosmology: the Compton Gamma Ray Observatory, the Chandra X-Ray Observatory, and the ground-breaking Hubble Space Telescope have revealed more about our universe than the previous century of ground-based observations. Indeed, the mystery of dark matter and dark energy would not be puzzling today's theoreticians without the observations from these magnificent instruments that were launched from the space shuttle. Shuttle crews were called on to repair manufacturing faults originally afflicting Hubble and to visit regularly to restore its capabilities to do cutting-edge science.

The shuttle also served as a platform for observing our home planet. Over time, shuttle crews documented and commented on the changing landscape below: volcanoes, hurricanes, floods, droughts, and glaciers received extensive study. Humanity's impact on the planet became clear: deforestation, pollution, and urban sprawl were all seen and photographed. Popular awareness and understanding of such changes, routinely imaged by shuttle crews and experiments, helped expand our environmental consciousness.

Less well known and sometimes still veiled by secrecy are the eleven flights the space shuttle flew to enhance the security of the United States. Medals awarded in secret for that work, and the stories behind them, may someday be declassified, adding to the shuttle's lengthy list of successes.

The shuttle's promise to carry hundreds of people into space was perhaps the most transformative to our vision of society's future in space. With its innate capacity to carry up to eight human beings up and down from Earth orbit—there were over a thousand seats available over three decades—the shuttle opened up space travel to a range of Americans and our partners. Three hundred fifty-five individual fliers got that opportunity, and many flew more than once, with the shuttle providing 852 rides in total. The mix of nationalities, genders, ethnicities, and backgrounds pointed toward a future where thousands of humans will routinely live and work in space. The shuttle showed for the first time that everyday people (albeit with thorough training), not just experimental test pilots, could go to space. The valuable lessons in international relations learned by multinational crews working smoothly together in

PPOSITE: **Mission Control at ohnson Space Center prepares or STS-135, the space shuttle rogram's final mission.**

BOVE: **Main engines ignite as *tlantis* prepares to leave Earth 1 2007 on the STS-117 mission.**

the space shuttle became the foundation for the successful multinational crews of the ISS.

Tragically, fourteen shuttle pioneers perished in two heart-wrenching accidents in 1986 and 2003. Completely avoidable, these accidents were rooted more in human error than the limitations of the space shuttle vehicle. The searing lesson that hubris and carelessness will exact a price in blood applies in space travel just as it does on terra firma. The stories of the accidents and the shuttle team's recovery and return to flight are found in many other volumes, and still carved into the hearts of those of us who lived them close at hand.

In the public mind, the space shuttle became the iconic transportation machine of the future. Dozens of feature films and popular novels employed the capabilities of the space shuttle—sometimes with artistic enhancements to the reality of spaceflight—and it came to permeate popular culture. For more than four decades, whenever human spaceflight was depicted in the media, the picture was always of the space shuttle. The image of Bruce McCandless floating in space while wearing the Manned Maneuvering Unit (MMU) became the classic depiction of a human being working and living in space. Bruce's lonely astronaut, drifting unencumbered against the blackness of space, is permanently etched in the public conception of space flight.

We who operated the space shuttle hope that the nation will develop a future vehicle which builds on our hard-won experience. America needs a vehicle, a system, that will possess great capabilities, but is distinctly safer, more economical to operate, and able to fly more frequently than the shuttle could. That future has not yet arrived. The space shuttle is still in a category by itself: the first reusable, winged spacecraft to carry people and supplies to and from the new frontier of space.

Those of us who had the privilege of working in the space shuttle program—the astronauts and flight controllers, and especially those who prepared it for flight and refurbished it after every mission—will never forget the vast promise and sturdy versatility that the space shuttle brought to the challenging but necessary task of pioneering the space frontier. *Space Shuttle Stories* celebrates that commitment and reminds us of what a team of dedicated Americans can achieve in extending our reach toward the stars.

Wayne Hale

Shuttle Flight Director, 1988–2003
Shuttle Deputy Program Manager, 2003–2005
Shuttle Program Manager, 2005–2008

Bruce McCandless flies the MMU 300 feet (91.4 meters) from *Challenger*.

NASA

Enterprise

Developing the Space Shuttle

The Apollo expeditions to the Moon were in full swing when NASA began pitching its idea of a follow-on program to the Nixon administration. The agency's vision was an ambitious progression of space exploration, beginning with an orbiting space station and continuing with a lunar base and eventually an expedition to Mars, but the plans depended on the creation of a more economical spacecraft than the costly, expendable Saturn V. Recognizing that its shrinking post-Apollo budget would not sustain continuing Saturn launches to the Moon and beyond, the space agency sought a reusable "shuttle" to low Earth orbit, which, the agency hoped, would free enough funds downstream to build its space station, the next step in humankind's outward journey. Early shuttle definition studies had begun in 1968, one year before Neil Armstrong, Buzz Aldrin, and Mike Collins completed the first lunar landing mission, Apollo 11.

NASA wanted a fully reusable, orbital space plane, lifted vertically atop a piloted, reusable, flyback booster which would lift the orbiter and return for a runway landing. Its orbital work done, the recovered space plane would be serviced, refueled, and launched again just a month after landing. With the reduced operations costs that reusability afforded, NASA figured the shuttles would be in such demand that they would fly fifty missions a year. But tight budgets and the agency's underestimation of the technical difficulties in designing a reusable space truck made it impossible ever to reach those rosy projections.

Budget reductions proved even more drastic than NASA had feared, leading it to propose its shuttle as a national space transportation system that would serve both agency needs and those of the US Air Force. But in exchange for Air Force and intelligence agency support in the budget fights to come, NASA had to size the payload bay of the new shuttle to carry the largest planned military reconnaissance satellites on the drawing boards, fixing the length of the bay at sixty feet (18.3 meters), with a fifteen-foot (4.6-meter) width to accommodate future space station modules. The agency also accepted the military's requirement for a craft that could range well left or right of its orbital track to reach a landing runway. This capability was driven by the Air Force's desire to launch from Vandenberg Air Force Base in California, place a reconnaissance satellite in polar orbit, and land back at Vandenberg after one orbit—even as Earth (and the runway) rotated 1,100 miles (1,770 kilometers) to the east.

[P]rototype orbiter *Enterprise* is [r]eadied for vibration testing at [t]he Marshall Space Flight Center [d]ynamic test stand, April 1978.

NASA's Max Faget, designer of the blunt-bottomed capsule shapes used in Mercury, Gemini, and Apollo, proposed a straight-winged orbiter riding atop a three-million pound (1.4 million-kilogram), piloted, reusable

TOP: **Orbiter *Columbia* under construction in 1978 at the Palmdale, California, manufacturing hangar.**

ABOVE: ***Columbia* rolls out of the Palmdale hangar for shipment to Kennedy Space Center, March 1979.**

ABOVE RIGHT: ***Enterprise* separates from its 747 carrier for its second free-flight approach and landing, September 13, 1977.**

booster, but industry studies showed a delta-winged craft would have much better hypersonic gliding capabilities, the key to achieving the cross-range needs of the Air Force.

By 1971, NASA narrowed its choices to these two orbiter concepts, each riding on a fully reusable fly-back booster to keep operations costs low. As budget pressures weighed more heavily on the agency, though, it abandoned the reusable booster concept to hold the line on development costs. The still-reusable orbiter would now ride a throwaway booster before separating at 6,000 miles (9,656 kilometers) per hour and rocketing on to orbit.

Even this design compromise wasn't enough. When the Office of Management and Budget told NASA that shuttle development costs would be capped at $5 billion over five years, with a first flight planned

in 1977, the agency was forced again to redesign the spacecraft. During 1971, NASA shrank the orbiter by moving its propellants into an external tank. The 122-foot (37-meter), reusable orbiter would ride piggy-back on this mammoth, 154-foot (46.9-meter) external propellant tank, which became the system's structural backbone. The tank supplied propellants to three reusable hydrogen-oxygen engines housed in the orbiter's tail; these space shuttle main engines would ignite at liftoff and run all the way to orbit. To reduce initial development costs, designers had discarded the liquid-fueled, fly-back booster and substituted strap-on, solid-propellant rockets, which powered the shuttle on its initial two-minute ride through the lower atmosphere. Just prior to reaching orbit, the shuttle would drop the empty tank back into the atmosphere and use smaller maneuvering engines to achieve orbital velocity. The new shuttle's design was set.

President Nixon approved the shuttle's development on January 5, 1972. In April, while Apollo 16's John Young and Charlie Duke explored the Moon's Descartes highlands, the US Congress approved the new Space Transportation System. Designed to carry large, heavy cargoes to orbit, repair and retrieve satellites, and reduce costs through reusability, the shuttle was to serve as the primary payload launcher for NASA, the Pentagon, the intelligence agencies, and domestic satellite companies. Offering as well the potential for scientific research in orbit the space shuttle was to be a versatile, capable, all-things-to-all-users marvel.

The shuttle's development required costly technological innovations. Manufacturing the heat shield's lightweight, silica tiles and producing high-performance yet reliable main engines were especially troublesome challenges. As that work slowly progressed through 1977 and '78, NASA used the prototype shuttle *Enterprise*, launched from the back of a converted Boeing 747 airliner, to prove that the orbiter could return after reentry to an unpowered, precision landing. The Approach and Landing Tests buoyed NASA's shuttle team, but repeated testing delays pushed the shuttle's orbital debut well past the planned 1977 first launch. The agency had to abandon its plan for a shuttle rescue of Skylab, America's first space station, launched in 1973. NASA had hoped the new shuttle could first salvage and then restore Skylab to useful service.

Skylab fell to a destructive reentry in July 1979 with the shuttle still untried. Engineers were further dismayed that year when *Columbia*, the first space-ready orbiter, lost thousands of heat shield tiles when ferried by air from its California factory to Florida. NASA had to develop stronger adhesives and laboriously pull-test each tile to ensure it would stay attached during launch and reentry. Because of its complexity —2.5 million parts, 230 miles (370 kilometers) of wire, 1,060 valves, and 1,440 circuit breakers—and the precision runway landing required, NASA decided that human judgment would be necessary to ensure the first flight's success, and accepted the risk of crewing the shuttle without a prior, unpiloted flight test. Astronauts would be at *Columbia*'s controls the first time up.

Technicians at Kennedy Space Center began stacking the shuttle in November 1979, first erecting the solid rocket boosters (SRB), then suspending the external tank between them. In November 1980, after 613 days of assembly and testing at Kennedy Space Center, orbiter *Columbia* was finally rolled to the Vehicle Assembly Building and mounted on the external tank. The final major milestone before launch was a successful twenty-second test firing of the three main engines at Pad 39A on February 20, 1981. *Columbia*'s first crew of astronauts now aimed for a historic first launch in April 1981. Their lives—and NASA's future—depended on a first-launch success.

Major orbiter components (top) and dimensions of the space shuttle stack, with external tank, solid rocket boosters, and orbiter.

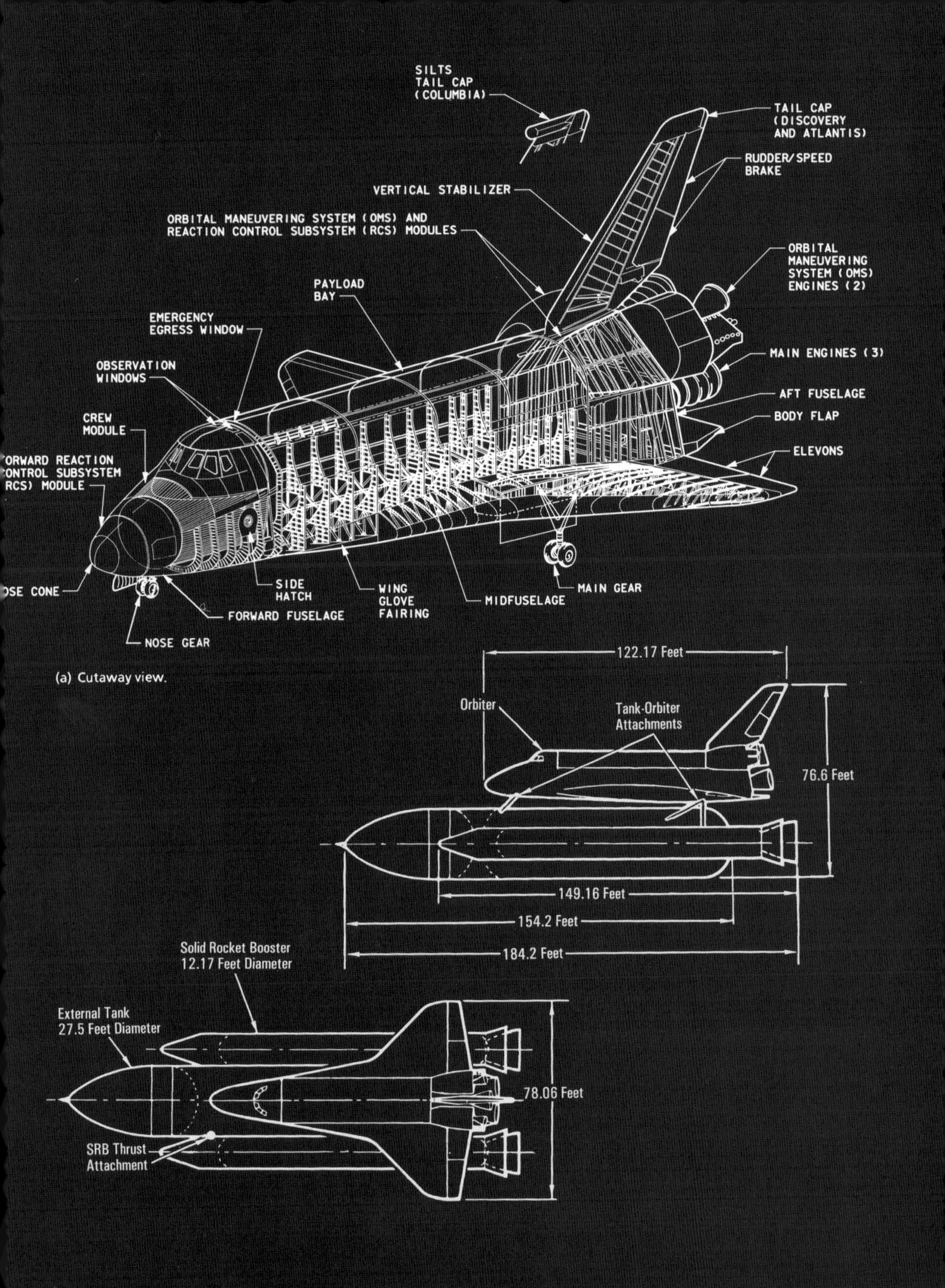

(a) Cutaway view.

List of Acronyms

ACCESS	Assembly Concept for Construction of Space Structures
ACS	Advanced Camera for Surveys
ACTS	Advanced Communications Technology Satellite
AMS	Alpha Magnetic Spectrometer
AOA	abort once around
AOS	Acquisition of Signal
APU	auxiliary power unit
ARIS	Active Rack Isolation System
ATLAS	Atmospheric Laboratory for Applications and Science
ATMOS	Atmospheric Trace Molecule Spectroscopy Experiment
ATO	abort to orbit
COSTAR	Corrective Optics Space Telescope Axial Replacement
CRISTA-SPAS	Cryogenic Infrared Spectrometers and Telescopes for the Atmosphere Shuttle Pallet Satellite
DAP	Digital Autopilot
Dextre	Special Purpose Dexterous Manipulator
DOD	Department of Defense
DSP	Defense Support Program
DTO	detailed test objective
EASE	EVA Assembly of Space Equipment
EDO	Extended Duration Orbiter
ELM-PS	Experiment Logistics Module, Pressurized Section
EMU	Extravehicular Mobility Unit
ERBS	Earth Radiation Budget Satellite
ESA	European Space Agency
ET	External Tank
EURECA	European Retrievable Carrier
EVA	Extravehicular Activity
FGS	Fine Guidance Sensor
FGB	Functional Cargo Block
GRO	Gamma Ray Observatory
HAC	Heading Alignment Cone
HST	Hubble Space Telescope
IAE	Inflatable Antenna Experiment
IBSS	Infrared Background Signature Survey
IEH	International Extreme Ultraviolet Hitchhiker
IML	International Microgravity Laboratory
IMU	Inertial Measurement Unit
ISS	International Space Station
IUS	Inertial Upper Stage
JEM	Japanese Experiment Module
JPL	Jet Propulsion Laboratory
LAGEOS	Laser Geodynamic Satellite
LDEF	Long Duration Exposure Facility
LITE	LIDAR In-Space Technology Experiment
LMS	Life and Microgravity Spacelab
MECO	Main Engine Cut Off
MFD	Manipulator Flight Demonstration
MLI	multi-layer insulation
MMU	Manned Maneuvering Unit
MPLM	Multi-Purpose Logistics Module
OASIS	Orbiter Experiments Autonomous Supporting Instrumentation System
OAST	Office of Aerospace Science and Technology
OBSS	orbiter boom sensor system
OMS	Orbital Maneuvering System
ORFEUS-SPAS	Orbiting and Retrievable Far and Extreme Ultraviolet Spectrometers Shuttle Pallet Satellite
PAM-D	Payload Assist Module
PCU	Power Control Unit
PMA	Pressurized Mating Adapter
RCC	reinforced carbon-carbon
RCS	Reaction Control System
RMS	Remote Manipulator Syste
RTG	Radioisotope thermoelectri generator
SAFER	Simplified Aid for EVA Rescue
SAREX	Shuttle Amateur Radio Experiment
SARJ	Solar Array Rotary Joint
SFU	Space Flyer Unit
SIR-B	Shuttle Imaging Radar-B
SLS	Spacelab Life Sciences
SPARTAN	Shuttle Point Autonomous Research Tool for Astronom
SPAS	Shuttle Pallet Satellite
SPIFEX	Space Shuttle Plume Impingement Flight Experiment
SRB	solid rocket booster
SRL	Space Radar Laboratory
STA	Shuttle Training Aircraft
STIS	Space Telescope Imaging Spectrograph
STS	Space Transportation System
TAS	Technology Applications an Science
TDRS	Tracking and Data Relay Satellite
TPAD	Trunion Pin Attachment Device
TRRJ	Thermal Radiator Rotary Joint
UARS	Upper Atmosphere Researc Satellite
UHF	ultra high frequency
USML	United States Microgravity Laboratory
USMP	US Microgravity Payload
WFPC	Wide Field and Planetary Camera

he *Atlantis* STS-79 stack atop
s mobile launcher rolls to the
ad, August 20, 1996.

Columbia's STS-61C liftoff lights up Kennedy Space Center on January 12, 1986.

Section 1

TESTING THE SPACE SHUTTLE

1981–1986

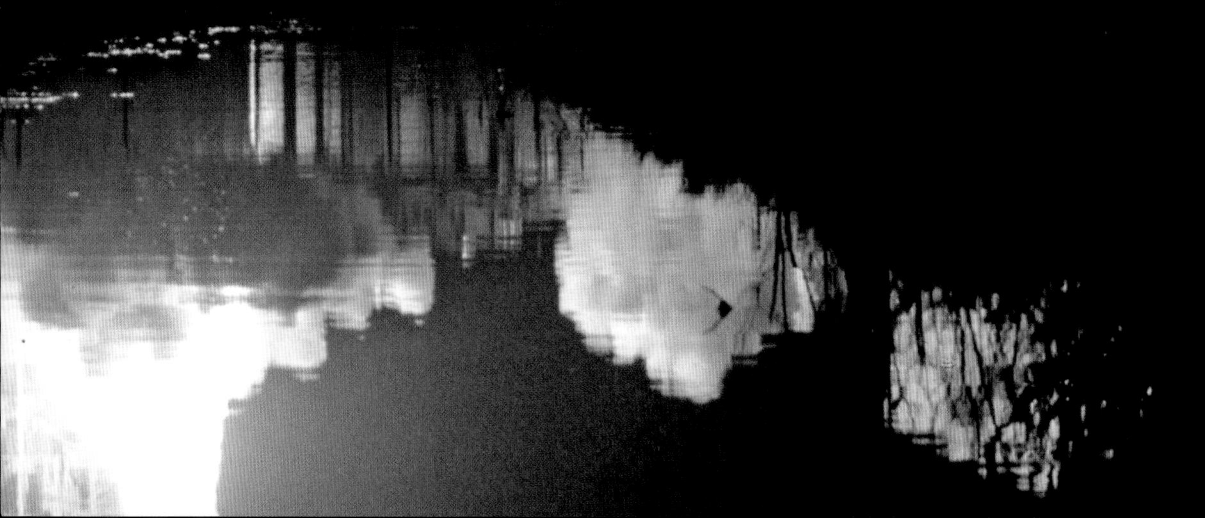

Starting with *Columbia's* first flight in April 1981, NASA embarked on an intensive, six-flight test program to prove the shuttle's capabilities and define its operational envelope during launch, orbital flight, reentry, and landing.

For each of these orbital test flights, the shuttle's three main engines would roar to life at six seconds before liftoff, producing 37 million horsepower (27.5 million kilowatts) and 1.2 million pounds (5.33 million Newtons) of thrust at full throttle. At zero, the twin 150-foot (46-meter) boosters ignited to produce an additional six million pounds (26.7 million Newtons) of thrust, consuming eleven tons (ten metric tons) of solid propellant every second. The 4.5-million-pound (2.05-million-kilogram) shuttle stack (boosters, tank, and orbiter) vaulted upward, cracking the sound barrier in just forty-five seconds. Two minutes after launch, the spent boosters dropped away, parachuting into the Atlantic for reuse. Just eight-and-a-half minutes after liftoff, external tank nearly empty, the main engines cut off just below orbital velocity at nearly five miles (eight kilometers) per second per second. Once separated from the tank, the orbiter fired two small maneuvering engines to achieve orbit while the tank burned up in the atmosphere over the Indian Ocean.

After completing its work in orbit, the orbiter used that same pair of maneuvering engines to reduce its velocity, dropping the orbiter back into the atmosphere to begin its hypersonic reentry. Enveloped in a searing, pink-white plasma, the spacecraft used thrusters and hydraulically powered aero surfaces to control its hypersonic glide toward a landing at Edwards Air Force Base in California, or at Kennedy Space Center in Florida. Just five minutes before touchdown, the commander disengaged the autopilot and guided the 110-ton (100-metric-ton), unpowered orbiter to a gentle landing.

President Ronald Reagan met the returning STS-4 crew at Edwards on July 4, 1982, ceremonially declaring the shuttle "operational" after just four test flights. Yet the shuttle always remained an experimental spaceplane, operating in an extreme environment at the limits of applied human technology. Design deficiencies and system vulnerabilities had yet to surface, and even as NASA adopted the "operational" label in its public communications, those who designed, built, and flew the shuttle knew that a machine of this complexity built by humans was still very much an experimental system.

Early flight crews of just two test pilots were followed quickly by missions carrying four or more astronauts. On the flight deck were the commander and pilot in the front seats. Behind them sat the flight engineer, with the Mission Specialist 2 in the center seat, and the Mission Specialist 1 to their right. Downstairs on the middeck, anywhere from one to four mission or payload specialists were strapped in for launch or landing. Although the crewmembers' light blue flight coveralls reinforced the perception of an airliner-style operation, and the media characterized shuttle flights as "routine," the spaceplane was still very much an experimental system. Crews lacked an escape system, and in an emergency their survival depended on the orbiter landing intact on a runway.

But the shuttle went to work, launching commercial communications satellites out of its fifteen-by-sixty-foot (five-by-eighteen-meter) cargo bay and collecting revenue from paying commercial customers. The Remote Manipulator System (RMS), the robot arm supplied by Canada, was nimble and rugged, serving to deftly snare spacecraft and hoist astronauts to their work sites. Spacesuited astronauts sortied out to retrieve satellites in jet-powered backpacks. The orbiter itself could pirouette elegantly with its thrusters and close to within inches of co-orbiting spacecraft for inspection or retrieval.

To boost satellites to their working orbits from the shuttle's low Earth orbit altitude of 100 to 300 nautical miles (185 to 555 kilometers), the shuttle employed a variety of solid-fueled upper stages attached to each satellite, like the Inertial Upper Stage (IUS) and Payload Assist Module (PAM). To send heavy satellites to geosynchronous orbit, or planetary probes on Earth-escape trajectories, NASA planned to employ the liquid-fueled Centaur upper stage, set to debut in 1986.

The space agency quickly ramped up the shuttle's launch pace, lofting nine missions in 1985. Although the science research results and commercial revenues were satisfying, nine was still a far cry from the fifty annual missions once touted by the agency. In fact, between missions the shuttle required intensive maintenance, closer in character to the experimental X-15 rocket plane than a familiar Boeing 747 airliner. The complex shuttles were far from trouble-free. Tiles were lost in flight or damaged by debris falling from the tank during ascent. Inspections of recovered boosters revealed disturbing evidence of heat damage and exhaust soot deposited in the field joints, where sections of the steel booster casings bolted together. This evidence of "blow-by" meant that 5,000-degree Fahrenheit (2,760-degree Celsius) exhaust gases were leaking past the joint seals. Engineers also found that greater blow-by damage appeared to correlate with colder launch temperatures. A jet of white-hot exhaust burning

With many heat-shield tiles still missing, *Columbia* rolls out of its manufacturing hangar in March 1979.

hrough a steel booster casing would be catastrophic to he shuttle and its crew.

Engineers working at solid-rocket-motor maker Thiokol n late 1985 proposed a change to the booster joint design o prevent the flexing at ignition that allowed hot exhaust as past the joint O-ring seals. NASA rejected the redesign ut asked for further analysis of its costs and benefits.

Just after the clear but chilly dawn of January 28, 1986, he crew of shuttle mission STS-51L *Challenger*—Dick Scobee, Mike Smith, Judy Resnik, Ron McNair, Greg arvis, Ellison Onizuka, and Christa McAuliffe—strapped n for launch. Sub-freezing overnight temperatures had raped the launchpad with icicles and frosted the skin of he external tank. But as a brilliant Sun warmed the spaceort, NASA decided the cold did not pose a safety risk and leared the shuttle for liftoff.

Challenger left Earth at 11:38 a.m. At ignition, a flexing ooster joint opened for a few milliseconds, long enough to allow superheated exhaust to escape past the O-ring seals. Acting like a blow torch, incandescent gases gradually cut a widening path through the damaged casing, spraying white-hot exhaust onto the side of the external tank. At seventy-two seconds into the flight, as the shuttle reached 1,977 miles (3,182 kilometers) per hour and 46,000 feet (14,020 meters) in altitude, the plume melted its way through a booster attachment strut and the base of the hydrogen tank. The detached booster pivoted into the upper tank, collapsing it instantly. In the blink of an eye, *Challenger* and its crew disappeared into a blazing white fireball of burning propellant. Supersonic air loads ripped the tumbling orbiter apart, and the seven astronauts rapidly lost consciousness. The incapacitated crew perished when their falling cabin struck the ocean two minutes and forty-five seconds later. The highly visible catastrophe and resulting national trauma caused many to wonder if *Challenger's* loss meant the end of NASA's human spaceflight program.

Mission no. **1**

STS

1

Orbiter	Columbia
Launch	April 12, 1981
Landing	April 14, 1981
Duration	2 days, 6 hrs., 20 mins., 53 secs.

John Young and Bob Crippen in *Columbia*'s cockpit during training

Crew	John W. Young, Robert L. Crippen

Mission: Unlike the first flights of all previous US-piloted spacecraft, astronauts were an essential element on STS-1: the orbiter could not be operated without a crew. Young and Crippen volunteered for "the boldest test flight in history." STS-1 proved the shuttle "stack" could be flown safely to orbit and make a hypersonic return to Earth.

Bob Crippen, Pilot:

My biggest worry was "Would we launch?" There were so many things that could go wrong with the vehicle. It was pretty complicated and I could see us going through several scrubs, but we only went through one. I was pleasantly surprised when we did lift off.

John said that if they were going to light off seven million pounds (31,137,551 Newtons) of thrust under you and you weren't a little bit excited, you didn't understand the situation. Both John and I understood the situation—we had worked on the design of the shuttle from its inception.

The first stage was quite a ride. The solid rockets were loud and produced a low-frequency vibration. We later learned that the acoustic shock of their ignition had reflected off the launch pad and actually damaged some of the structural components in the nose of the vehicle.

Two minutes later, the solids expended their fuel and separated from the vehicle with a fiery blast from the jettison rockets. Then I felt my first moment of alarm. It seemed too quiet. The thrust level had changed dramatically—I thought our main engines had quit—but a quick check of the instruments verified that everything was fine. We had reached an altitude so high that not enough atmosphere remained to allow engine noise to reach the cockpit.

The biggest surprise was how everything worked well. The first time I had a chance to look out at the Earth from orbit was right after orbit insertion. I was busy because we had an orbital maneuvering system burn shortly after main-engine cutoff to get into the orbit we wanted. I thought I knew what it would look like, to look out at this gorgeous Earth we live on, but it was about a hundred times better.

I've been asked what the most memorable moment on the mission was. I like to use John's expression: "The part between takeoff and landing." It was all marvelous—from ascent, to floating around in weightlessness in orbit, to looking out at the Earth, to coming in on reentry, to landing on the runway—it was all a fantastic experience.

We landed at Edwards Air Force Base and John took over from the autopilot at about Mach 1 as we normally do, and we did a big left bank to circle around to the dry lakebed runway. I looked out his window and could see these thousands of people out there on the lakebed. *Columbia* really got people's attention. I was just hoping that none of them were out there on the runway!

I'd started working on the shuttle in 1972. We worked hard to get to the first flight, and it cost quite a bit of money—more than we anticipated—but the fact that we got there, we flew it, and it worked was important. STS-1 was the first space shuttle mission that went into orbit, and its success was extraordinarily important to NASA and to the country. The excitement of flying STS-1 didn't really hit me until the launch countdown passed one minute to go. Then I turned to John and said, "I think we might do it."

Columbia STS-1 lifts off from Pad 39A on the shuttle's maiden flight, April 12, 1981.

Mission no. **2**

STS 2

Orbiter	Columbia
Launch	November 12, 1981
Landing	November 14, 1981
Duration	2 days, 6 hrs., 13 mins., 12 secs.

Dick Truly and Joe Engle wave to reporters after a scrubbed STS-2 launch attempt.

Crew	Joe H. Engle, Richard H. Truly

Mission: STS-2, *Columbia*'s second flight, demonstrated the first reuse of a crewed space vehicle. Astronauts conducted the first test of the Canadian RMS, the mechanical arm for grappling payloads, and operated the OSTA-1 payload, including the Earth-scanning SIR-A radar. Engle and Truly flew a manual reentry profile on a mission shortened by a faulty fuel cell. STS-2 was also the only time an all-rookie crew flew a shuttle mission.

Dick Truly, Pilot:

All the hype about the shuttle was that it was reusable, but NASA hadn't proven it until we flew. I think that history will show that the first reusable flight sent us on our way.

Joe was a great pilot; he and I had a lot of laughs. We had flown *Enterprise*, the space shuttle atmospheric test vehicle, together, then immediately backed up John Young and Bob Crippen, and then trained for STS-2. So for several years, Joe and I flew together, traveled together, simulated together—it was almost like we were married. The "all-rookie" designation was good for stories, but we didn't think of ourselves that way: Joe had flown the X-15 and I'd been a military and then a NASA astronaut since 1965.

I had waited a long time to fly STS-2, so it was a thrill. The flight was scheduled three or four weeks earlier, but on the launch pad, we had a leak in the forward RCS that washed off some of the tile surface coatings. The delay just happened to schedule the launch for my forty-fourth birthday—it was great!

After we were in orbit for about five or six hours, a fuel cell failed and we had to shut it down. With one fuel cell down, we were fine, but if we had lost another one, that would have created an emergency, and losing a third would have been "lights out."

Joe and I were working like hell during the shortened flight to get all our on-orbit testing done. I was the principal RMS flier, and we had planned for several hours over two separate days to systematically test it. We were thrown into "hurry up." But we got it all done, and we went through the whole checklist with the RMS, and luckily it performed beautifully. We checked out all the modes, joints, and grapple functions, took some video of ourselves, and all that.

We were airborne about fifty-two hours, and after the flight I looked back at my crew activity plan. Because of the alarms and the notes in the flight plan, I only slept continuously for about two or three hours. Houston was continually changing the flight plan to figure out how to get five days of work into two-and-a-half days. When they signed off during our rest period, which was also our sleep period, there was Japan and Italy out the window, so it was hard to sleep!

Joe made sure I got some time piloting the craft both in the *Enterprise* flights and on STS-2. During reentry we would come out of automatic control and fly it manually. Naturally he did most of it, but we would go back into auto and I would then fly. *Enterprise* trained us to land, so that was the easiest part of the experience.

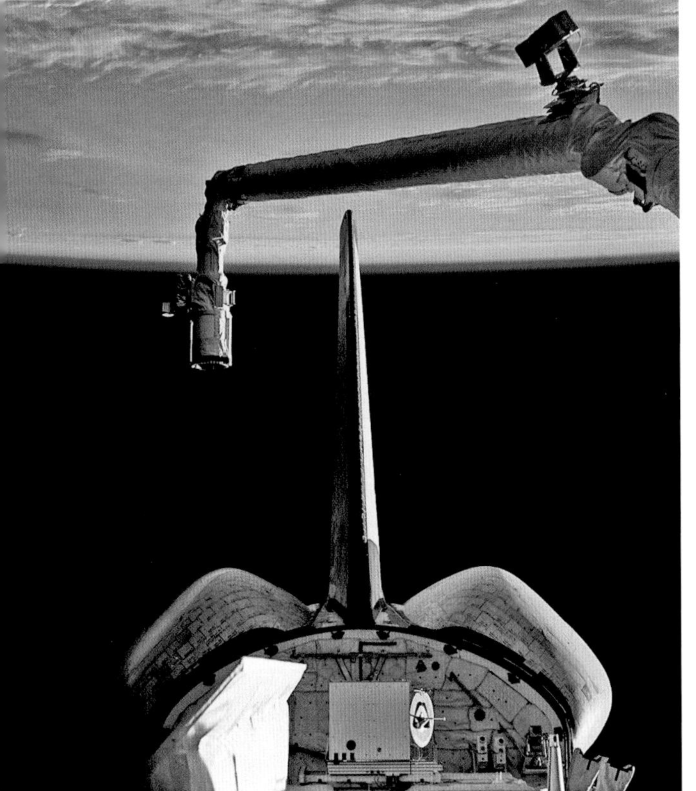

ABOVE: **President Ronald Reagan speaks to the STS-2 crew from Mission Control at Johnson Space Center.**

LEFT: **STS-2's crew unlimbered Canadarm, the shuttle's Remote Manipulator System, for the first time.**

BELOW: ***Columbia*'s wingtips generate contrails during the return to Edwards Air Force Base, California.**

Mission no. **3**

STS

3

Orbiter	Columbia
Launch	March 22, 1982
Landing	March 30, 1982
Duration	8 days, 0 hrs., 4 mins., 45 secs.

Gordo Fullerton and Jack Lousma walk toward Pad 39A for a countdown rehearsal on February 19, 1982.

Crew	Jack R. Lousma, C. Gordon "Gordo" Fullerton

Mission: This developmental flight test of the shuttle system was the shuttle's third mission, designed to wring out the environmental limitations of the orbiter's systems during a long stay in space. The crew operated a suite of scientific experiments in the payload bay and expanded operations of the Canadian RMS.

Jack Lousma, Commander:

STS-3 is probably remembered most by space historians for being the only space shuttle mission to land at White Sands Missile Range in New Mexico. We also accomplished at least 98 percent of our objectives, tripled the shuttle flight duration—from fifty-four hours to 184 hours—in a single mission, and set the pace for working in space.

We had fifteen scientific experiments, some in the cockpit and some outside. We deliberately exposed the orbiter systems to the extremes of the harsh space environment. For example, we spent a day each with the nose or tail pointed at the Sun to see if the thermal control system could keep the warm end cool, and the cool end warm.

We planned to land on the lakebed at Edwards Air Force Base where multiple, very long runways were available. But rain on the Edwards' lakebed meant we might have to land on the brand new 15,000-foot-long (4,572-meters-long), hard-surfaced runway at the Cape (with alligators at both ends). Because the precision landing required at the Cape seemed to me premature for the shuttle at the time, I suggested landing on the lakebed at White Sands Missile Range instead. That's where we did most of our approach training in both the Shuttle Training Aircraft (STA) and T-38 jets.

We put *Columbia* on autopilot for a test of an automatic final approach to the lakebed, requiring a manual takeover for landing once we were stabilized on the inner glideslope, descending through 1,000 feet (305 meters). That approach did not go as planned.

We were lower and faster than planned when the automatic approach system established us on the shallow, 1.5-degree inner glideslope. When I took over manually to make the landing, I didn't have much time to make corrections. We were approaching about 500 feet (152 meters) above the runway but still had to slow to get the landing gear down, so I asked Gordo, "What's the speed?", and he said, "275." So, I said, "Put 'em down!" This apparent late, low-altitude deployment of the landing gear was off-nominal, or abnormal, due to our inner glideslope overspeed, and it understandably worried most of the folks on the ground.

Columbia flying at lower speeds with the landing gear extended felt much different than flying it "clean" at 285 knots (146 meters per second) before the approach. Just before landing, this "glider" now felt more like a 220,000-pound (99,790-kilogram) freight train. This awkward, very late manual takeover out of a fully automatic approach was one of the reasons an orbiter autoland capability was never certified: We didn't want to force a crew into a late manual takeover if the autopilot failed.

It also appeared that we were targeted for touchdown about halfway down the runway. In order to get on the runway sooner to begin decelerating, our touchdown was faster and not as smooth as I had trained for. But all landing parameters were well within structural limits, and we had ample runway to make the rollout and evaluate the nose wheel steering, our final detailed test objective (DTO).

As we rolled to a stop on the lakebed, a wave of gratitude to God and of professional reward like no other swept over me. We were still alive, we had accomplished all of our planned objectives, and we had presented our whole team with the successful outcome for which so many had worked so hard for so long. It doesn't get any better than that.

LEFT: **The crew maneuvers the RMS, the shuttle's robotic arm, with the plasma experiment package attached.**

ABOVE: **Jack Lousma prepares a strawberry drink pouch on *Columbia*'s middeck.**

BELOW: **Lousma flares *Columbia* for landing on the dry lakebed at the White Sands Missile Range in New Mexico.**

Mission no. **4**

STS

Orbiter	Columbia
Launch	June 27, 1982
Landing	July 4, 1982
Duration	7 days, 1 hr., 9 mins., 40 secs.

Hank Hartsfield and Ken Mattingly in their STS-4 preflight portrait

Crew	Thomas K. "Ken" Mattingly, Henry W. "Hank" Hartsfield Jr.

Mission: STS-4's developmental flight test launched with a then-classified strategic missile defense sensor experiment, nine student experiments, a shuttle contamination sensor package, and flight test instrumentation and recorders. After a week in orbit, *Columbia* returned to its first concrete runway landing.

Ken Mattingly, Commander:

Flight 4 was the last mission of the flight-test program. One of the most significant things at this point was the capability to turn the vehicle around from one flight to the next in about three months.

The shuttle lift-off is not noisy. It doesn't shake and rattle, it just goes. It was a totally different experience than the Saturn V launch. Compared to the Saturn, the shuttle was like electric propulsion.

After working on this device for ten years, the most magical thing was the experience of seeing Earth. We got into orbit and opened the payload bay doors for the first time towards Earth, and all of a sudden, it was like you pulled the shades back on a bay window—Earth appeared. Because of the orbiter windows, we could now see the world in cinemascope.

I don't care how long you're up there, I can't imagine anyone ever getting tired of it. So Hank would say, "You know, we probably ought to get some sleep here." I'd say, "Yeah, yeah, yeah. You're right. We've got another day's work tomorrow." So Hank would sleep in the middeck, and I slept on the flight deck. And after we turned in, I told the ground I went to sleep so they wouldn't bother me, but I'd sit there looking out, having a wonderful time.

We came home, and of course we were all bug-eyed about this reentry and being able to see out the windows and see things that we had never seen before as we sailed down over the west coast. In Apollo, the reentry gave you a healthy dose of deceleration as you came through the atmosphere. On the shuttle, you can stand up through reentry—I'm sure others have.

We set up for approach and landing, and I was still having this inner-ear, vestibular sensation of motion that was unusual, but once we got on the glide slope, the picture was normal. I would listen to Hank tell me what the altitude was, and I'd make little adjustments to position the nose.

He was calling off air speeds and altitude, and I was just staring at the horizon, and I had no idea what it was going to feel like to land. When I would shoot touch-and-go's in the big KC-135 tanker, a NASA training aircraft, there was never any doubt when we landed. You could always tell.

So I was expecting "bang, crash, squeak"—something. But nothing. Finally Hank said, "You'd better put the nose down." "Oh," I said, "All right." So I put it down, and I was sure we were still in the air. I thought, "We can't be very far off the ground." But we were already on the ground, and neither one of us knew it. I had never done that before. And I've never been able to do it again in any airplane.

This business of learning about space flight and what goes into it will never be boring. Every day you take one more little step to find out something that's really, really cool.

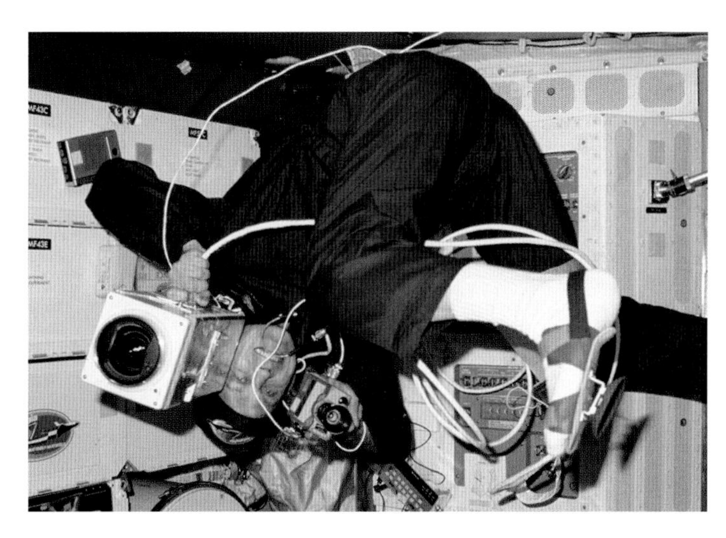

LEFT: **The shuttle robotic arm holds the Induced Environment Contaminant Monitor above *Columbia*'s tail.**

ABOVE: **Ken Mattingly operates a handheld TV camera on *Columbia*'s middeck.**

BELOW: **On July 4, 1982, Nancy Reagan and President Ronald Reagan greet the STS-4 crew after landing.**

Mission no. **5**

STS

5

Orbiter	Columbia
Launch	November 11, 1982
Landing	November 16, 1982
Duration	5 days, 2 hrs., 14 mins., 26 secs.

Ace Satellite Launchers (clockwise from bottom left) Vance Brand, Bill Lenoir, Bob Overmyer, and Joe Allen pose in the middeck.

Crew	Vance D. Brand, Robert F. Overmyer, Joseph P. Allen IV, William B. Lenoir

Mission: STS-5 was billed as the shuttle's first operational flight, deploying a pair of Hughes 376 communications satellites, ANIK-C3 and SBS-3, from the payload bay. *Columbia* carried the first American crew of four astronauts, who flew for the first time without pressure suits.

Vance Brand, Commander:

STS-5 was billed as the first operational flight, but in a sense, the shuttle was experimental until the very end.

I had an exceptional crew. STS-5 was the first time mission specialists flew on the shuttle, with Joe Allen and Bill Lenoir. It was the largest crew up to that point; Bob Overmyer, Bill, and Joe waited more than thirteen years for their first flight. These three were around during the long development of the shuttle. The ground team was really very experienced by this time and very good at their job. The motto of the flight was "We Deliver."

We only had three seats upstairs. We split it up so that Bill sat upstairs for ascent, and Joe sat upstairs for reentry, in order to spread the wealth. We decided to disable the two ejection seats, because we couldn't use them: If the guys in the front had ejected, the one in the middle seat would have gotten the blast, and there was one guy downstairs who wouldn't have gotten out at all.

The Hughes 376 satellite that went to geosynchronous orbit was deployed for the first time with its Payload Assist Module (PAM-D) by a mechanical deployment system, which would spin the satellite up and then push it off from the shuttle with spring action. For forty-five minutes, the satellite drifted away from the shuttle, then the PAM-D engine fired to send the satellite to geosynchronous orbit. We maneuvered to put the bottom of the shuttle facing the PAM-D engine burn so that if any high-speed particles reached the shuttle, we'd have some protection, albeit weak. The shuttle was a mile or two away during the burn.

We worked our butts off. I think the first day we worked probably seventeen or eighteen hours. The deployments came early in the mission. Extravehicular Activities (EVA), or work on the outside of the vehicle, were to be done later, but there were two EVA suit failures, one in Bill's and one in Joe's suit. Ground wasn't too keen on doing the EVA, and at that point I wasn't either, so we cancelled it.

On reentry, I had a chance at Mach 18 to do a push-over, pull-up maneuver. It wasn't a big maneuver—from 40 degrees angle-of-attack down to 35 degrees, then a pull up of about 15 degrees, followed by a push back down to 40 degrees alpha again. The whole thing didn't take more than fifteen or twenty seconds, but the three of us were pulling this off while I was maneuvering with the control stick. At that point in the reentry, Mach 18, there was a big white shock wave coming off the shuttle's nose. When I pushed over, the shock wave came up and sat on the windscreen. The crew's eyes got pretty big!

That white shock wave was hot. Of course we had a three-pane front windscreen, but it made everybody nervous. When I finished by pushing over to 40 degrees again, the shock wave walked back to the nose where it should have been. It was an interesting experience to hand-fly the shuttle at Mach 18 for a short time and hear the crew's "oohs" and "ahhs." It didn't hurt the windscreen at all, but after landing, technicians found "slumped," or deformed, tiles around the nose wheel well. After that, engineers installed panels there of reinforced carbon-carbon (RCC), a more heat-resistant material, to replace the tiles that felt that heat.

STS-5 rises through scattered clouds as photographed from a NASA T-38 trainer.

Mission no. **6**

STS

6

Orbiter	Challenger
Launch	April 4, 1983
Landing	April 9, 1983
Duration	5 days, 0 hrs., 23 mins., 42 secs.

Donald Peterson, Paul Weitz, Story Musgrave, and Bo Bobko in their STS-6 preflight portrait.

Crew	Paul J. "P. J." Weitz, Karol J. "Bo" Bobko, F. Story Musgrave, Donald H. Peterson

Mission: *Challenger*'s first flight carried the first of the TDRS, designed to provide nearly continuous communications between Mission Control, the shuttle orbiters, and scientific satellites. Story Musgrave and Don Peterson wore the new shuttle Extravehicular Mobility Units (EMU) during a four-hour, seventeen-minute spacewalk.

Story Musgrave, Mission Specialist:

The 1982 rollout of the new *Challenger* at Palmdale was exciting and beautiful. But, nine months later when we got STS-6 into space, we had an incredibly bad day. Nothing was checking out on the IUS—the booster for the Tracking and Data Relay Satellite (TDRS). Later, technicians found out that they'd mis-wired two star trackers on the IUS. The wires that were supposed to go to star tracker 1 were going to star tracker 2, and the ones to 2 were going to 1. But I didn't know that at the time. So we were trying like crazy in this eight-minute Acquisition of Signal (AOS) to solve the problem and get the IUS into a good configuration for release with TDRS. We got to the last AOS before the planned deployment, and nothing was working. With one minute to go, Mission Control said, "Story, you're GO for deploy!"

I'm talking to myself, saying "we can't do this," and I said to Mission Control, "You don't mean that, sir." "Yes, Story, GO for deploy." That was the last communication before we lost signal, so I sent it off—IUS/TDRS. "There you go, baby." I wasn't feeling good about any of this. Once I deployed it, I couldn't touch it. It was gone.

The IUS second-stage rocket had a nozzle failure. So we lost control of the TDRS satellite, stuck about 9,000 miles (14,500 kilometers) short of the altitude required for its service orbit. Over a period of months, controllers at the NASA Goddard Space Flight Center in Greenbelt, Maryland, eased it up there using little one-pound (4.44 Newtons) thrusters. After all that, it was a phenomenal success. That TDRS was supposed to have a fifteen-year lifetime, but it lasted twenty-seven years.

On the mission spacewalk, everything went the way it should. That was a great experience. I had helped design the EVA suit, so simply to check out the systems, the ergonomics—where you can be, where you can reach, how you can maneuver—was very satisfying. I did some tether dynamics: If you ever got loose from the handrail, what it would be like to grab the tether, and haul yourself back?

One really tough part of that EVA was that after we completed the planned spacewalk work, which took about four hours, Mission Control asked us to hang out a while longer because they wanted to get more data. How does this suit and the backpack perform after, say, seven hours? What's our rate of consumable use? That's why we were out there, to test stuff. So they asked us to continue the EVA, but P. J. Weitz, the commander, said, "No, sir, they're coming in now."

P. J. overruled the Mission Control Center. I was a communicator in Mission Control for twenty-five missions. Crews are supposed to work with them—instead of overriding them, you thank them. "Gee whiz, yeah, we get to do some *more* of this nice stuff." But not P. J. Nope. "We are coming in now." That was a rough point.

But we had a good time out there. My heart rate never reached sixty. If anything was hard work, I rechoreographed it. If I was short of breath, or if something was taking a little muscle, I said, "Don't do that, Story." If challenged, thinking about it as choreography for a ballerina or a figure skater was a nice way to make it work. Save the muscle and the breath.

ABOVE: **Story Musgrave and Donald Peterson traverse the payload bay on the first shuttle EVA.**

LEFT: ***Challenger* lifts off in this beach view from north of Pad 39A.**

RIGHT: **During the EVA, Musgrave works at *Challenger*'s forward bulkhead.**

Mission no. **7**

STS 7

Orbiter	Challenger
Launch	June 18, 1983
Landing	June 24, 1983
Duration	6 days, 2 hrs., 23 mins., 59 secs.

The STS-7 crew on the flight deck: Norm Thagard, Bob Crippen, Rick Hauck, Sally Ride, and John Fabian

Crew	Robert L. Crippen, Frederick H. "Rick" Hauck, John M. Fabian, Sally K. Ride, Norman E. Thagard

Mission: STS-7 carried five astronauts, the largest crew so far. Sally Ride became the first American woman to fly in space as the crew deployed the ANIK C-2 and PALAPA-B1 communications satellites. They launched and retrieved with the RMS the German-built SPAS, which carried ten microgravity research and remote-sensing experiments.

John Fabian, Mission Specialist:

The number one thing in media coverage of the mission had nothing to do with communications satellites or our German free-flier. It was about Sally Ride being the first American woman in space. That was enormously important and long overdue, and Sally was exactly the right person to be selected.

One of my favorite jokes is that flying with Sally Ride made me famous. Sally was a very hard-charging individual and extraordinarily bright. I've never met anyone who was as good a natural pilot. She was easy to work with, and yet like everybody else, there would be times when her attitude would swing, and you could see it: this mental exercise when she'd been given something that she wasn't really happy about. In a matter of thirty seconds, she would be back. It was just amazing.

One other thing about the mission that sets it apart is that it truly was the first international flight of the space shuttle era. We launched two communications satellites—one for Canada and one for Indonesia—and we had the German-made Shuttle Pallet Satellite (SPAS).

Sally Ride, Mission Specialist:

As we were about to step into the orbiter, right before launch, I think we were all a little bit apprehensive about what was about to happen. But more than apprehension, I think we all felt excitement. You're the only ones on the launch pad. You've just gone all the way up this elevator to the top, and there are just thre or four people up there. They're going to strap you into the orbite and then they're going to leave. It's a very lonely feeling as soon as they close the hatch and wav goodbye to you and leave the pad.

Circling 200 miles (322 kilometers) above the Earth, I saw the atmosphere—just a thin, blue band above the horizon—separating Earth from the blackness o space. I watched a huge hurrican swirling in the Atlantic Ocean, anc an enormous dust storm spreading across the entire Sahara Desert. Some nights, I saw lightning light up the clouds below, and jump from cloud to cloud across the sky. One night, as the shuttle passed over Florida, I watched all the cities from Miami to New York twinkle in the darkness.

There's a sense of camaraderie while you're up there. You're part of a team and you're great friends You've been training with them for a year; it's just like a family in a new experience. There are funny things that happen, almost like slapstick comedy. The whole effect of being weightless and bumping into people and spinnin around and spilling your food keeps you laughing for quite a bit of the time.

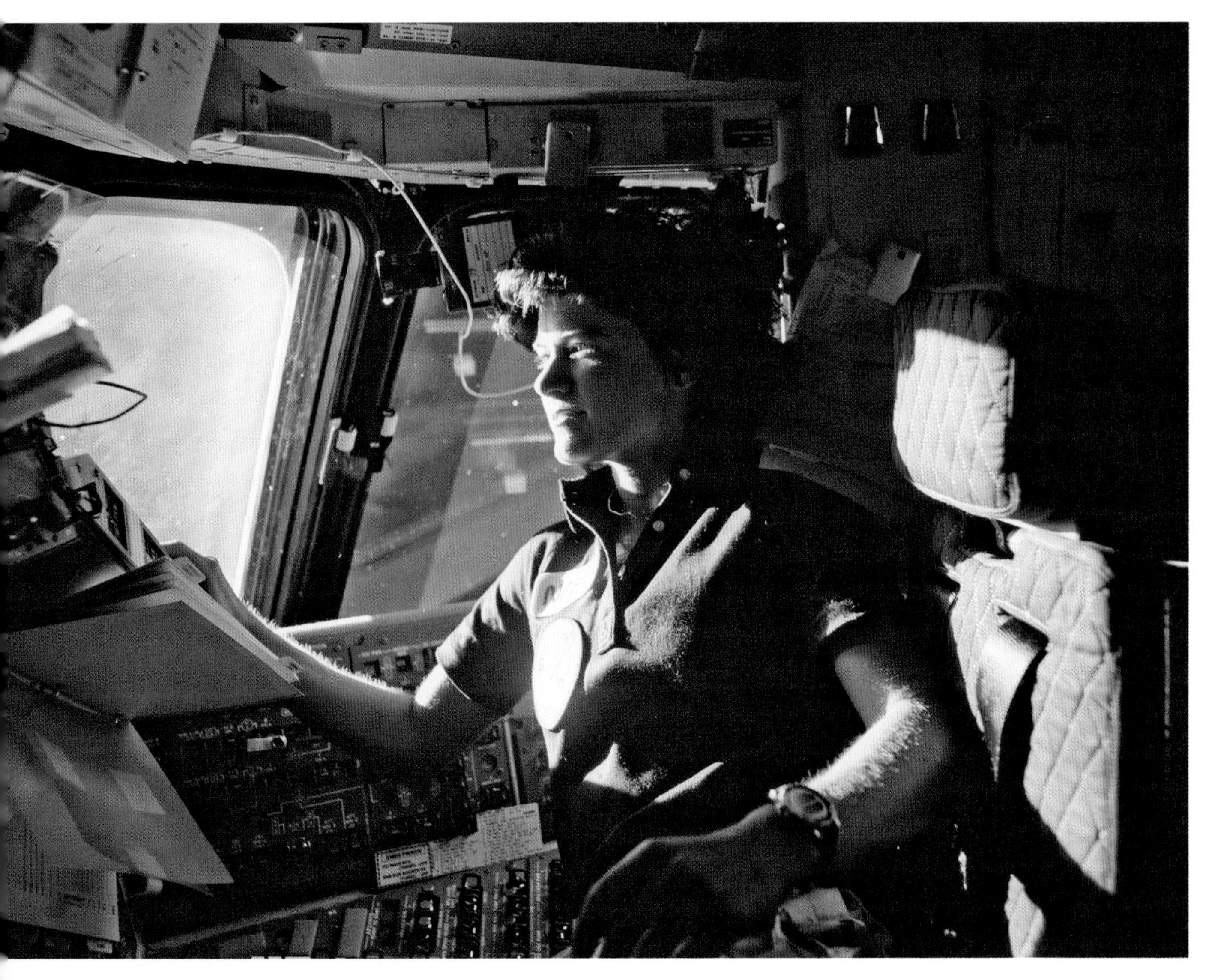

ABOVE: **Sally Ride enjoys Earth views from *Challenger*'s pilot seat during STS-7.**

LEFT: ***Challenger* as imaged by the SPAS satellite during its research flight.**

BELOW: **Bob Crippen flies *Challenger* during reentry as glowing plasma flows past the windows.**

Mission no. **8**

STS

8

Orbiter	Challenger
Launch	August 30, 1983
Landing	September 5, 1983
Duration	6 days, 1 hr., 8 mins., 43 secs.

The STS-8 crew at Pad 39A: Dale Gardner, Guy Bluford, Bill Thornton, Dan Brandenstein, and Dick Truly

Crew	Richard H. Truly, Daniel C. Brandenstein, Guion S. "Guy" Bluford Jr., Dale A. Gardner, William E. Thornton

Mission: *Challenger*'s crew made the first shuttle night launch and landing, deployed the INSAT-1B communications satellite, and tested the Canadian RMS by maneuvering a 7,350-pound (3,333-kilogram) test satellite above the payload bay. Guy Bluford became the first African American to fly in space.

Guy Bluford, Mission Specialist:

There were three African Americans selected in the 1978 astronaut class. They included Major Fred Gregory, an Air Force test pilot; Ron McNair, a PhD physicist, and me. I was an Air Force combat fighter pilot and a PhD aerospace engineer. I flew in space four times, Ron flew once and was killed on his second spaceflight in the *Challenger* accident, and Fred flew three times in space and became the first African American to command a spacecraft.

STS-8 attracted a lot of public attention. Our primary payload was the Indian weather and communication satellite called INSAT-1B, which was going to be put in a geosynchronous orbit above the equator over the Indian Ocean. In order to reach the targeted deployment point on time and inject the satellite into its ultimate orbit, we had to launch the space shuttle at night.

The first surprise on liftoff was the amount of light generated by the SRBs. We had darkened the cockpit in order to maintain our night vision, just in case we had to make an emergency landing at Kennedy Space Center. The cockpit was lit up like daylight as we rose off the pad and rotated to go down range. It looked like we were inside a fireball with the bright glow flooding through the windows. Our 500-foot-(152-meter) long fiery tail could be seen as far as 170 miles (273 kilometers) away in Fort Myers and 220 miles (354 kilometers) away in Miami.

The ride up on the SRBs and the three main engines of the shuttle was thrilling—very noisy and bumpy as the G-forces increased to about 2.5 g's, like we were driving a pickup truck over railroad ties. I remember thinking: "This thing moves! Like riding the simulator but it *moves*!" The next surprise was the brilliant flash of light just after two minutes from the firing of the pyros at SRB separation. This was *not* modeled in the simulator.

It took me several hours to get my "space legs" as I floated clumsily around the cockpit. Once in orbit, I had lunch and was ready to go to work. As part of Bill Thornton's studies, I found that due to zero-g, I was almost two inches (five centimeters) taller in space, and both of my thighs were one inch (two and a half centimeters) thinner due to the fluid shift in my body.

The crew operated the Canadian robotic arm with a large payload attached and put the middeck electrophoresis experiment through its paces. We gathered data on human adaptation to microgravity and were ready after six days for a night landing at Edwards Air Force Base.

After the deorbit burn, we maneuvered *Challenger* to the reentry attitude over Australia. As we descended, the hot plasma glow surrounding us became a 3,000-degree Fahrenheit (1,649-degree Celsius) blowtorch that danced on the nose and upper sides of the orbiter. At 50,000 feet (15,240 meters), Dick Truly took manual control and flew us in a wide sweeping turn to line up on final descent diving towards the end of the runway. For ground observers, the only indication of our arrival was the double sonic boom created by *Challenger* as we circled over Edwards, followed by our sudden appearance in the floodlights as we crossed the runway threshold for a smooth landing.

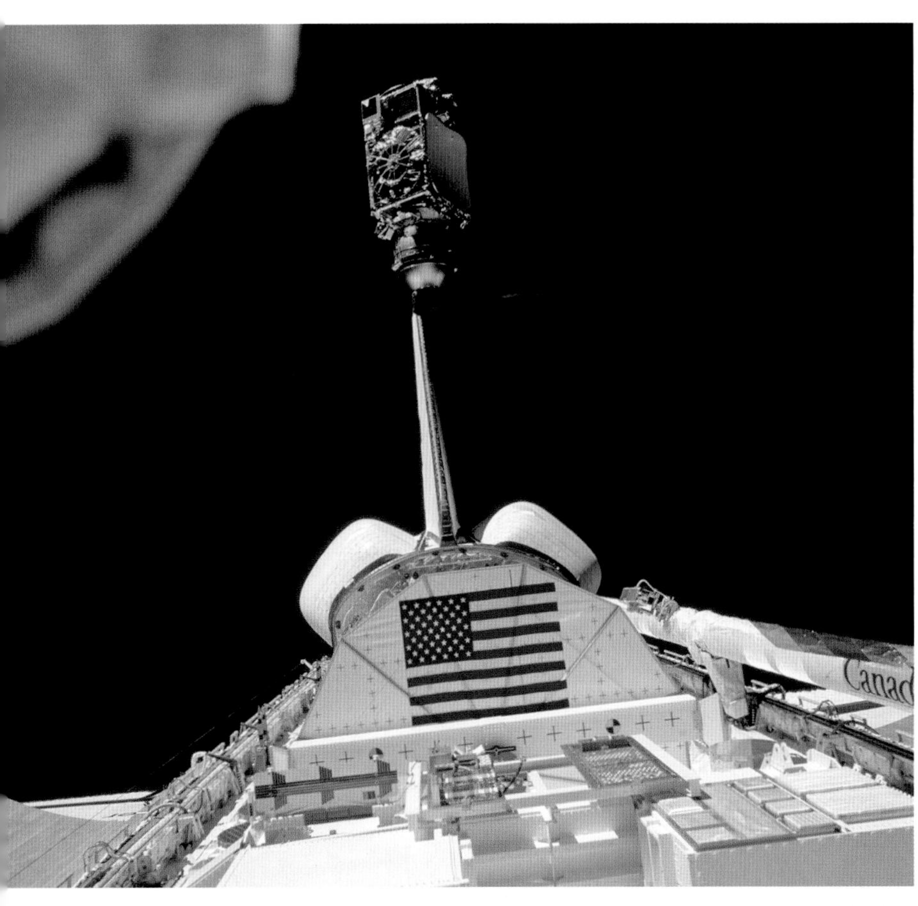

LEFT: **The INSAT satellite spins out of the payload bay on its way to geosynchronous orbit.**

ABOVE: **The STS-8 crew maneuvered a 4.5-ton (4.1 metric ton) practice payload during RMS testing.**

BELOW: ***Challenger* sweeps in for the first night orbiter touchdown at Edwards Air Force Base, California.**

Mission no. **9**

STS

Orbiter	Columbia
Launch	November 28, 1983
Landing	December 8, 1983
Duration	10 days, 7 hrs., 47 mins., 24 secs.

The STS-9 crew on the middeck: (clockwise from left) Brewster Shaw, Byron Lichtenberg, Bob Parker, John Young, Ulf Merbold, and Owen Garriott

Crew	John W. Young, Brewster H. Shaw Jr., Owen K. Garriott, Robert A. R. Parker, Byron K. Lichtenberg, Ulf D. Merbold

Mission: Six astronauts worked in the ESA-built, pressurized Spacelab module. Their research—seventy-three experiments in a dozen science fields—showed the potential of a future space station. Two of the flight computers failed, and two of the three auxiliary power units caught fire during reentry.

Brewster Shaw, Pilot:

On Spacelab 1, there were ninety-four verification flight tests conducted, and they were to demonstrate the capability of doing scientific research in that environment in that laboratory in low Earth orbit. The whole Spacelab project was a precursor to what people thought we would do on the ISS.

It was a wonderful experience just to be mentored by John Young. In the simulator they would throw us a failure signature, and I would react to that and do the proper procedure, and John would look at me and say, "Why did you do that?" And I would say because of this, this, and this. He'd say, "Yeah. But that's fake. Look here. Here's the real information you should have paid attention to. That stuff was just to suck you in. Make sure you look at more information before you decide that something is going on." And in the usual John, "aw, shucks" fashion, he never got upset and was always very calm.

SRB ignition surprised the hell out of me. I didn't think it would be as sharp: It felt like somebody hit the back of my chair with a sledgehammer. First-stage dynamics were also a shock: We hadn't seen any cameras at that time mounted in the crew module, and nobody told me it was going to shake like that.

Owen Garriott was a very even, steady, smart guy and immediately after entering the Spacelab module, Owen's body remembered what zero-g was like from his experience on Skylab, the first US space station that operated in 1973–74. He immediately tucked his toes under a handrail and he was just fixed there. Everyone else was bouncing off the walls in their "What in the hell do you do in zero-g?" mode.

What an experience to be able to see our planet from that perspective! There were 4.7 billion people on Earth then, and I bet every one of them would love to have been where I was and have that experience of seeing Earth. I was just lucky as hell.

Bob Parker, Mission Specialist:

This was the first mission to have payload specialists or "semi-astronauts," as some called them before we flew. There were four; two who flew and two backups. One was German, Ulf Merbold, and Byron Lichtenberg was from the US. Their training was focused almost entirely on the payloads, as they had actually helped design each principal experiment.

Meanwhile, Owen and I were training to be the mission specialists.

The Spacelab was just narrow enough that you could reach from one handrail across to the other wall. You could push off from one end and float gently to the other, and we had fun launching from the front tunnel hatch and zooming back to Spacelab.

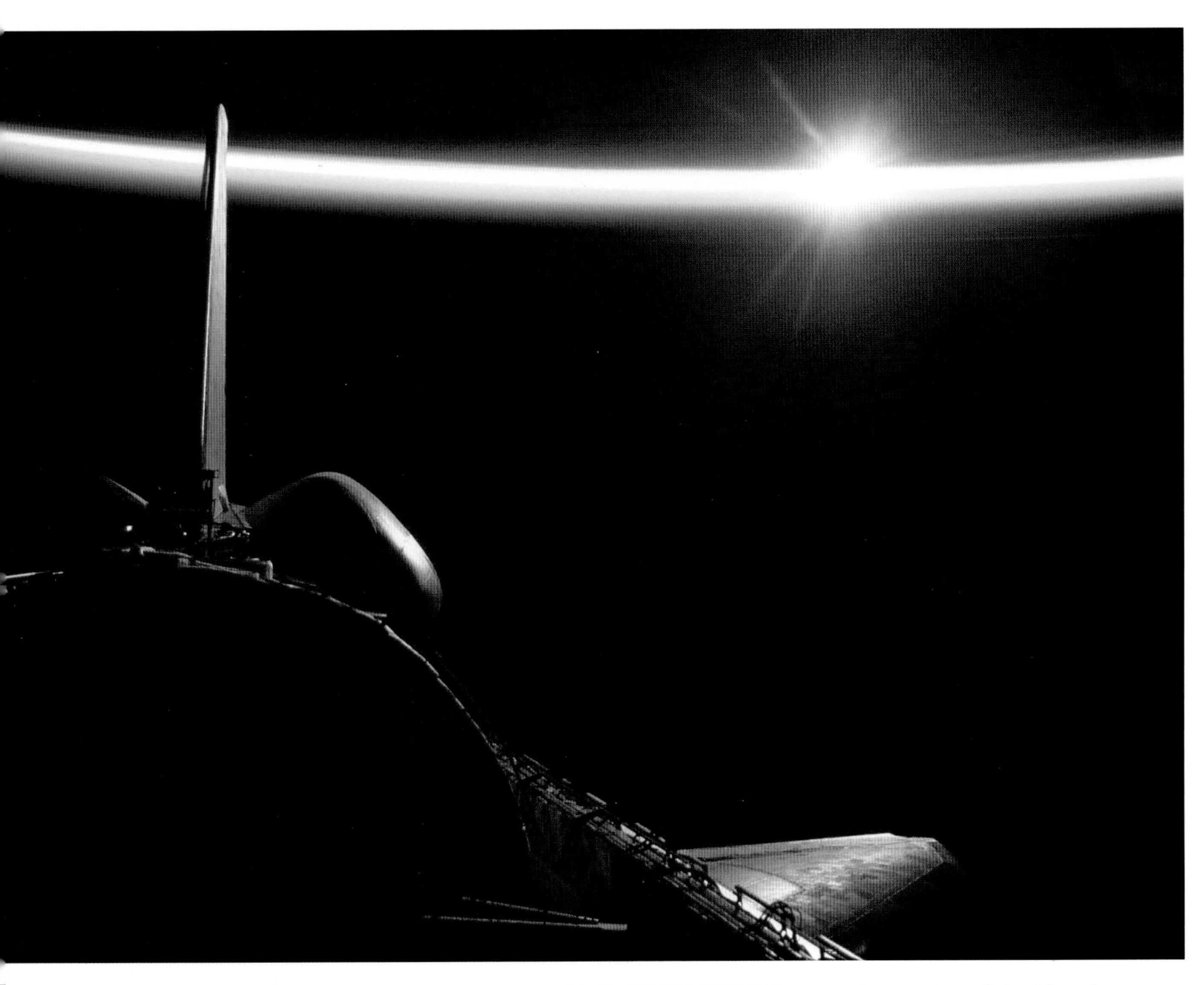

ABOVE: **Sunrise above *Columbia* and the shadowed Spacelab 1 module.**

LEFT: **Mission and payload specialists (from left) Bob Parker, Byron Lichtenberg, Owen Garriot, and Ulf Merbold in Spacelab.**

BELOW: **Spacelab 1 module nestled in *Columbia*'s payload bay.**

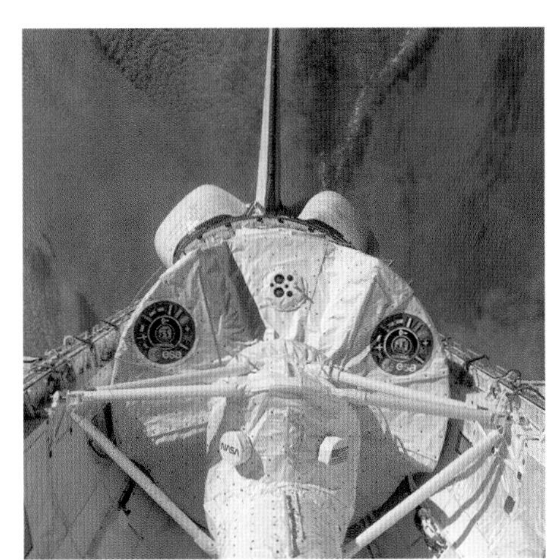

Mission no. **10**

STS

41B

Orbiter	Challenger
Launch	February 3, 1984
Landing	February 11, 1984
Duration	7 days, 23 hrs., 15 mins., 55 secs.

The STS-41B crew on the middeck: (clockwise from top left) Hoot Gibson, Vance Brand, Bob Stewart, Bruce McCandless, and Ron McNair

Crew	Vance D. Brand, Robert L. "Hoot" Gibson, Bruce McCandless II, Ronald E. McNair, Robert L. Stewart

Mission: *Challenger*'s crew deployed the WESTAR-VI and PALAPA-B2 communications satellites toward geosynchronous orbit, but both booster motor nozzles failed, stranding the pair in low Earth orbit. Astronauts tested the MMU on the first untethered spacewalks in history, creating unforgettable images of free-flying astronauts suspended against the cosmos.

Bruce McCandless, Mission Specialist 1:

Our concept for the MMU on the space shuttle was mounting the unit in a flight support rack in the orbiter's payload bay in vacuum. Bob Stewart and I would don our spacesuits and come out of the airlock, get in a foot plate in the payload bay, remove some launch-restraint bolts, check out the MMU, service it, and back our suits into it. And I emphasize that there was no connection between us and the MMU other than a mechanical latch and a seat belt—no electrical or gas connection between the pressure suit, whose integrity we wanted to preserve, and the MMU.

After four days on orbit, we came out, went through the routine, set out, and flew the unit uneventfully. We put a great deal of effort into testing it. I successfully went out about one hundred meters, which was the plan, and came back in.

When I got out away from the shuttle, I became very cold. Away from the payload bay, I wasn't getting any heat reflected back into the pressure suit. We had a big knob on the front of the suit's display and control module for temperature control. It was marked "C" on one end, with several gradations, and "H" on the other one. Naively, I thought that "H" stood for "hot," but "H" really stood for "not quite so cold." The suit was designed to support someone doing honest physical work in the space environment, and flying the MMU was a matter of using your fingertips. So my metabolic load was extremely low and I got cold.

Bob Stewart, Mission Specialist 2

I can't tell you how easy it was to fly the MMU. It was just a delight. You could thread a needle with that sucker. As Newton taught us, force equals mass times acceleration, and you controlled all the forces! It was a delightful machine; therefore NASA scrapped it. Why isn't it up there on the ISS right now?

Of course, when you were drifting out there in the MMU, you had plenty of time to look around, as long as you paid attention to keeping the MMU at the same orbital altitude as the orbiter. While I was out there looking at the orbiter and at Earth, I got to thinking, "What would it be like to be the only person in the universe?" So I turned the MMU to where I couldn't see the Earth, Moon, or Sun. I could see only the blackness of space. I lasted about fifteen seconds, and I thought, "Well, let's just turn around and make sure everything's still there." That was an interesting feeling that I did not expect.

We did the first landing at the Cape. On the approach, coming inbound toward Tampa, my thought was: "We ain't going to the Cape; we're going to end up in Portugal."

But we dug into the atmosphere coming over the Tampa coast of Florida and landed. Vance's landing was just incredibly accurate. I think we were tenths of a knot off in airspeed, and I think we were within one hundred feet (thirty meters) of the touchdown point that we'd planned three years before.

ob Stewart free-flies the MMU
94 miles (312 kilometers)
bove Earth.

Mission no. **11**

STS

41C

Orbiter	Challenger
Launch	April 6, 1984
Landing	April 13, 1984
Duration	6 days, 23 hrs., 40 mins., 7 secs.

The STS-41C crew of "Ace Satellite Repairers" on the flight deck: Dick Scobee, Pinky Nelson, Ox van Hoften, Terry Hart, and Bob Crippen

Crew	Robert L. Crippen, Francis R. "Dick" Scobee, George D. "Pinky" Nelson, James D. A. "Ox" van Hoften, Terry J. Hart

Mission: STS-41C deployed the bus-size LDEF satellite with fifty-seven materials science experiments, then rendezvoused with the Solar Max satellite. Spacewalkers Ox van Hoften and Pinky Nelson repaired the satellite's electronics using MMUs. Astronauts then returned Solar Max to orbit.

James van Hoften, Mission Specialist:

We were the eleventh shuttle flight. It was originally designated STS-13 back when they were still numbering flights. NASA, however, was upset by the number 13 and comparisons to Apollo 13 so the agency renamed all future flights and we became STS-41C. That designation meant that the flight launched in fiscal year 1984, from launch pad 1 at Kennedy Space Center, and the "C" meant that it was the third scheduled flight that year. Very confusing.

We had a great crew, with Bob Crippen as an experienced commander, and four astronauts from the large class of thirty-five selected in 1978 who became known as the Thirty-Five New Guys (TFNGs). On our mission, the TFNGs were Terry Hart on the RMS, then Pinky and me on EVA, and Dick Scobee as the rookie pilot.

What was important about STS-41C was that this would be the first shuttle flight to demonstrate the value of humans in space and the unique utility of the shuttle. Besides launching the Long Duration Exposure Facility (LDEF) satellite, we were scheduled to repair the Solar Max satellite which had lost its primary gyro system and suffered a few other technical issues. Luckily, Solar Max was the first satellite designed with the idea of on-orbit service.

The MMU was designed for just this type of work. It was propelled by compressed nitrogen and steered by forty-four reaction control nozzles. To attach the MMU astronaut to the Solar Max, we were supposed to use a device called the TPAD—the Trunion Pin Attachment Device. It was designed to dock with a grapple fixture on the satellite similar to the ones used for attachment by the RMS, the shuttle's Canadian-built robot arm. We trained over and over for this retrieval, but when we got in orbit and Pinky tried to dock with the Solar Max, the TPAD wouldn't lock on the trunnion pin. He made several attempts but only managed to bump the satellite into a tumble. Pinky did try to fly out to the satellite's solar array and manually stabilize the motion but was unsuccessful. We later were told that there was a thermal blanket, not mocked up in the simulator, which prevented the docking.

We terminated the EVA on that first day and went back inside with our tails between our legs. From inside we could see the Solar Max slowly tumbling nearby. Luckily that evening the ground controllers figured out how to stabilize the motion, and we flew close by so that Terry Hart could use the shuttle arm to grapple and place the satellite on the orbiter platform. We were then go for the EVA repairs the next day.

We went out two days after the first try and were out for about eight hours. We performed the repairs in about a quarter of the time we thought it would take. Of course our training was planned around all the things that could go wrong, but the repair went without any issues. We got the gyros on it in nothing flat and then removed and replaced the computer board.

Given the problems that we had, the fact that we were able to complete the repairs successfully made it just a spectacular flight. I recall having one selfish thought on the flight back after landing at Edwards: "You can't take this experience away from me."

ABOVE: **Ox van Hoften test flies the MMU away from Solar Max in the payload bay.**

LEFT: **The LDEF satellite grasped by the RMS, ready for release.**

BELOW: ***Challenger* touches down on Edwards Air Force Base lakebed Runway 17.**

Mission no. **12**

STS

41D

Orbiter	Discovery
Launch	August 30, 1984
Landing	September 5, 1984
Duration	6 days, 0 hrs., 56 mins., 4 secs.

The STS-41D crew: (front row) Mike Mullane, Steve Hawley, Hank Hartsfield, and Mike Coats; (back row) Charlie Walker and Judy Resnik

Crew	Henry W. "Hank" Hartsfield Jr., Michael L. Coats, Judith A. Resnik, Steven A. Hawley, Richard M. "Mike" Mullane, Charles D. Walker.

Mission: An engine problem aborted *Discovery*'s first launch attempt, sparking a nearly invisible hydrogen fire on the pad. After two months of repairs, Discovery reached orbit successfully and the crew deployed SBS-4, Syncom IV-2, and Telstar 302, then extended a 102-foot-tall (thirty-one-meter-tall) solar array for the future space station.

Steve Hawley, Mission Specialist:

No one knew the orbiter was on fire. There was still some hydrogen at the aft end of the vehicle. Until the fire spread to something they could see burning, they didn't know. Only when the heat-shield tile gap-filler fabric started burning did they realize they had a fire, but it had probably been burning since the moment of the abort.

Onboard we were talking about whether we should make a run for it. We decided we should wait and let them tell us we should evacuate. It's good we didn't as we might have inadvertently run through places where there was fire. They got the fire under control before we got out, but we got soaked running along the orbiter access arm.

When we finally launched, we were the first mission to deploy three satellites, an important milestone because it was one of the capabilities we claimed for the shuttle. One of our payloads that is underappreciated was the Office of Aerospace Science and Technology (OAST-1) demonstration. It was a mockup of a space station solar array. It didn't have real solar cells on it, just solar cell simulators to make the mass and resulting dynamics appropriate. The goal was to show whether the 102-foot (thirty-one-meter) array would behave dynamically as the models had predicted. It led to what are now the solar arrays on the ISS. In terms of long-term consequences, that was probably more important than launching the satellites.

As part of the first flight of *Discovery*, there were some changes to the fleet, including insulation changes—some of which were good and some not. They put heat-resistant blankets in places where they had had tile on *Challenger* and *Columbia*. They changed the insulation around the supply and wastewater dump nozzles. When we did a supply water dump, the heaters weren't able to protect the nozzle and the spray formed an icicle. That created a couple of problems: one, the ground was worried about the icicle being there and what it would do during reentry. The other was that we couldn't do any more water dumps, including wastewater dumps.

In those days, we didn't have a backup for the toilet. We had plastic Apollo bags but nothing else. That was a stressor for the crew: How do you get through the next three or four days without being able to use the toilet? We wanted to get rid of that icicle.

That's when ground controllers came up with the idea of using the RMS to knock the icicle off. We had to violate a flight rule, which was to not operate the RMS in a position where you can't see the end effector. But they came up with a plan where if they could monitor the position with the closed-circuit TV system, we would know the starting position and calculate the distance we could move without hitting the orbiter structure. Even though we couldn't see the end of the arm, that was safe enough to attempt. It was safer than the EVA they were planning to fix the problem. Henry was the arm operator, and after the arm maneuver, we actually saw the icicle floating away.

One of my other fondest memories was of my crewmate Judy Resnik. She was just a delight to work with and very capable. We lost Judy on *Challenger* in 1986. The day we were assigned in February 1983, the other guys all had families to go home to. She and I were single, and so we went out for dinner and a beer. We couldn't tell anybody we'd been assigned, but we celebrated together.

LEFT: **Three satellites ready for launch in *Discovery*'s payload bay on the pad.**

RIGHT: **The OAST-1 solar array towers above the payload bay.**

BELOW: **Judy Resnik sends greetings to her family from the middeck.**

Mission no. **13**

STS

41G

Orbiter	Challenger
Launch	October 5, 1984
Landing	October 13, 1984
Duration	8 days, 5 hrs., 23 mins., 38 secs.

The STS-41G crew: (front row) Jon McBride, Sally Ride, Kathy Sullivan, and Dave Leestma; (back row) Paul Scully-Power, Bob Crippen, and Marc Garneau

Crew	Robert L. Crippen, Jon A. McBride, Kathryn D. Sullivan, Sally K. Ride, David C. Leestma, Marc Garneau, Paul D. Scully-Power

Mission: *Challenger*'s crew launched the Earth Radiation Budget Satellite (ERBS), measuring the Sun's energy output, and mapped Earth at optical and radar wavelengths. Dave Leestma and Kathy Sullivan demonstrated satellite refueling during a three-and-a-half-hour spacewalk, the first by an American woman.

Kathryn D. Sullivan, Mission Specialist:

It was the first mission with two women, and I'll observe that we didn't crash *and* we met the mission objectives, so that worked out well. It was the first substantial space shuttle mission dedicated to looking back at Earth. I think the ethos of the astronaut office, embodied by Bob Crippen as commander, was: "Know your stuff. Be ready to do your stuff. You gotta deliver." We were all striving to live up to that.

On launch, I was in the right rear seat on the flight deck. The only thing you have to do in that seat is juggle checklists for the flight engineer, Mission Specialist 2, during shuttle mission simulator ascent runs. I was staring at the gauges all the way uphill because I couldn't quite believe that this was not going to be another messy ascent where we don't get to orbit. But of course we did, and eight and a half minutes into the flight, the main engines cut off, and I finally lifted my gaze from the front panel up towards the windows. I looked out over Jon McBride's shoulder and saw this gorgeous arc of Earth—we were up over the northeast Atlantic, soon to be over Britain—blue, white, and gorgeous. The view literally pulled the words out of me, "Wow! Look at that!"

Of course, right away, Crippen waved his hand over the C3 panels and said, "No, no, not yet! We're busy. We've got checklists to do." So, it was my first flight and I got blown away by the sight of Earth, and I got demerits from the commander just eight and a half minutes into the mission.

Getting an American female outside was a first. During the spacewalk, as I was making my way across the pallet, Dave Leestma was on the port sill when the pallet was, and Jon was going to shoot the scene from the cabin with the IMAX camera. Jon wasn't quite ready, so he ordered us to stop for a second, and that brief pause while he was getting the IMAX camera ready was the only real moment I had to take in Earth and where we were. I remember looking at my hands on this little instrument shelf with my feet pointed up out of the payload bay. My first thought was to chuckle that the Shuttle Imaging Radar-B (SIR-B) guys would have killed me if I'd told them before launch that I was going to do handstands on their instrument shelf amid all their precious gear. But there I was doing a handstand on their instrument shelf. Then I looked away from my hands, out level, then down towards my feet, and instantly I felt that I was hanging from a tree limb and looking down at the ground below me. We happened to be over the southern Caribbean, and I watched the very distinctive Maracaibo peninsula of Venezuela go right between my boots.

I felt as completely secure around the spacecraft as I was in my own house—I didn't need to wrack my brain to remember where the light switch or thermostat was. It was delightful to realize, "I live here. I know this place. I've got this."

ABOVE: **Kathy Sullivan (front) handstands on the SIR-B instrument shelf while Dave Leestma moves down the sill.**

BELOW LEFT: **Sullivan and Sally Ride corral clips and fasteners for a sleep restraint on the flight deck.**

BELOW RIGHT: **The crew deploys ERBS into its Earth/Sun observing orbit.**

Mission no. **14**

STS

51A

Orbiter	Discovery
Launch	November 8, 1984
Landing	November 16, 1984
Duration	7 days, 23 hrs., 44 mins., 56 secs.

STS-51A's Dave Walker, Dale Gardner, Anna Fisher, Rick Hauck, and Joe Allen hold signs celebrating their launch of two satellites and "repossession" of two more.

Crew	Frederick H. "Rick" Hauck, David M. Walker, Anna L. Fisher, Dale A. Gardner, Joseph P. Allen

Mission: After *Discovery*'s crew deployed Canada's Telesat-H (ANIK) and LEASAT-1 communications satellites, Dale Gardner and Joe Allen undertook two EVAs with MMUs to retrieve the crippled PALAPA and WESTAR satellites deployed on STS-41B. The crew secured them in the payload bay for Earth return, refurbishment, and relaunch.

Anna L. Fisher, Mission Specialist: No one had ever brought orbiting hardware back from space, especially items as large as communication satellites, and that set the stage for what we were later able to do on the ISS.

The morning of launch went smoothly, like most everything on the flight. Near launch, I thought a little bit about my husband Bill and my daughter Kristin. She was only fourteen months old. That morning I'd called to talk to them again. This was pre-cell phone, so I had to call while still at crew quarters. She said, "I-La," which at the time was her way of saying "I love you."

The crew went out there and we had the usual sort of banter. Then when the clock got inside a minute I thought, "Remind me again why I want to do this?" Then inside thirty seconds I thought, "Whatever's going to happen is going to happen. You're along for the ride."

PALAPA and WESTAR-VI were not designed to be retrieved, and we'd never handled anything that large in orbit. Rick Hauck tells the story of how Dale came into the office one day and drew on a piece of paper the idea of a "stinger," the long retrieval probe he eventually used to insert into the nozzle of an engine on the WESTAR-VI satellite. Of course, Dale didn't design it, but he came up with the idea of how to do it and then worked to develop it with the engineers.

The one thing we were most worried about was the piece of hardware that fit over the top of the satellite. The spacewalkers would put this bracket on top, over the antenna, at the opposite end from the nozzle.

After Joe had captured PALAPA using the MMU and stinger, we finally got to the part where I used the shuttle arm to maneuver the satellite down into the payload bay. Dale and Joe were trying to attach the adapter across the top when Dale said, "It doesn't fit."

We asked if he could try to make it fit, but he said, "No, it's too short." There was mechanical interference from a stud atop the satellite. Rick asked the guys how they'd feel about going to the backup plan that we'd discussed. We were all in.

We were in uncharted territory: The hardest thing was that we'd never practiced with Joe holding the satellite while we directed him how to berth it. PALAPA was surfaced with solar arrays, and if he'd damaged them, then what was the point of doing the retrieval?

So Joe had this big satellite in his face, and we were trying to give him verbal guidance as to where to maneuver it. Joe and Dale outside did the berthing from the arm and foot restraints, with us just talking to help them. When it was all over I thought, "Wow! Can't believe that really worked."

They came back inside, and I'll never forget that smell when they first came in. It's on the spacesuits, and lasts just a very short time, maybe five minutes—just after we reopen the inner hatch. It's kind of like an electrical, or ozone, smell.

When they came back in, and we had deployed the two satellites successfully, and the two retrieved satellites were sitting in the payload bay, it was just the most amazing feeling. No one had ever done that. Then we had a steak dinner! The only bad part? There was no wine.

ABOVE: **Dale Gardner, flying the MMU, closes in on WESTAR VI.**

LEFT: **Gardner, photographed by Joe Allen, jokingly offers two retrieved comsats for sale.**

Mission no. **15**

STS

51C

Orbiter	Discovery
Launch	January 24, 1985
Landing	January 27, 1985
Duration	3 days, 1 hr., 33 mins., 23 secs.

The STS-51C crew: Gary Payton, Loren Shriver, Ken Mattingly, Jim Buchli, and Ellison Onizuka

Crew	Thomas K. "Ken" Mattingly II, Loren J. Shriver, Ellison S. Onizuka, James F. Buchli, Gary E. Payton

Mission: The successful three-day Department of Defense (DOD) mission deployed a classified payload, boosted into its final orbit by an IUS. With a liftoff temperature of 54 degrees Fahrenheit (12 degrees Celsius), the SRB O-rings experienced serious charring, evidence of a joint design flaw that presaged *Challenger*'s catastrophic booster failure.

Loren Shriver, Pilot:

STS-51C was the first dedicated DOD mission. This particular mission was highly classified and still is today. NASA and DOD had to admit that the mission involved the use of an IUS, but that's about as far as I can go in discussing anything about the mission. We couldn't even take pictures inside the crew module on orbit. In contrast to the usual NASA environment where the agency wanted all the publicity they could get, we weren't able to talk about anything. To this day, my wife doesn't know what we did on that mission.

The "customer" thought it was essential that we have an all-DOD crew. Our payload specialist, Gary Payton, was part of the Manned Spaceflight Engineer Program within the Air Force.

Ascent was the most eye-opening part of the mission. We'd watch the early guys come back and listen to their debriefings. Being the macho people that they are, they'd come back and say, "That first stage is pretty dynamic." So, you'd say to yourself, "Okay, I know what dynamic is." But the shaking and the rattling caught me by surprise. I got my own definition of what "first stage is dynamic" meant. In second stage, it's the opposite: Once you get rid of the boosters, it's like electric drive.

O-ring erosion was not a hot topic in those days. After we flew the mission, I was not even aware that there was considerable O-ring erosion. After *Challenger*, we learned that STS-51C was the next-most-severe erosion of any of the missions up to that point. The temperature and conditions were almost identical to *Challenger.*

After the seven years I'd waited to get into orbit, I was so happy to be there that I was giddy with it. It was three packed days, and it helped that the main mission went off without any hitches. We thought we had a fourth day, but Mission Control started telling us on Flight Day 3 that we would be coming home really early on the fourth day.

Ellison Onizuka flew our mission and then just a year later was on *Challenger*. He was a second-generation Japanese American, and was highly respected by the Japanese population on Hawaii's Big Island, his boyhood home. He had a very austere upbringing, and that humble background went a long way toward him making so many friends. Having grown up on a coffee farm, he was very intent on all of us enjoying the Kona coffee "experience." He was a hard worker in everything he did, and very technically capable. He also had a personality that let him get along with everybody.

ABOVE: **The booster ignition vaults *Discovery* from Pad 39A.**

BELOW: **Orbital view of the entire Florida peninsula with Cape Canaveral at right center.**

Mission no. **16**

STS

51D

Orbiter	Discovery
Launch	April 12, 1985
Landing	April 19, 1985
Duration	6 days, 23 hrs., 55 mins., 23 secs.

The STS-51D crew holding their "flyswatters": Jake Garn, Jeff Hoffman, Don Williams, Rhea Seddon, Bo Bobko, Dave Griggs, and Charlie Walker

Crew	Karol J. "Bo" Bobko, Donald E. Williams, M. Rhea Seddon, Jeffrey A. Hoffman, S. David Griggs, Charles D. Walker, E. Jake Garn

Mission: *Discovery*'s crew launched two comsats, but one failed to activate. The EVA crew attached snares to the RMS to trigger the satellite's activation lever, but the satellite failed to respond. Senator Jake Garn became the first congressional space flier.

Bo Bobko, Commander:

STS-51D was important because it demonstrated, once again, the capability of people in space and the work they might be able to do there.

We had gone through the countdown and had gotten to the end of the launch window. I told the crew "Hey, I see us at the end of our window, and so it doesn't look like we're going to go today." We lay there there for a couple of minutes, and then the ground informed us that our launch window had been extended. By that time Dave Griggs had swung around in his seat and was relieving himself. He was Mission Specialist 2, between the two pilots. So I said, "Goddammit, Dave, get back in that seat! Get your straps on!"

We had trained to deploy and retrieve an astronomical observatory that the shuttle could launch and retrieve, called SPARTAN. But the spacecraft was held for a later mission that year. So we launched without any rendezvous checklists aboard. Facing a rendezvous with the faulty Syncom satellite, Mission Control had to get the information up to me. They teleprinted the rendezvous and proximity operations procedures, and Don Williams and Dave cut it all up and taped it into book form. We had practiced all those things for SPARTAN, and didn't have to start from square one when doing the rendezvous and prox ops.

Rhea Seddon and Jake Garn made what we called the "fly swatter," an improvised device that we hoped would trip an activation switch on the Syncom satellite. Mission Control engineers had built a prototype on the ground and sent the pictures and assembly procedures up to us. Once we built it, they said: if anything, ours looked better than theirs. Dave and Jeff Hoffman then went on an EVA and installed the fly swatter and another snare we called the "lacrosse stick" on the robot arm.

Rhea Seddon, Mission Specialist:

Controlling the arm, I worked hard to align the fly swatter with the switch as the Syncom rotated slowly above us. It was like trying to touch something on a merry-go-round as it swung by.

On the next pass we slowly moved the swatter into position. Three, two, one: Time to go for it! Whoosh—as predicted, the switch caught on the plastic rungs of the device and tore through, tugging the switch with the right amount of force.

We let the satellite go around again and tried to catch the switch with the wire loop on the lacrosse stick. Yes! It caught, and we could tell that it yanked pretty hard on the switch. If we succeeded, the satellite would spin up. We couldn't see the antenna on top but knew it should also pop up. Neither happened.

Bo Bobko:

Even though Syncom was still dead, we knew we had done everything we had been asked to do on the mission, and it was more than I ever had expected.

ABOVE: **Syncom activation snares strapped to the shuttle arm's end effector.**

BELOW LEFT: **Dave Griggs and Jeff Hoffman attach improvised "flyswatters" to the RMS in the payload bay.**

BELOW RIGHT: **Rhea Seddon maneuvers the shuttle arm snares close to a spinning Syncom satellite.**

Mission no. **17**

STS

51B

Orbiter	Challenger
Launch	April 29, 1985
Landing	May 6, 1985
Duration	7 days, 0 hrs., 8 mins., 46 secs.

The STS-51B crew in the Spacelab before flight: Don Lind, Bob Overmyer, Taylor Wang, Norm Thagard, Bill Thornton, Fred Gregory, and Lodewijk van den Berg

Crew	Robert F. Overmyer, Frederick D. Gregory, Don L. Lind, Norman E. Thagard, William E. Thornton, Lodewijk van den Berg, Taylor G. Wang

Mission: Spacelab 3 was the first operational flight of the ESA orbital laboratory. *Challenger*'s crew operated fifteen primary Spacelab microgravity experiments in materials processing and fluid physics. In the life sciences field, researchers studied laboratory animals and their response to orbital free fall.

Norm Thagard, Mission Specialist:

I loved it when I went on a spaceflight, because when I was a kid, I was so obsessive-compulsive that if I was assigned anything I immediately did it. As I got into puberty and began to notice girls, that attitude partially went away, but every time I went into space it came back in spades. I was not satisfied if I couldn't do the stuff I was supposed to do. I didn't feel like it was work. It was something that needed to be done and I was supposed to do it.

Our mission was the first one to fly animals on the shuttle. We had two squirrel monkeys and twenty-four rats. There was a bit of controversy because the research animal holding facility, which was supposed to channel airflow only from the crew cabin to the Spacelab into the cages, went the other way.

We didn't do anything to the animals during the flight. It was a test flight of the cage system itself. I had my suspicion preflight that we were going to have a problem because the air was clearly coming out of the cages, not going into them. There was a breeze blowing into my face standing in front of it during the ground test. My suspicion was that it had to do with the enclosure access or maybe seals on the cage openings.

Liquid and materials were leaking out of the animal cages into the crew cabin. I'm not sure if it was fecal material or food but you could see stuff coming out and it floated around. Both rat and monkey enclosures probably leaked.

There was a nationally published editorial page cartoon and it was clearly Bob Overmyer in it, because Bob had been up on the flight deck and a particle of fecal material from the cage out in the Spacelab had come all the way up on the flight deck and went by Bob's eyes. Bob's ending comment in the cartoon conversation was "No, I'm just happy we don't have elephants on board."

I didn't learn until much later that an O-ring on one of the SRB's nozzle joints failed to seal. We were just a few seconds away from having a problem with that. Crippen, who by then was the deputy director of flight crew operations, knew about it right after we flew, but I didn't.

It was a very successful flight, with folks onboard that were well trained in what they were doing. In fact, two of them, Lodewijk van den Berg and Taylor Wang, were flying their own experiments in which they had already invested years of research.

ABOVE: **Lodewijk van den Berg floats feet first into the Spacelab module.**

BELOW LEFT: **A lab-grown mercuric iodide crystal inside the Spacelab's furnace.**

BELOW RIGHT: **A newspaper cartoon highlights the leaky animal enclosures that plagued the mission.**

Mission no. **18**

STS

51G

Orbiter	Discovery
Launch	June 17, 1985
Landing	June 24, 1985
Duration	7 days, 1 hr., 38 mins., 52 secs.

The STS-51G crew on the flight deck: Salman Al-Saud, J. O. Creighton, Steve Nagel, Shannon Lucid, John Fabian, Patrick Baudry, and Dan Brandenstein

Crew	Daniel C. Brandenstein, John O. "J. O." Creighton, Shannon W. Lucid, John M. Fabian, Steven R. Nagel, Patrick Baudry, Sultan Salman Al-Saud

Mission: *Discovery*'s crew deployed the MORELOS-A, ARABSAT-A, and TELSTAR-3D comsats, then used the RMS to deploy the SPARTAN astronomy satellite. The crew then retrieved SPARTAN for reuse. Prince Sultan was the first Arab and first royal family member to fly into space.

John O. Creighton, Pilot:

The engines start to come alive at about six and a half seconds, and the whole vehicle starts rumbling and shaking. You can't believe that these big bolts are still holding you down. It feels like the shuttle's trying to rip itself off the ground.

The count goes—although you can't hear it—down to zero and they blow the bolts, ignite the SRBs, and it's a giant kick in the rear as this vehicle leaps off the ground.

We cleared the tower very rapidly. I could see a little bit of the gantry from my seat before launch, but that quickly disappeared and the vehicle whipped around, did a roll. In this first flight, it was a 28.5-degree inclination orbit, which meant that the shuttle executed a 90-degree roll off the pad and headed due east from the launch site.

I couldn't believe how fast it got dark. At an altitude of about 100,000 feet (30,480 meters), there's not enough air to scatter the light. You can still look at the bright Sun out there, but there's no surrounding air to diffuse the light. That was kind of a surprise to me.

Until you experience it for the first time, you don't really appreciate the tremendous brute force pinning you in the seat, with all the vibration on top of it. That lasts for about two minutes and eleven seconds, and then the SRBs shut down and I was sort of thrown forward in the straps.

Then the bolts that are holding the SRBs to the tank blew, and rockets of about one thousand pounds (4,448 Newtons) of thrust each ignited—four of them in the nose and four of them in the tail of each—and they totally engulfed the windscreen in flame. It sounded like World War III was going on right outside the windows as these things fired to push the solids away from the shuttle.

The design of the shuttle limits to about three g's of acceleration. That's what we experienced while the SRBs were burning, and then it dropped off to about a g and a half, and then gradually built back up over the next several minutes, until eventually we got back up to about three g's in about seven minutes.

On all three of my flights, when the main engines shut down, there was a spontaneous cheer from the crew. You get five astronauts in a room, and you'll get six opinions. But most people, at least up until the time of the *Columbia* tragedy, would agree that you just survived the most hazardous part of a spaceflight, which is the first eight and a half minutes. So, I think we all shared a collective sigh of relief

ARABSAT 1 spins out of *Discovery*'s payload bay toward geosynchronous orbit.

Mission no. **19**

STS

51F

Orbiter	Challenger
Launch	July 29, 1985
Landing	August 6, 1985
Duration	7 days, 22 hrs., 45 mins., 26 secs.

The STS-51F crew in the middeck: (clockwise from upper left) J. D. Bartoe, Gordon Fullerton, Tony England, Karl Henize, Roy Bridges, and Loren Acton, with Story Musgrave at center

Crew	C. Gordon "Gordo" Fullerton, Roy D. Bridges Jr., F. Story Musgrave, Anthony W. England, Karl G. Henize, Loren W. Acton, John-David F. Bartoe

Mission: The STS-51F crew operated Spacelab 2, a three-pallet cargo bay laboratory with experiments in life sciences, astronomy, astrophysics, technology, plasma, solar, and atmospheric physics. A faulty sensor shut down one of *Challenger*'s main engines during the ascent, forcing a tense abort-to-orbit. Yet the Spacelab 2 work achieved broad success, fulfilling all scientific and technology objectives.

Roy D. Bridges Jr., Pilot:

Spacelab 2 was the first pallet-only Spacelab mission. We operated eleven major science experiments in the payload bay from the flight deck, and two additional ones were in the crew compartment. Those experiments covered seven scientific disciplines. There were four solar physics experiments mounted on the ESA instrument pointing system, which could keep those telescopes pointed accurately enough to focus on a quarter from three-quarters of a mile (.4 to 1.2 kilometers) away.

Because we operated twenty-four hours a day for the full eight-day mission, everyone had to be extremely competent in shuttle operations and housekeeping. Of course, I did most of the front-end-of-the-orbiter work. Despite initial problems with some of the payloads and the orbiter, we were able to get all of our mission objectives accomplished.

We were the only mission to lose an engine on ascent. We lost the engine at five minutes and forty-five seconds into the flight. We missed setting a transatlantic speed record by thirty seconds—because if we'd lost it thirty seconds earlier, we would not have reached orbit, and would have had to abort to a landing at Zaragoza, Spain. That Atlantic crossing from Florida to Spain would have taken us just thirty-seven minutes. I was happy to leave that speed record for someone else.

I was looking at some overhead switches under two g's, which was what we were pulling at the time. I put my hand behind my head to make sure I could reach them under that g-load—you dor get to simulate that—when I felt the g's relax, and all the bells and whistles went off. We got the call right away, "*Challenger*, Houston, abort to orbit!" So we had to pres the abort button. Then we did a fuel dump from our Orbital Maneuvering System (OMS) to lighte the weight and give us a better chance of getting uphill.

Then a few minutes later, we g a call: "*Challenger*, Houston, limit to inhibit." We had never simulate that, ever. That's a switch on the center console, and it turns off al the safety systems in the remaining two engines. Jenny Howard, who was the Booster console operator, had noticed the same failure mode in one of the remaining two engines that had caused the shutdown on the first. She ha asked Flight to get Capcom to call us on that switch throw. She probably saved our lives.

I do remember that we were all anxiously looking at the Mach meter, because that's going to tell us whether we're ATO, abort to orbit, or AOA, abort once around—that is, after a single orbit. It turned out we had enoug velocity, although we ended up ir a lower orbit than planned.

I felt very lucky that God had given me the opportunity, among the billions down below, to fly in space. And I was lucky, too, to have survived not only the incidents going uphill, but also th many things I had to do to even become an astronaut: fighter pilo test pilot. There were lots of close calls along the way.

ABOVE: **STS-51F lifts off, the only shuttle to lose an engine during ascent.**

LEFT: **The Instrument Pointing System aims telescopes from the payload bay.**

BELOW: **Story Musgrave works on a vitamin D blood sample experiment in the middeck.**

Mission no. **20**

STS

51I

Orbiter	Discovery
Launch	August 27, 1985
Landing	September 3, 1985
Duration	7 days, 2 hrs., 17 mins., 42 secs.

The STS-51I crew on flight deck: Ox van Hoften, Bill Fisher, Joe Engle, Dick Covey, and Mike Lounge

Crew	Joe H. Engle, Richard O. Covey, James D. A. "Ox" van Hoften, John M. "Mike" Lounge, William F. "Fish" Fisher

Mission: *Discovery*'s crew deployed the ASC-1, AUSSAT-1, and Syncom IV-4 comsats, then rendezvoused with the inert Syncom IV-3, which had launched on STS-51D. Spacewalkers grasped the satellite, rewired its control systems, and later launched it into orbit.

Joe H. Engle, Commander:

We had three satellites, so the three mission specialists were each given the responsibility to launch one of those. Ox van Hoften and Fish Fisher were designated as the two who would perform the EVA to repair the satellite. Mike Lounge was the remote arm operator, and all three of them played an important and demanding role in the capture, repair, and redeployment of Syncom IV-3.

Dick Covey and I got to ride in the front seats, and I was going to get to land the vehicle, so I gave Dick the responsibility of doing the rendezvous with the Syncom. He trained and practiced in the simulators and did an absolutely superb job. When he had completed the rendezvous maneuver and had stabilized, he said something like, "Okay, boss, it's all yours."

Syncom was a huge satellite—about 15,000 pounds (6,800 kilograms) and 14 feet (4.3 meters) in diameter—and the separation-joint surface, we were warned, might have sharp edges. It was just kind of randomly tumbling in space. As we approached the satellite, it became obvious to us that the EVA grappling tool was not going to be of any use at all. Ox was going to have to just grab it with his hands. With his feet connected to the shuttle arm, he grabbed it by the edge of the sunshield to stop it and to slowly get it rotated around where he could grab these trunnion bolts and stop them right in front of him. It was a tremendous job of real-time adaptation on orbit to capture that satellite.

The piloting task was keeping Ox in position all the time to do this, by flying the orbiter and keeping him positioned. Because of a degraded RMS elbow joint, Mike Lounge had to fly the arm manually in its "single-joint" mode using just one joint at a time. He couldn't use the hand controllers to easily fly Ox around and keep him in position.

It took a little more time to orient and position the Syncom, presenting the surfaces to Ox and Fish, where they could remove panels, hotwire the avionics to the battery bus with a wiring harness that we had taken up, and make the electrical repair to the satellite.

The redeploy was not as taxing, nor as stressful, as capturing and securing Syncom, but it was a matter of positioning Ox again on the arm and letting him give some pushes on the grab bar—we called it a "towel rod"—that he had left on. He started the Syncom spinning, and we flew Ox back up and kept him close in, so that when the towel rod came by again, he could grab it and give it another push to spin it a little bit faster.

It was a phenomenal sight. As it was spinning and moving away, we flew the shuttle down around to line Ox up with the satellite so that we could get a picture of him and the satellite in the background. He gave it a salute and then the Charles Atlas bodybuilder pose.

We all worked together. That was the story of the crew the whole way through—just a totally dedicated, totally prepared, very competent crew.

ABOVE: **Bill Fisher on the starboard sill of *Discovery*, photographed by Ox van Hoften.**

BELOW: **Fisher, in a foot restraint, positions LEASAT 3 for release.**

Mission no. **21**

STS

51J

Orbiter	Atlantis
Launch	October 3, 1985
Landing	October 7, 1985
Duration	4 days, 1 hr., 44 mins., 38 secs.

The STS-51G preflight portrait: Bob Stewart, Dave Hilmers, Bo Bobko, Bill Pailes, and Ron Grabe

Crew	Karol J. "Bo" Bobko, Ronald J. Grabe, David C. Hilmers, Robert L. Stewart, William A. Pailes

Mission: This first, four-day, DOD flight of *Atlantis* carried two Defense Satellite Communication System III satellites, boosted by an IUS deployed successfully from the payload bay. The Air Force declassified the payload in September 1994.

Dave Hilmers, Mission Specialist: It was quite a memorable mission for me, not only for my first mission, but to work on the first flight of a new orbiter. The year 1985 was the busiest we ever had with the space shuttle. We flew nine missions that year, and shortly thereafter in 1986 we had the *Challenger* accident. Whether that launch pace contributed to it or not, I don't know. Certainly, in some of those missions around that time we were starting to get evidence of something going on with the SRBs and maybe with the tiles. Yet it was full speed ahead. We're not going to curtail the launch pace with these problems—instead, we'll work on them in real time. That attitude led to disaster.

We all had the confidence that NASA was always right, and if they said we were ready to launch, then we'd launch. That kind of naivete disappeared after *Challenger*. I remember telling my family that this is going to be all right. We've flown twenty times and nothing's happened, so we're going to be fine. I think that was comforting for them, but looking back on it, we should have had more respect for the vehicle and its attendant risks.

This mission gave me a view of what Earth looks like from a very high altitude—278 nautical miles (514 kilometers)—higher than the ISS is today. It was sunset over Dakar in west Africa by the time I got to look outside. This was something I'd never experienced. The beauty of the sunset was like a nuclear explosion. I turned to B and said, "Look at that!" And he said, "Well, it's just a sunset."

I don't think I'll ever live this one down. I was scheduled to cook breakfast. Maybe I was in a hurry, but I took the freeze-dried sausage the one you were supposed to rehy drate with four ounces of water, an I forgot to do that. I'll never forget Ron Grabe taking a bite out of it an just spitting it out: "Who's serving us up a cinder block?" I think I was banned after that. I got off galley duty for the rest of the mission.

Like some of the other military missions—and I was on another one—this flight is often overlooked. But it was extremely important to national security, and I think that these military missions (to my knowledge all were successful) went a long way toward making our country safe and giving us capabilities to keep us secure that we don't really appreciate. It's not that we should get special thanks for it—we were doing our job.

tlantis **powers toward orbit**
n its first flight.

Mission no. **22**

STS 61A

Orbiter	Challenger
Launch	October 30, 1985
Landing	November 6, 1985
Duration	7 days, 0 hrs., 44 mins., 51 secs.

The STS-61A crew on the middeck: (front row) Ernst Messerschmid, Wubbo Ockels, Steve Nagel, and Guy Bluford; (back row) Hank Hartsfield, Bonnie Dunbar, Jim Buchli, and Reinhard Furrer

Crew	Henry W. "Hank" Hartsfield Jr., Steven R. Nagel, James F. Buchli, Guion S. "Guy" Bluford, Bonnie J. Dunbar, Reinhard Furrer, Ernst Messerschmid, Wubbo J. Ockels

Mission: *Challenger's* seven-day flight, the orbiter's last successful mission, saw the crew operate the German-funded Spacelab D1 laboratory research mission, with seventy-five numbered experiments in microgravity science, materials science, life sciences, technology, communications, and navigation.

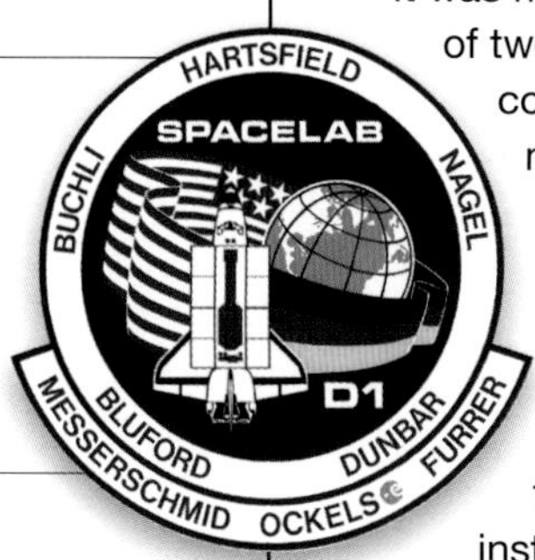

Bonnie Dunbar, Mission Specialist:

It was my first flight and the first of two dedicated German commercial flights. Germany paid roughly $60 million to the US government for the flight. It was the first time that our payload training facility was outside the United States—instead of at Marshall, it was at DLR, the German Space Agency headquarters in Bonn and Cologne, Germany. And it was the first time that the payload itself was controlled outside the United States, so the payload operations control center was actually at Oberpfaffenhofen, Germany, near Munich.

It gave me the opportunity to really become part of their team, and I can't remember a time when we weren't working together. We knew who the customer was, and my job was not to be the hard-nosed astronaut—"Do it my way"—but to make sure they had mission success, safely. My role was to make sure the Spacelab operated properly and that the payload specialists all understood that when one of the computers sounded a master alarm, they had to inform the flight deck crew. Hank said more than once, "What the hell's going on down there?"

It was a seven-day flight, with round-the-clock operations, but very successful with a lot packed in. We had about a hundred experiments altogether: We had a small centrifuge in one of the lockers, we had fruit flies, and we worked many experiments in the glove box.

To minimize shuttle disturbances to the Spacelab environment, we were actually in a gravity-gradient attitude, tail down. We were doing that for several experiment reasons—one was microgravity—and I think we also had a German navigation experiment in a Getaway Special can, a self-contained experiment canister in the payload bay. But I tell you, flying in that attitude was like being on the Starship *Enterprise*: You can imagine being on the flight deck with the overhead windows facing the horizon. You must mentally shift your legs to being toward the tail, and you're just floating there and seeing Earth. That view out the overhead windows into the velocity vector and seeing no obstructions out there—Captain Kirk had nothing on us. I remember being at the terminator and looking to my right and seeing the aurora australis over the South Pole.

There were eight people on our crew, so the training folks had this running joke about Snow White and the Seven Dwarfs. Among ourselves, I was not Snow White—it was Hank. We'd play music over the intercom into the Spacelab and Hank brought the theme music for *Snow White.* What we didn't realize was that our German colleagues had never heard that Disney music before. It was a cultural gap, but we thought it was pretty funny. I'm not sure that Hank ever knew that he was Snow White.

We developed a common culture, shared values, and a love of flight, and I think that's what made the mission successful. I think there was a bigger cultural divide in the beginning than with Russia, due to all this culture and history, and we bridged that as individuals because we had a common goal.

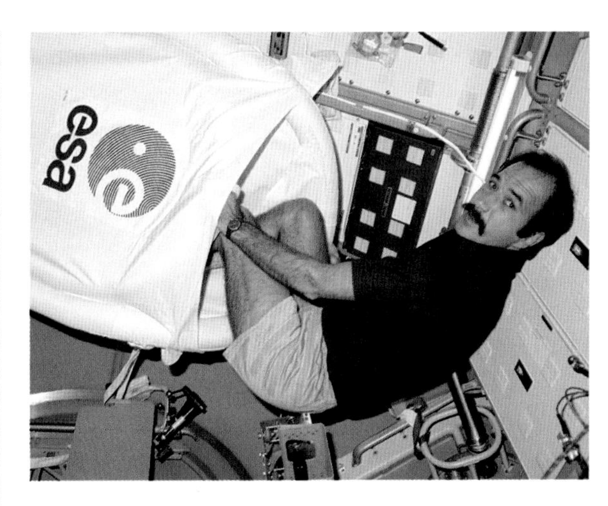

LEFT: ***Challenger*'s ascent viewed from a NASA chase aircraft.**

ABOVE: **ESA payload specialist Wubbo Ockels slips into his sleep restraint in Spacelab.**

BELOW: **Guy Bluford, Reinhard Furrer, and Ernst Messerschmid at work in Spacelab D1.**

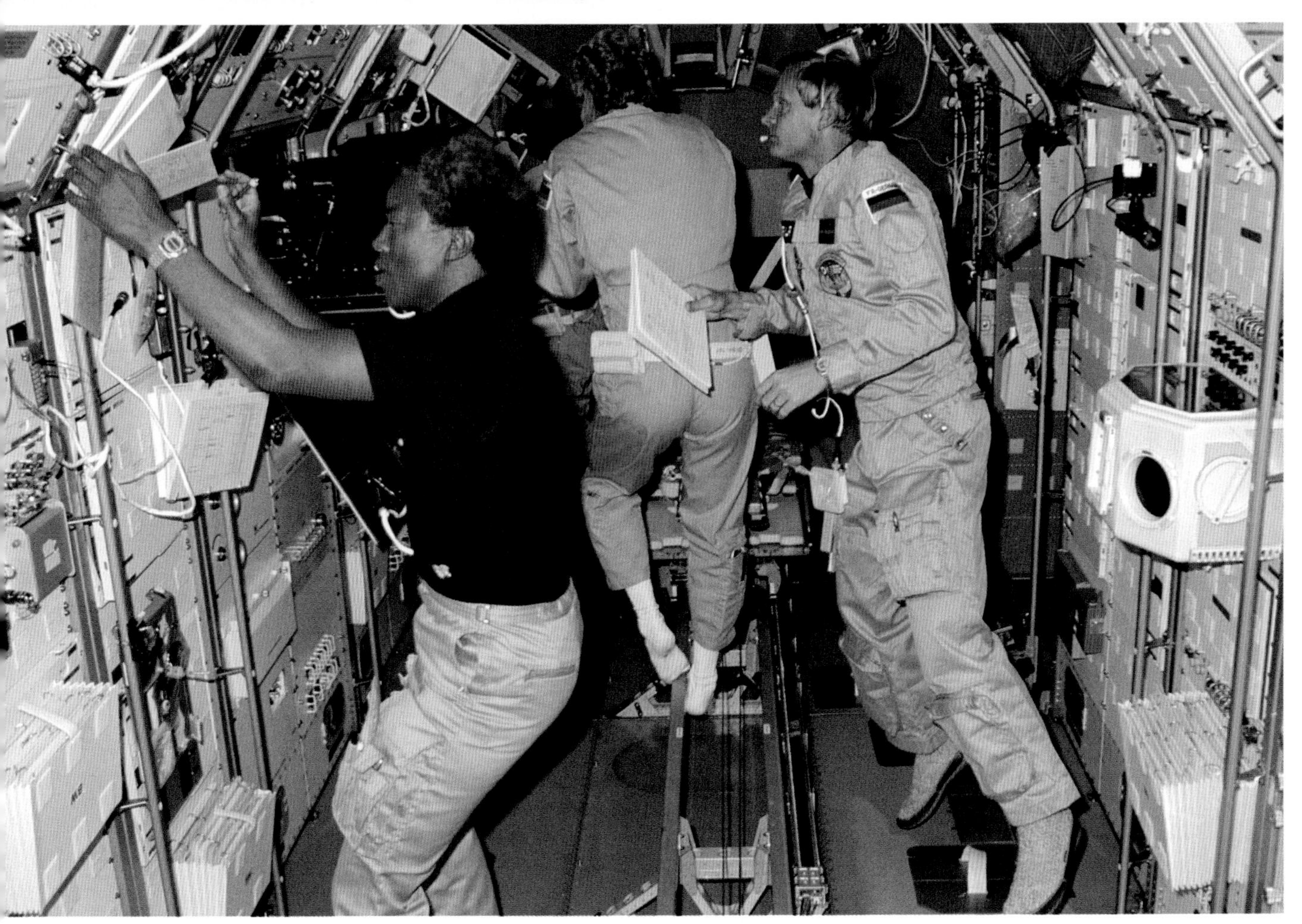

Mission no. **23**

STS

61B

Orbiter	Atlantis
Launch	November 26, 1985
Landing	December 3, 1985
Duration	6 days, 21 hrs., 4 mins., 49 secs.

The STS-61B crew on the flight deck: Jerry Ross, Charlie Walker, Brewster Shaw, Mary Cleave, Bryan O'Connor, Woody Spring, and Rodolfo Neri Vela

Crew	Brewster H. Shaw Jr., Bryan D. O'Connor, Mary L. Cleave, Sherwood C. "Woody" Spring, Jerry L. Ross, Rodolfo Neri Vela, and Charles D. Walker

Mission: The *Atlantis* crew deployed the MORELOS-B, AUSSAT-2, and Satcom KU-2 comsats into orbit, then conducted two spacewalks to test techniques for assembling large space-station-type structures in orbit. Astronauts also operated a suite of middeck and cargo bay experiments.

Woody Spring, Mission Specialist:

Except for Brewster Shaw and Charlie Walker, we were a complete rookie crew. Our paying payload was three communications satellites—MORELOS, AUSSAT, and RCA Satcom—which gave eight to ten years of communications and TV to Australia, New Zealand, Mexico, and the United States. However, I believe that the most important and longest lasting thing we did were the two EVAs. Looking to the future, we had to get the entire EVA community ready to repair the Hubble Space Telescope and build the ISS, which we thought would require ten to twelve EVAs per week.

The two spacewalks were basically a time and motion study: How hard can we make astronauts work? Jerry and I did an EVA run in the water tank every week for the six months prior to launch, and those were a really a good workout getting ready for the mission.

Our two EVAs were for building both the EVA Assembly of Space Equipment (EASE) and the Assembly Concept for Construction of Space Structures (ACCESS). EASE was an inverted tetrahedron made of six beams that were six inches (fifteen centimeters) in diameter, each ten feet (three meters) long and weighing sixty pounds (twenty-seven kilograms) on Earth. I had to hold myself in position with one hand while I maneuvered the beam with the other. Not a good way to work in orbit, but a good exercise with a recorded learning curve. We also did the same with the ACCESS structure, but for ACCESS, both of us worked out of foot restraints and pulled the struts and nodes from canisters within easy reach. We were able to build a rigid forty-five-foot-tall (fourteen-meters-tall) structure in about thirty minutes. We built those multiple times. For both these experiments, the correlation between space work and the water tank was almost perfect.

There were two times on orbit that I was momentarily afraid: One was when I was up on top of the EASE tetrahedron when night fell. It was like going to parachute jump school—all of a sudden it felt like a good idea to hold on to that structure. I remember hugging it for a second until I got my visor up, and it takes your eyes about five seconds to adjust to the darkness and the lights in the payload bay. I do remember hugging that thing, thinking, "This is not cool." Five seconds later I was fine: "OK, now we're back to cool again."

I still remember when I got a moment to just lean back in my foot restraints watching Earth go by—a very relaxing view going down the entire west coast of Africa. We launched just when the monsoons were breaking, so Africa was an absolute fireworks show. Lightning flashes all over the place, behind the clouds, on top of the clouds—what a gorgeous view. Just a quarter inch (half a centimeter) of Lexan between you and infinity.

Woody Spring (left) and Jerry Ross salute the flag on the assembled ACESS truss.

Mission no. **24**

STS

61c

Orbiter	Columbia
Launch	January 12, 1986
Landing	January 18, 1986
Duration	6 days, 2 hrs., 3 mins., 51 secs.

The STS-61C crew on *Columbia*'s middeck: (bottom) Steve Hawley, Franklin Chang-Diaz, Pinky Nelson, and Hoot Gibson; (top) Bob Cenker, Bill Nelson, and Charlie Bolden

Crew	Robert L. "Hoot" Gibson, Charles F. Bolden Jr., Franklin R. Chang-Diaz, Steven A. Hawley, George D. "Pinky" Nelson, Robert J. "Bob" Cenker, Clarence William "Bill" Nelson II

Mission: Nearly four weeks of technical and weather delays preceded *Columbia's* liftoff. The crew launched the RCA-built Satcom K1 then operated a varied complement of experiments. Representative Bill Nelson was the second legislator to fly in space.

Bob Cenker, Payload Specialist:

STS-61C was the last commercial flight of the space shuttle and deployed its last commercial payload, a communications satellite. After the *Challenger* accident—the mission right after this one—shuttle payloads were either science or government.

I never bought into the idea that the shuttle was going to act like a truck, because picking things up and dropping them off in space, due to orbital dynamics, is not like driving a truck. But you could drop off three, four, five communications satellites without any problem at all with the shuttle. Considering at the time the boom in commercial communications satellites, that was expected to really help drive that sector of the commercial world.

I've met a lot of the other astronauts since that mission, and I have a great deal of respect for all of them, but I can't imagine having worked with a better crew. Here was this army of people looking out for me. There were people at NASA who would have given their right arm to fly—engineers and so on—and they were taking care of me. I didn't get any angst or anything; they took care of me because I was part of the flight crew. It was an amazing experience.

The most satisfying moment was watching my satellite be deployed. When you work on something for two years, and you get embedded in this business, the satellite is yours. It was worth probably $100 million.

We delayed landings twice due to weather. For the first attempt, we got everything packed up, but we got stuck in space. I decided to stay up all night and look out the window. Imagine eight hours of uninterrupted window time. I'd like to stay in space long enough to find out when you get bored looking out the window.

The night after I stayed up all night, I was just wiped. I didn't even want to unroll a sleeping bag. I took some Velcro strips, hung myself on the wall, and fell asleep. At some point in the night I pulled away from the wall, and I went floating around the cabin, sound asleep. The air pulled me to wherever it was that Bill Nelson had tied up his sleeping bag. So I floated into him, and he pushed me away, and the air pushed me back. Apparently after two or three cycles of this, Bill took out the roll of gray tape and secured me to the wall. That's how I found myself the next morning, duct-taped to the wall.

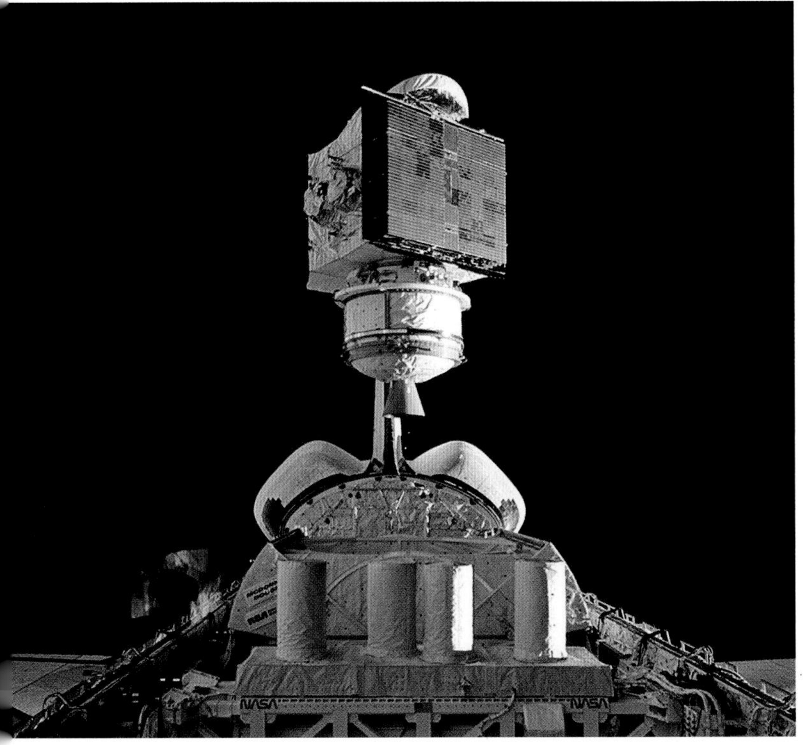

ABOVE: **Payload specialists Gerry Magilton, Christa McAuliffe, Bill Nelson, Barbara Morgan, and Bob Cenker enjoy free-fall training in NASA's KC-135 aircraft.**

LEFT: **Satcom K1 spins out of the *Columbia* payload bay.**

BELOW: **Protein crystals precipitate from solution in *Columbia*'s middeck PCAM experiment.**

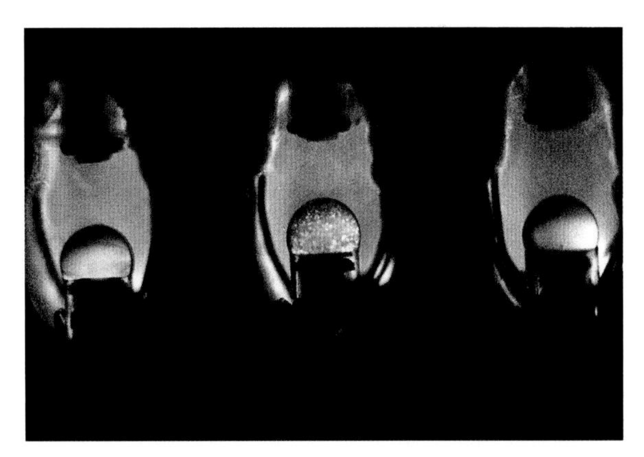

Mission no. **25**

STS

51L

Orbiter	Challenger
Launch	January 28, 1986
Landing	*Challenger* broke up seventy-three seconds into ascent.
Duration	1 minute, 13 secs.
Crew	Francis R. "Dick" Scobee, Michael J. Smith, Judith A. Resnik, Ronald E. McNair, Ellison S. Onizuka, S. Christa McAuliffe, Gregory B. Jarvis

Mission: An SRB field joint failure enabled hot exhaust gases to breach the booster casing. At seventy-three seconds into flight, at Mach 1.92 and 46,000 feet, the leak caused a structural failure that exposed *Challenger* and its crew to destructive aerodynamic forces. The ensuing orbiter breakup and ocean impact resulted in the loss of all seven crewmembers.

Author's Note: The following excerpts are from press events, interviews, speeches, and personal writings in which the STS-51L crew spoke about their hopes for the upcoming *Challenger* mission.

Dick Scobee, Commander:

Persistence is the key to reach your goal, but once you arrive at your destination, you realize the joy was in the journey along the way. We have whole planets to explore.

And if only a tiny fraction of the human race reaches out toward space, the work they do there will totally change the lives of all the billions of humans who remain on Earth.

Mike Smith, Pilot:

It's good to be flying a vehicle that we know many folks have worked very hard on. I'm one of the three people on board who will be having their first opportunity to fly, with Christa McAuliffe and Greg Jarvis. We're all looking forward to getting on orbit and getting the secret handshake.

Ellison Onizuka, Mission Specialist:

Every generation has the obligation to free men's minds for a look at new worlds—to look out from a higher plateau than the last generation. Your vision is not limited by what your eye can see, but by what your mind can imagine. Make your life count—and the world will be a better place because you tried.

Judy Resnik, Mission Specialist:

The best part of the astronaut job technically is the very well-rounded approach to science and technology. We get to do a little bit of everything in the state of the art. It's always a challenge.

Ron McNair, Mission Specialist:

The road between South Carolina and spaceflight is not a very simple one, nor is it one filled with guarantees. In fact, the only guarantees to be found are those that reside in the unchallenged depths of one's own determination. The true courage of spaceflight is not strapping into one's seat prior to liftoff. It is not sitting aboard seven million pounds (3.2 million kilograms) of fire and thunder as one rockets oneself away from the planet. But the true courage comes in enduring the preparation and believing in oneself.

Greg Jarvis, Payload Specialist:

It's a great pleasure finally to get this far. I'm very proud to be part of the program that NASA and the Hughes Aircraft Company have put together, and I'm glad to be representing Hughes as their payload specialist.

Christa McAuliffe, Payload Specialist:

Space is for everybody. It's a new world out there, a new frontier, and a lot of people who we have in our classrooms are going to be living and working in space. I feel a great responsibility and excitement that I'm representing my profession. I am hoping that this is going to elevate the teaching profession in the eyes of the public and of potential teachers out there.

When you think about the future and those people flying in space, it's going to be those kids in our classrooms. It's a wonderful thought.

The STS-51L preflight portrait: Ellison Onizuka, Mike Smith, Christa McAuliffe, Dick Scobee, Greg Jarvis, Ron McNair, and Judy Resnik

The shuttle's breakup created a cloud of burning hydrogen as the boosters spiraled away.

Astronauts Carl Meade (top) and Mark Lee test the SAFER rescue jetpack on STS-64.

Section 2

RETURN TO FLIGHT

The Space Shuttle as Science Platform

1986–1998

The loss of *Challenger* and seven astronauts shocked the nation and the world, but the shuttle program had had plenty of early warning that the faulty joint design in the SRBs could cause an accident. NASA decided to continue with launches while studying the flaw; there was no fleet-wide grounding. The night before *Challenger*'s launch, when Thiokol engineers recommended waiting for warmer conditions, NASA booster managers dismissed that advice and pushed for a liftoff. The investigating Rogers Commission not only pinpointed the faulty joint design, but also cited serious communications, cultural, and organizational failures that led to the fatal launch decision.

NASA's shuttle managers, for example, pushing to increase the flight rate to as many as twenty-four launches per year, subconsciously elevated schedule and budget concerns above flight safety. During discussions with Thiokol engineers about the wisdom of flying the SRBs at colder temperatures, agency managers exhibited a "convince me that the system is unsafe to fly" attitude, instead of the proven, conservative posture of "prove to us that the system is safe and ready for launch."

While NASA dealt with its institutional deficiencies, it moved swiftly to correct the SRB joint design. It adopted Thiokol's recommendation to add a third O-ring to keep hot gas from reaching the joint, and a metal capture feature to keep the joint from opening under the bending forces caused by motor ignition.

Congress approved construction of a replacement orbiter using structural spares manufactured during the initial shuttle build. *Endeavour* would incorporate upgraded computers, many avionics upgrades, and structural modifications suggested by shuttle flight experience. The new orbiter also packed a forty-foot (twelve-meter) drag chute to reduce loads on its new carbon-carbon brakes, along with main engine and propellant line improvements.

Astronauts assisted in the design and testing of a partial escape system, giving the crew a limited bail-out option. If a gliding orbiter were unable to reach a runway, the crew would engage the autopilot, jettison the side hatch, and extend a telescoping steel pole into the slipstream. Each astronaut in turn would attach their parachute harness to the pole, then roll through the open hatch, with the pole guiding them safely below the orbiter left wing. Importantly, the escape system wouldn't work during powered ascent or at altitudes above 40,000 feet (12,192 meters). By September 1988, the return-to-flight orbiter *Discovery* was ready on the pad, equipped with nearly all of the new modifications and upgrades.

The cost of the accident and recovery was gut-wrenching $2 billion (almost $5 billion in today's dollars) just to build *Endeavour.* The added expense of payload delays, the restart of expendable rocket production lines to lift commercial satellites—henceforth banned from the shuttle—and purchasing the rockets and adapting grounded spacecraft to fly on them approached nearly $12 billion by the time of *Endeavour's* 1992 maiden flight.

The shuttle would return to space, but only as a launch vehicle for scientific spacecraft and a science platform for Spacelab and payload bay experiments. Commercial payloads and most Pentagon spacecraft would shift to expendables like Titan, Atlas, and Delta, avoiding any risk to humans except where astronaut skill and intervention were justified by scientific return. Another focus was developing science operations for the future space station.

First priority for launch aboard the shuttle went to the backlog of TDRSs for communication between spacecraft and ground stations, classified Department of Defense payloads, planetary probes, and the Hubble Space Telescope—all delayed by the stand-down after the *Challenger* tragedy. The high-energy Centaur upper stage, meant to loft heavy military and planetary spacecraft and fueled by volatile, cryogenic hydrogen and oxygen, was correctly deemed too dangerous to launch on the shuttle (astronauts had termed the "Death Star"). Its cancellation forced the *Galileo* planetary spacecraft to use the reliable but less-powerful Inertial Upper Stage; its new gravity-assisted trajectory took six years to reach Jupiter instead of the two that a Centaur boost would have offered.

Returning to service in 1988, shuttle crews deployed high-value military spacecraft and a variety of Earth observation and space science satellites, many of them retrievable for subsequent missions. Microgravity research craft were placed in orbit, then retrieved for sample analysis back on Earth. The orbiters hosted the European Space Agency Spacelab module and later the commercially built SPACEHAB, full of biomedical, microgravity, and space technology experiments. The shuttle itself served as an observing platform for integrated suites of Earth science or astronomical instruments, like Space Radar Lab and the ASTRO telescope cluster.

Orbiter *Atlantis* deployed the Hubble Space Telescope in 1990, but its main mirror was flawed during manufacture, embarrassing NASA. On STS-61 in 1993, astronauts installed corrective optics to restore Hubble's performance, beginning a series of servicing and upgrade missions that kept the observatory at astronomy's cutting edge.

Jay Apt captured crewmates Curt Brown (upside down), Jan Davis, Mamoru Mohri, Mark Lee, Hoot Gibson, and Mae Jemison troubleshooting an experiment in Spacelab-J.

To maximize experiment and observing time in orbit, ASA pushed the shuttle mission duration from about a eek to longer stays of two weeks or more. Providing the ectricity and life support for longer flights required extra xygen and hydrogen tanks, a regenerative CO_2 scrub- ing system, and more cabin stowage for equipment and ood. Worried that pilot flying skills might degrade after everal weeks in free fall, the program briefly considered ght-testing an orbiter "autoland" system, but decided the sks of a low-altitude component failure outweighed any cremental loss of pilot capability. Shuttle commanders efreshed their flying skills in orbit by using a laptop-based pproach-and-landing simulator.

Although the post-*Challenger* flight rate never qualed 1985's nine missions, the 1990s program amped up to a steady six to eight missions annually, flying about forty astronauts every year. With incorpo- ration of post-Soviet Russia as a major partner in the ISS program in 1993, the shuttle's job jar soon included flights to the Mir space station as Phase I of ISS. These missions delivered and returned American astronauts to and from Mir for months-long stays and led directly to the first ISS assembly flights, beginning with STS-88 in late 1998.

This impressive yet intimidating assembly effort would soon fill the entire shuttle launch manifest, save for a few space research and Hubble servicing missions. The space shuttle in its final decade would ultimately build the Space Station envisioned at its inception. But well before Station completion, many of the same leadership, institutional, and engineering flaws that led to *Challenger* reappeared to threaten another shuttle with catastrophe.

Mission no. **26**

STS

26R

Orbiter	Discovery
Launch	September 29, 1988
Landing	October 3, 1988
Duration	4 days, 1 hr., 0 mins., 11 secs.

The STS-26 "Loud and Proud" crew on the middeck: (front) Mike Lounge and Dick Covey; (back) Dave Hilmers and Pinky Nelson; (center) Rick Hauck

Crew	Frederick H. "Rick" Hauck, Richard O. Covey, John M. "Mike" Lounge, George D. "Pinky" Nelson, David C. Hilmers

Mission: STS-26R (for Return to Flight) verified the engineering changes made to the shuttle after *Challenger* and launched the geosynchronous TDRS-C satellite to replace one lost on STS-51L. The crew ran the OASIS study of hydrodynamic flow in free fall.

Rick Hauck, Commander:

What made this mission important and memorable was the personal aspect of recovering from tragedy. The families of our friends who were killed on the *Challenger* were on our minds.

We were focused on a clean launch, orbit, and return, with a relatively simple timeline, because it was generally conceded that if we had a major problem with STS-26, it could spell the end of the human spaceflight program.

I think most of us in the office had a sense of confidence in NASA. We believed they really knew how to do this business of spaceflight. So you say, "Well, I don't really have to worry about that. I have to worry about whether I've trained well enough so that I don't make mistakes." But honestly, I did not worry about going into space with these crewmembers. I told my family that this flight will be the safest flight that NASA has flown to date. What I did not say to them, of course, was that it was guaranteed that I was coming home, because you just can't have that guarantee.

I went into orbit with a refreshed, renewed sense of wonder at what I saw outside and inside. This is the most complex machine made by human beings, and as we accelerated through Mach 16 on the way uphill, I said to myself, "I hope it doesn't blow up." I had that feeling, then I thought, "That's not productive. You should be looking at the gauges." So I just tossed it out of my thinking.

We had a great crew. Everyone had a sense of humor. That's a management style, I think—letting people inject humor as long as it's backed up by professionalism. Dick Covey was a low-key guy and Mike Lounge was our representative following the TDRS. He was like a bulldog; once he got his teeth into a subject, he'd pull and pull on it. Every now and then I'd have to tell him, "They hear you. Just dial it down a notch." And Dave Hilmers was the class genius, who could memorize procedures with such facility that I had to ask him at one point, "Don't use your memory, use the printed pad." That worked from then on.

We decided that each one of us would compose and read a section of a tribute to the *Challenger* crew. So all those words were spoken individually, and they were all from the heart. Without exception, the families thanked us for what we did there. I said, "Today, up here where the blue sky turns to black, we can say at long last to Dick, Mike, Judy, to Ron and El, and to Christa and Greg: Dear friends, we have resumed the journey that we promised to continue for you. Dear friends, your loss has meant that we could confidently begin anew. Dear friends, your spirit and your dream are still alive in our hearts."

ABOVE: ***Discovery* lifts off from Pad 39A on its Return to Flight mission on September 29, 1988.**

BELOW LEFT: **The TDRS/IUS stack moves away from *Discovery* on its way to geosynchronous orbit.**

BELOW RIGHT: ***Discovery*'s tail backlit by an orbital sunrise.**

Mission no. **27**

STS 27R

Orbiter	Atlantis
Launch	December 2, 1988
Landing	December 6, 1988
Duration	4 days, 9 hrs., 5 mins., 37 secs.

The STS-27 crew on the flight deck: Hoot Gibson, Mike Mullane, Jerry Ross, Bill Shepherd, and Guy Gardner. The football shown here was later presented to the NFL during the 1989 Super Bowl.

Crew	Robert L. "Hoot" Gibson, Guy S. Gardner, Richard M. "Mike" Mullane, Jerry L. Ross, William M. Shepherd

Mission: This DOD mission's payload was the Lacrosse 1 radar surveillance satellite; its sponsors were the US National Reconnaissance Office and the CIA. Debris shed during ascent from the right SRB nose cap damaged hundreds of heat shield tiles. Beneath one tile, the exposed orbiter skin nearly melted through.

Hoot Gibson, Commander:

We didn't know it, but during the launch, something broke loose. The nose cap ablative coating of the right-hand SRB disintegrated during launch, and it showered the right wing of *Atlantis* with debris. The day after we got to space, Mission Control called: "Guys, we'd like you to take the shuttle's robot arm and hang it over the side of the orbiter and look at your right wing, and tell us what you see."

I'll never forget—we looked at the right wing, and when we first brought up the arm's end effector camera, I said to myself, "We are going to die." I was looking at more than seven hundred shredded tiles on the right wing of *Atlantis*.

Guy Gardner, Pilot:

They had us take the arm and look under the belly. You could see the little white streaks of the damage on the black tiles. I didn't even know Hoot was feeling that concerned—that is a compliment to Hoot. Ground told us it wasn't any concern.

Hoot as commander was phenomenal in his knowledge of shuttle ascent and reentry. I could tell during reentry he was closely watching these small differences from what we'd seen in the simulation, because of the tile damage.

But as for the rest of the crew, there was no doom or gloom. It wasn't like Apollo 13 coming back from the Moon and wondering if you're going to make it or not. I was enjoying the ride, and not worrying about tiles. I think what made this a successful mission was the dedication and the teamwork. All of those thousands of people across the country—their efforts came together to result in successful spaceflight.

But even more sensual, emotional, and impressive than being weightless is looking down at Earth from low Earth orbit. The diversity of our planet is awesome. Oceans have different colors. Mountains look different. If you look down at the Andes Mountains, it's the Andes. It's not the Rockies, not the Himalayas. Deserts look unique in their coloring and texture. You look at places around the planet and you say, "Oh! I know where I am—those are those funny gumdrop mountains in south China. Those are the Galapagos Islands."

Hoot Gibson, Commander:

When we landed at Edwards and got out of the orbiter, there was already a crowd of people gathered outside *Atlantis*, looking at our right wing, looking at all the shredded tiles. Even more significant was the one entirely missing tile which probably burned up during reentry, and the melted metal that we had on the surface of the orbiter. We were fortunate because there was a large steel plate in that area, and the steel plate lasted a lot longer in the heating region than aluminum would have. The plasma was working on the steel when we successfully made it through the reentry heating region.

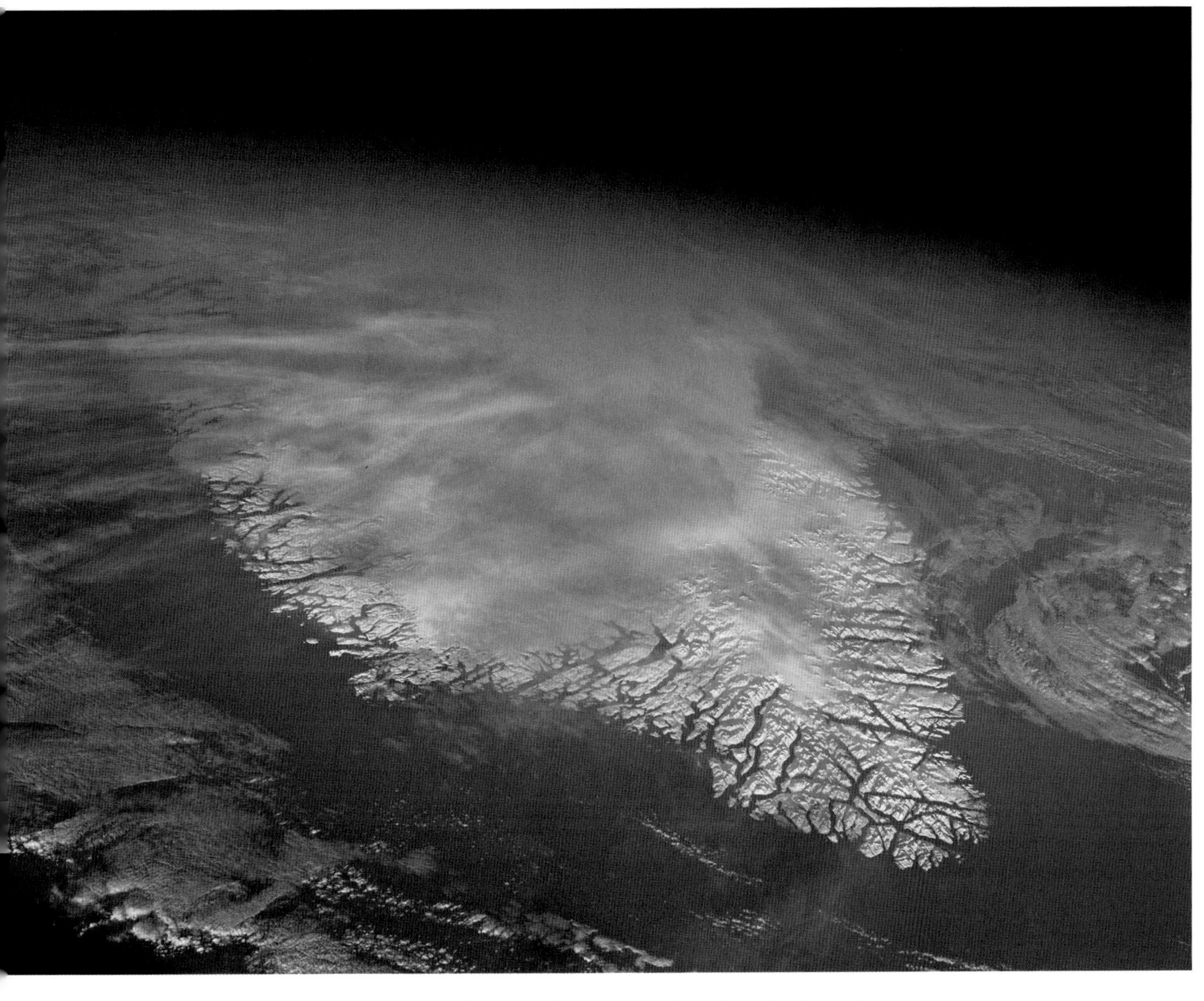

ABOVE: ***Atlantis*'s high-inclination orbit gave the crew this view of southern Greenland's icy fjords.**

BELOW LEFT: ***Atlantis* rolls to a stop on the lakebed Runway 17 at Edwards Air Force Base, California.**

BELOW RIGHT: **Reentry heating partially melted the orbiter's skin, which was left exposed by a missing tile.**

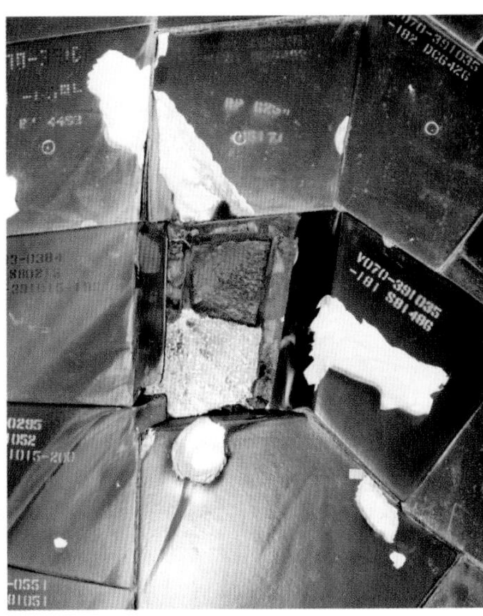

Mission no. **28**

STS 29R

Orbiter	Discovery
Launch	March 13, 1989
Landing	March 18, 1989
Duration	4 days, 23 hrs., 38 mins., 50 secs.

The STS-29 crew on the simulator flight deck: (front) John Blaha and Mike Coats; (middle) Jim Bagian and Jim Buchli; (back) Bob Springer

Crew	Michael L. Coats, John E. Blaha, James P. Bagian, James F. Buchli, Robert C. Springer

Mission: *Discovery's* crew deployed the TDRS-4 relay satellite atop its IUS booster. Within a day of liftoff, TDRS arrived successfully in geosynchronous orbit. Astronauts operated protein crystal growth and plant cell division experiments in the middeck and a potential space station cooling system in the cargo bay.

Mike Coats, Commander:

This flight deployed one in a series of shuttle-launched TDRS. STS-29 was short, sweet, and successful. We deployed the TDRS on the first day, and other than a small cryogenic reactant failure that led to a minor power down, we had no significant failures to cope with.

This was a fun crew to fly with. We had two Marine colonels, Jim Buchli and Bob Springer, who enjoyed giving the Air Force colonel, John Blaha, a hard time. John did an amazing job, but his incredible ability to focus on the task and be oblivious to his surroundings got him a steady ribbing. Mission Specialists Jim and Bob had to occasionally remind Pilot John that "MS" stood for "mission specialist," not "manservant."

Jim Buchli was on his third flight and that extensive experience had him ahead of all of us. I'd mention to him that he was cleared to perform the next item on the crew activity plan, and he'd usually reply, "Yeah, I just finished that."

Jim Bagian kept us entertained through those months of training and long hours in the sim with stories from his medical career, especially tales from his service in the emergency room at a Philadelphia hospital. His experiences there made us realize that being an astronaut was a comparatively easy job.

Bob Springer had big responsibilities in deploying the TDRS right away on Flight Day 1. Bob was right at home immediately and acted like a veteran, even though this was his first spaceflight. Watching the TDRS float off majestically was a huge relief since it was the primary purpose of the mission.

One neat thing on the mission was the chance to maneuver the orbiter to help our ailing heat pipe experiment, which was meant to demonstrate a cooling technology of potential use to the upcoming space station. We were asked to tumble the orbiter—rotate it in pitch, nose over tail—to assist the fluid transfer inside the long heat pipe tubing in the payload bay. I would hold a constant pitch rate, about 2 degrees per second, as we flipped the orbiter end over end. That was a rare opportunity to manually fly the shuttle on orbit in a way that few crews had experienced.

Because of the spectacular views that tumbling afforded out the cabin windows, the IMAX film team at the last minute asked us to film Earth scenes while we had this unique vantage point. One horizon-to-horizon scene captured as we tumbled for five to ten minutes made it into the opening shot of the film *Blue Planet*.

At a White House dinner after the flight, I was seated with President George H. W. Bush. I'd made the comment—this was the third flight after *Challenger*—how wonderful it was for the crew and families to have a tour of the White House after all the tension and with all the families being under a lot of stress. I said it was a shame every crew couldn't tour the White House with their spouses after their mission. The president asked "Well, why can't they?" All I could think of to say was, "Well, you're the president, sir." He smiled and said, "Why, yes I am." During his four years as president every crew got to tour the White House.

ABOVE: ***Discovery*** **departs Pad 39B, as seen from the Merritt Island shore at Kennedy Space Center, Florida.**

BELOW LEFT: **TDRS 4 and its IUS await deployment from** ***Discovery*****'s payload bay.**

BELOW RIGHT: ***Discovery*** **nears touchdown on Runway 22 at Edwards Air Force Base, California.**

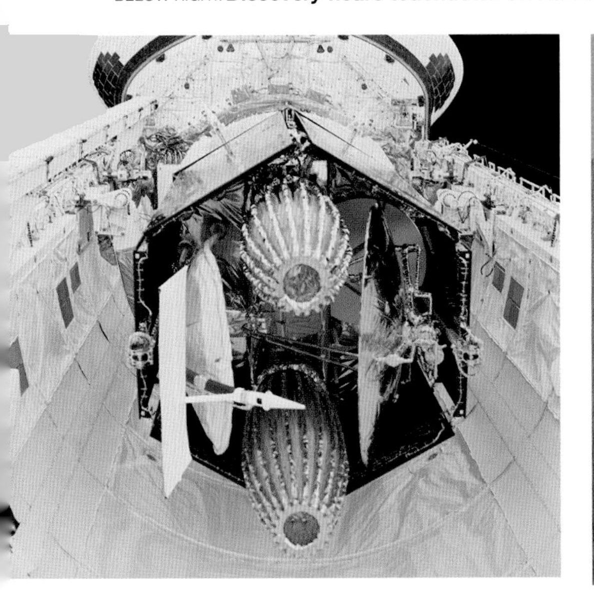

Mission no. **29**

STS

30R

Orbiter	Atlantis
Launch	May 4, 1989
Landing	May 8, 1989
Duration	4 days, 0 hrs., 56 mins., 27 secs.

The STS-30 crew on the flight deck: Norm Thagard, Ron Grabe, Mary Cleave, Dave Walker, and Mark Lee

Crew	David M. Walker, Ronald J. Grabe, Norman E. Thagard, Mary L. Cleave, Mark C. Lee.

Mission: On Flight Day 1, the crew deployed the Magellan radar mapper atop its IUS, which then propelled the spacecraft toward orbit around Venus. Magellan obtained high-resolution radar mapping of Venus's cloud-shrouded surface, imaging 98 percent of the planet by mission end in October 1994.

Mary Cleave, Mission Specialist:

Magellan was the first planetary probe to be deployed from the shuttle. It was big, and we had a lot of work to do at the Jet Propulsion Laboratory (JPL) in California. Mark Lee and I were the deploy team, so we spent a lot of time out at JPL.

This was my second launch. My only instrument was an altimeter, right in front of me on the forward lockers, so I'd know when to deploy the side-hatch escape pole if we had to bail out. I was down on the middeck all by myself, hootin' and hollerin'. When you're not working, it's a great ride!

I remember right after launch turning heads down, shaking during the first two minutes and ten seconds from the solids, and then the solids coming off and the smooth ride all the way up. Then the main engines cut off and I felt like I was upside down! I was fighting with myself. My job is to be the first one out of the seat to get everybody set up with communications. I had to fight with myself to even undo my seatbelt. Even though I'd been in space before, I knew I was going to fall on my head. It was the funniest thing. When I undid my seatbelt, I didn't fall on my head, so it was okay.

Magellan got our attention because it was a big sucker and once deployed, it came right over the cabin. We pulled away a bit and waited for the solar arrays to deploy, and made sure it was oka before we fully pulled back.

We were told we wouldn't be able to see Venus—it was going to be setting too fast. But we spotted it near the horizon and got all excited, because that was Magellan's destination. I got my "Wow!" moments out of looking up, not looking down. There's too much going on out there—we're not alone.

In terms of the crew, I knew Red Flash—that's what we called Dave Walker—well as he had rented the house next door to me. Mark Lee was great, we'd fly all over the place together because as an Air Force pilot he was qualified to fly i the T-38 front seat. So going out to JPL worked well and I enjoyed flying with him. Normie Thagard was a real character, and it's nice to have a doc on a mission. It's hand if you need it. Ron Grabe was goo to fly with; I got along with him as we grew up very close, geographically, on Long Island.

The most satisfying part for me was returning to flight and having a clean mission. Quietly I was working on a lot of stuff as I was the first woman to fly after the accident. It took effort to control that in the press because I didn't want to make a big deal out of it. Every time somebody would say something, I'd tamp it down: "A lc of women have flown and we're going to keep flying." Boom—that's it.

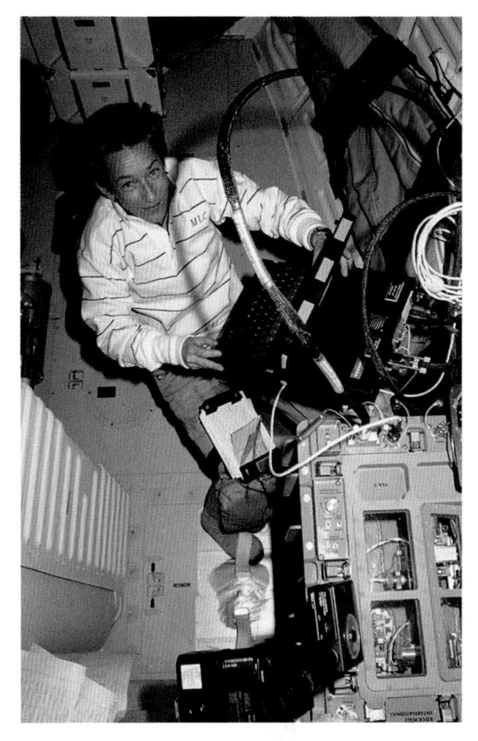

LEFT: **Magellan and its Venus-scanning radar dish are ready for deployment atop the IUS.**

ABOVE: **On the *Atlantis* starboard middeck, Mary Cleave operates the Fluids Experiment Apparatus.**

BELOW: **Magellan's target, Venus, hovers just above the atmosphere after sunset.**

Mission no. **30**

STS

28R

Orbiter	Columbia
Launch	August 8, 1989
Landing	August 13, 1989
Duration	5 days, 1 hr., 0 mins., 9 secs.

The STS-28 crew on the middeck: (clockwise from left) Brewster Shaw, Jim Adamson, Dave Leestma, Mark Brown, and Dick Richards

Crew	Brewster H. Shaw Jr., Richard N. Richards, James C. Adamson, David C. Leestma, Mark N. Brown

Mission: *Columbia's* DOD payload is still classified. A middeck locker carried a human skull studded with dosimeters to record radiation dosage at orbital altitude. During reentry, an infrared camera mounted on the tip of the tail measured temperatures on the orbiter's port wing and tail.

Jim Adamson, Mission Specialist:

When *Challenger* happened, about half a dozen payloads that could only go on the shuttle had accumulated and stayed grounded during the suspension of operations post-accident. After we started up again in 1988, NASA was trying to fly off these satellites that could not go out another way. All these DOD flights were part of a process to launch payloads that the country really wanted to get into space.

We had two national security assets that were among the most significant of any in the Cold War, and yet to this day we're not free to talk about them. It was truly amazing technology for that time. Today we've advanced way beyond what was possible then, but for those times it was just incredible what those things could see and do in orbit. I was gobsmacked at the capability that we had in our country that I never knew about.

Our crew was named to fly right after *Challenger*—we were originally scheduled for March 1986. All of us were military, but I was the only Army guy. Brewster Shaw and Mark Brown were Air Force, and Dave Leestma and Dick Richards were Navy. We had an interesting crew because we were all from similar backgrounds and were together a long time, so by the time we were ready to fly we felt like five fingers on the same hand. We were the closest crew I've ever been a part of and even observed at NASA.

The secondary job of this mission was a whole list of military objectives aimed at Earth observation. This turned out to be one of the most satisfying parts of the mission for all of us, because it was actually our job to stare at Earth and take a lot of photographs. We had some impressive DOD optics tools onboard to play with.

It was a very impressive first look at Earth for me. You can see lakes, rivers, and mountains really well, and long straight lines. You can spot the cities because they'r gray. We got to see Alaska and th Aleutian chain, Vancouver, and th Canadian Rockies. August was a great month to fly, because of all the months it was relatively cloud-free. Places on the Earth that are ordinarily hard to photograph, like Moscow, we could get pictures of with no clouds. There are a few areas on Earth that you almost never catch cloud-free, like the Panama Canal. My favorite photo turned out to be the Malaspina Glacier in Alaska.

I think the question of what value humans afford in space is still one that we don't have a grea answer for. I think there is, though some power that human beings have to process information in real time, and to collect valuable knowledge and act on it. I think that's why humans ought to explore, to participate in going to th Moon, Mars, and beyond.

ABOVE: ***Columbia*'s crew photographed Alaska's tidewater Malaspina Glacier on August 10, 1989.**

BELOW: **During reentry, a tail-mounted infrared camera imaged the hottest areas (red) of *Columbia*'s left wing.**

RIGHT: **Jim Adamson used a Hasselblad 250mm telephoto lens to photograph Earth from the flight deck.**

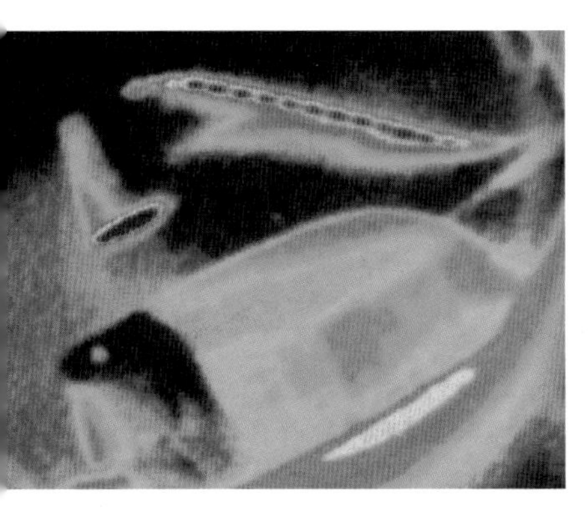

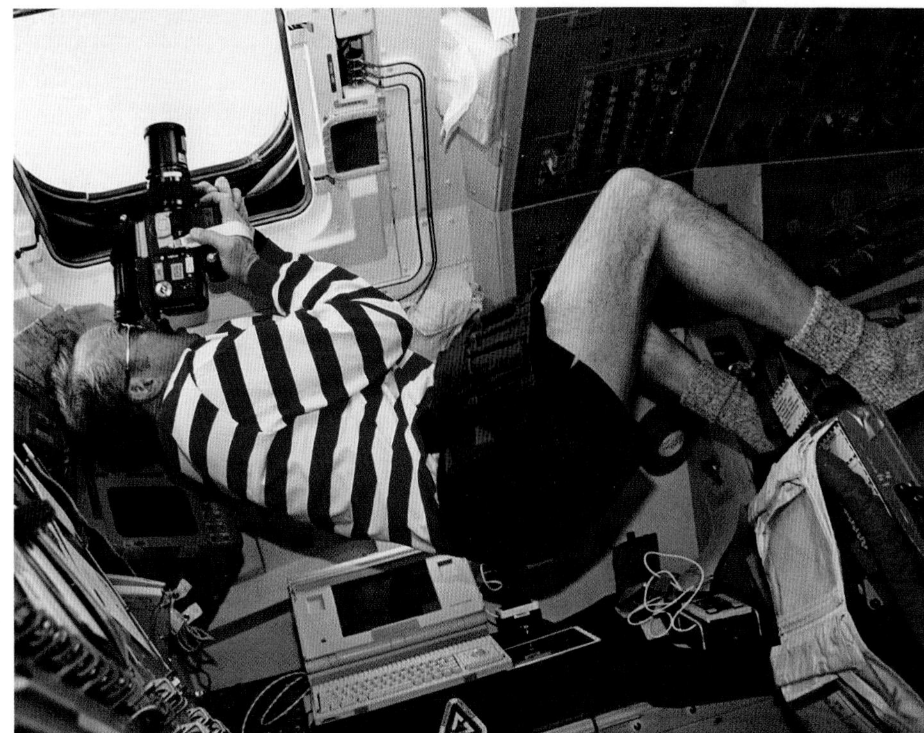

Mission no. **31**

STS

34R

Orbiter	Atlantis
Launch	October 18, 1989
Landing	October 23, 1989
Duration	4 days, 23 hrs., 39 mins., 20 secs.

The STS-34 crew on the flight deck: Don Williams, Ellen Baker, Franklin Chang-Diaz, Shannon Lucid, and Mike McCulley

Crew	Donald E. Williams, Michael J. McCulley, Franklin R. Chang-Diaz, Shannon W. Lucid, Ellen S. Baker

Mission: The *Atlantis* crew deployed the Galileo Jupiter probe atop its IUS six hours and thirty minutes into the flight. The probe reached Jupiter on December 7, 1995, beginning a nearly eight-year exploration of its atmosphere and moons. Galileo was the first spacecraft to orbit an outer planet.

Mike McCulley, Pilot:

The mission was fourteen years long, from the time we started until the time the Jupiter probe Galileo ran out of fuel, reentered, and burned up. It was successful in so many ways, and with so many new discoveries, it was the perfect mission to be part of as a one-time flier.

The first stage was far more violent—the rumbling, rocking, and rolling—than I expected. By that point, I had 400 carrier catapult launches, so the thrust, power, and energy weren't a surprise, but it shook a lot more than I'd anticipated. Then when the SRBs separated it was like a bomb going off in front of your window, rather than this little flash cube.

Getting the crew together was extremely satisfying; we trained and launched together. Of course, in the simulators you die all the time, but on launch day you just go fly. We were a few seconds into the launch when the master alarm went off, and we had a runaway, overspeeding auxiliary power unit (APU) turbine. "Bang, bang, bang, bang," and we looked down at the computer display on the front instrument panel and verified that we had an "APU Speed High." Don Williams and I just walked through that. I went over to the side to the switches, we verified it on the computer screen, and just worked our way through the procedure and shut the APU down. Then we looked at each other and realized we'd never said a word to Houston. It was one of those "Uh oh" moments.

We got back to the debrief, and were going through what kind of emergencies we'd had, and the APU came up. They properly chewed our asses out and really gave us a hard time about not calling, not checking in, not doing everything. We just put our tails between our legs and said, "Yes, sir." They did their job by yelling at us, but then they stopped and said, "However." There was a long pause. "It was a real, high-speed runaway, and if you hadn't done that, it was ready to blow up." Afterwards they made procedural changes for some of those situations that really needed immediate attention, where you didn't want to spend any time calling Houston.

Once in orbit, Galileo deployed right on time. I was in the pilot's seat for deployment, Franklin Chang-Diaz was running an IMAX camera out the back during the operation, and Ellen Baker and Shannon Lucid were working the deployment procedure. Like most crews, we just practiced and practiced, and so we just walked right through it.

As all of us know when we fly, you're the tip of the spear with thousands of people behind you. The JPL folks were just absolutely delightful. They had been working on the probe, they'd gone through two or three different launch and trajectory scenarios. We had a wonderful time working with their program manager and their engineers.

LEFT: **Galileo, atop its IUS, leaves its launch support ring on its glide out of *Atlantis*'s payload bay.**

ABOVE: **At the southern limit of *Atlantis*'s ground track, the aurora australis glows green and red.**

BELOW: **Next step on its long journey to Jupiter: Galileo, with its IUS, separates from *Atlantis* before an initial engine firing.**

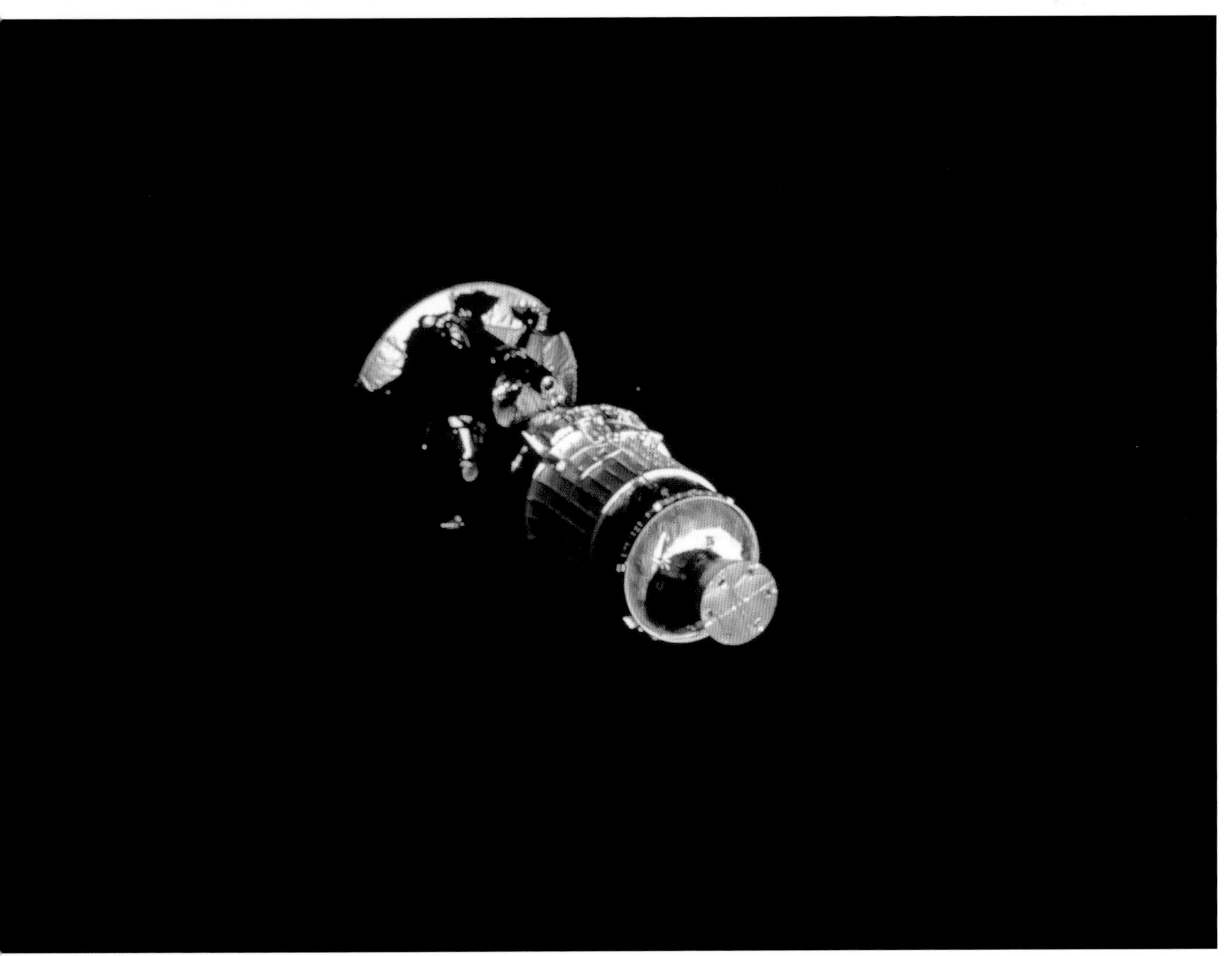

Mission no. **32**

STS

33R

Orbiter	Discovery
Launch	November 22, 1989
Landing	November 27, 1989
Duration	5 days, 0 hrs., 6 mins., 49 secs.

The STS-33 crew on the middeck: (clockwise from bottom left) Fred Gregory, Kathy Thornton, John Blaha, Sonny Carter, and Story Musgrave

Crew	Frederick D. Gregory, John E. Blaha, F. Story Musgrave, Manley L. "Sonny" Carter Jr., Kathryn C. Thornton

Mission: *Discovery's* crew flew the fifth classified DOD mission, deploying a satellite boosted to geosynchronous orbit by an IUS. The astronauts later positioned the orbiter so its optical and infrared signature could be measured by ground instruments in support of missile defense research.

Fred Gregory, Commander:

STS-33R was a DOD mission, associated with the National Reconnaissance Office. To this day, I believe, it is still classified.

I can tell you, however, that NASA and DOD chose an excellent crew to fly it. The crew originally included Dave Griggs as the pilot. Unfortunately, Dave was killed in an airplane accident, and so I had the unique opportunity to select the pilot. I thought seriously about it for about five minutes and then chose John Blaha, who had just completed a mission. We were pretty far along in our training when Dave was killed, and Blaha stepped in like a champ, and was a vital part of the success of the mission.

I remember our relationship with our training team: We always said they were attempting to kill us, and we were attempting to make fools of them. It wasn't adversarial, but there was this competition between us and the training team, like "Bring it!"

On landing, as soon as we came through Mach 1, I looked over at John and said, "Okay, you've got it." He flew 30 to 45 degrees of the spiral down the heading alignment cone (HAC), and then I took over, and we continued around. We were on glide path and everything was just working out great, but I think that was just because of the kind of training that we had been given in these 1,500 or 2,000 approaches that we had made ou at White Sands in preparation.

We touched down and I forgot I was in the shuttle. I thought I was in a T-38, and when landing at Edwards you always turned of at midfield. So when we touched down, I got on the brakes, and w slowed at twelve to thirteen feet-per-second-squared (three and a half to four meters-per-second-squared), the greatest braking deceleration they had ever seen. I suspect that we had the shortest hard-surface rollout distance without a drag chute of any orbiter.

When I took over on the HAC, flying this glider to touchdown, everybody on the crew was looking at me. I know they were a thinking, "Don't screw this up." S when we touched down and cam to wheels stop, I took in a huge breath. It was a fantastic time.

This was a very interesting fligh and I wish I could talk about it. After the flight, the crew was invit ed to CIA headquarters. The direc tor of the CIA, William H. Webste was there. We went in and they presented us with medals for the work that we had done. And then they took them immediately away We were gratified that we'd been acknowledged and recognized, but disappointed because we couldn't take this medal with us. We came in with nothing, we cam out with nothing.

Author's note: Later, the entire crew did receive their declassifie National Intelligence Medals of Achievement.

Discovery blasts off from Pad 39B amid a towering plume of exhaust steam.

Mission no. **33**

STS 32R

Orbiter	Columbia
Launch	January 9, 1990
Landing	January 20, 1990
Duration	10 days, 21 hrs., 0 mins., 36 secs.

The STS-32 crew on middeck: (front row) Marsha Ivins, Bonnie Dunbar, and G. David Low; (back row) Jim Wetherbee and Dan Brandenstein

Crew	Daniel C. Brandenstein, James D. Wetherbee, Bonnie J. Dunbar, G. David Low, Marsha S. Ivins

Mission: *Columbia*'s crew launched the Syncom IV-5 on Flight Day 2, then retrieved the LDEF materials exposure satellite, just weeks before atmospheric drag would have caused its uncontrolled reentry. Insights gained from its nearly six years in space informed materials choices for the exterior of the ISS.

Dan Brandenstein, Commander:

One of the main objectives was to use this mission as a stepping stone to longer-duration flights—some were concerned about the pilots' ability to reenter and land the shuttle after extended periods in free fall. We had a suite of experiments to determine the effects of longer-duration missions and remedies for those effects.

Astronaut Sonny Carter, not assigned to the mission but a former Navy flight surgeon, always had creative ideas. He convinced our crew to volunteer for muscle biopsies. Before we launched, each of us had a chunk of muscle removed from our thigh, and then when we got home, we got to go through it all again. Everybody signed up for it, but I think a few of the crew in hindsight wished they hadn't. Not a very pleasant experience, but you do it for science.

We were kind of on the bubble to get this flight off before the LDEF reentered. We did the standard rendezvous approach. It was a little different in that the LDEF was drifting in a gravity-gradient, standing-on-end attitude, and we weren't sure where on the circumference of the LDEF the grapple fixture was going to be. So we flew all the way around to the top of the R-bar, above LDEF. Then we could just do a yaw maneuver to align the arm with the grapple fixture, and we did our final approach from there.

We grappled it on the side, and then we went through the photo documentation procedure. They wanted a picture taken of every panel, so we had the spacecraft on the end of the arm and then maneuvered it all over the place and took a ton of pictures before we actually brought it back and stowed it in the payload bay.

The design of the orbiter was amazing. It was very easy to fly. You had the ultimate in control precision, accurate to a matter of inches. We didn't do it in flight, b[ut] in training, there were a few times where I could fly the arm over the grapple fixture and all Bonnie Dun[bar] had to do was hit the capture button. That's just an indication o[f] how well that vehicle was designed and how well it flew.

Marsha Ivins was the expert o[n] cameras and crew equipment. He[r] greatest thrill on a mission was doing that sort of stuff. Bonnie go[t] the LDEF captured and got it into the payload bay in fine style. And Jim Wetherbee was phenomenal; he knew every system frontwards and backwards.

I never knew G. David Low's dad, but from what everybody else told me about him, G. David was a chip off the old block. His father was George Low, a legend in the Apollo program, chiefly responsible for bringing the agency back after the Apollo 1 tragedy. G. David was a very dedicated, smart guy. He was the first perso[n] to run around the world; he got on the treadmill one day and ran for an orbit. Then we had a big piece of teleprinter paper that we held across the middeck, which he broke through as he complete[d] the run. He was a good guy, only fifty-two when he died of colon cancer, way before his time.

Typically, none of these missions would be successful withou[t] the tremendous team across the country that pulls them together and makes them successful, from programs to training to equipmen[t]. Getting to meet all these people was always my greatest joy abou[t] working in the astronaut office.

ABOVE: ***Columbia*'s crew grapples the LDEF satellite for berthing and return to Earth.**

RIGHT: **G. David Low breaks the "tape" as he completes a globe-circling treadmill run.**

BELOW: **Syncom IV-5 rolls slowly from *Columbia*'s payload bay, headed for geosynchronous orbit.**

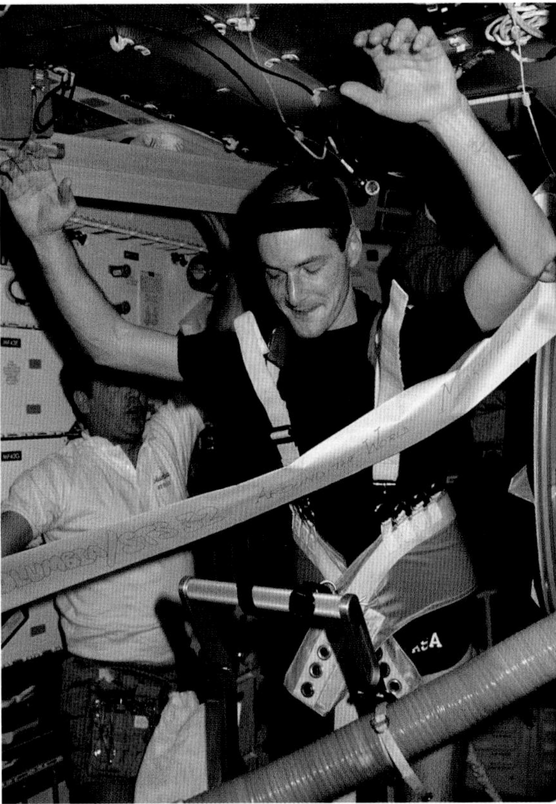

Mission no. **34**

STS

36

Orbiter	Atlantis
Launch	February 28, 1990
Landing	March 4, 1990
Duration	4 days, 10 hrs., 18 mins., 22 secs.

The STS-36 crew on the flight deck: J. O. Creighton, Dave Hilmers, Mike Mullane, Pierre Thuot, and John Casper

Crew	John O. "J. O." Creighton, John H. Casper, Richard M. "Mike" Mullane, David C. Hilmers, Pierre J. "Pepe" Thuot

Mission: On this sixth classified DOD mission, the *Atlantis* crew deployed a single satellite into orbit. Their orbital inclination to the equator of 62 degrees was the shuttle's highest ever, with a ground track reaching nearly to the Arctic and Antarctic Circles.

Mike Mullane, Mission Specialist:

The joint NASA-Air Force team was marvelous. Everybody was totally dedicated to the mission. We didn't wear our military uniforms, but it felt great to be back in "warrior" mode, doing something in the defense of the country.

While in our Kennedy Space Center crew quarantine, J. O. Creighton came down with some wicked upper respiratory illness that ultimately infected John Casper and Dave Hilmers. I came down with it a couple days after getting back from the flight. I think Pepe was the only person on the crew who never caught it. At any rate, the flight surgeon grounded the flight until everybody got well, resulting in a long delay that we waited out in the crew quarters. Even then, when we finally did launch, there's no doubt J. O. and John Casper were still sick. It was a long delay made worse by a miserable launch time—3:50 am—resulting in a crazy sleep cycle. We were all totally exhausted by the time we did fly.

We were at 62 degrees inclination, but also at low altitude—about 130 nautical miles (241 kilometers). What was really amazing to me was the atomic oxygen glow. It wrapped the whole payload bay in a yards-deep, foggy, greenish glow. I'd seen the oxygen glow on earlier missions, but it was just a very thin, gossamer glow. This time, it was very thick. I joked that it was as if we were in a ghost ship.

Becoming an astronaut was the completion of a life dream, and that made the mission moving for me. Early in my life, I was building pipe bomb rockets and launching them off in the deserts near my home in Albuquerque. When Sputnik was launched, I was out in the desert trying to see it pass overhead with the rest of the fearful population, worried about what the Russkies were up to. There was a big push to get kids interested in rocketry. I aimed my life toward pursuing a career in aviation and space.

The most emotional moments in my career were those times in orbit when I'd fly over Albuquerque and remember as a child standing in the desert and looking up and dreaming of someday being in space, and now there I was, up there and looking down on that desert, the birthplace of my dream.

And on this five-day flight, we were flying over parts of the Soviet Union where Sputnik was launched and the space race was born. It pains me to say we owe the Russians for anything, but we would never have gotten to the Moon without Sputnik. On STS-36, I lived the remarkable story of seeing both the area where Sputnik was launched and the deserts of Albuquerque, where I was inspired by Sputnik to become an astronaut.

Even though I can't talk about them, I feel very proud about those DOD missions. They had a significant impact on America's security, and I was part of it.

ABOVE: **Time exposure of the STS-36 ascent, photographed from near the Vehicle Assembly Building.**

LEFT: **The eastern half of a snow-bound Iceland as seen from STS-36.**

BELOW: **Dave Hilmers uses a large-format AeroLinhof Earth observations camera on the flight deck.**

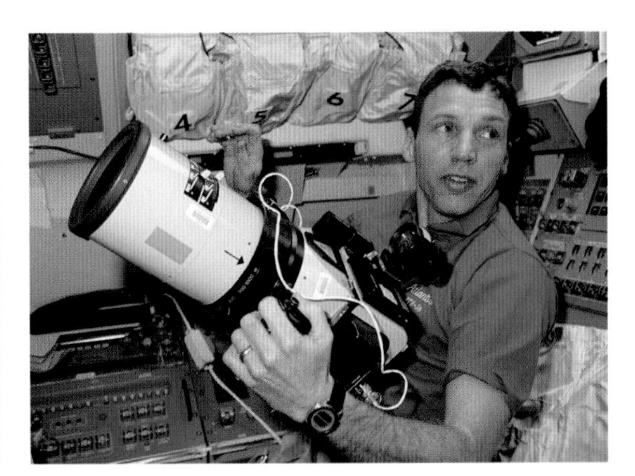

Mission no. **35**

STS 31

Orbiter	Discovery
Launch	April 24, 1990
Landing	April 29, 1990
Duration	5 days, 1 hr., 16 mins., 6 secs.

The STS-31 crew on the middeck: Charlie Bolden (top), Loren Shriver, Kathy Sullivan, Bruce McCandless, and Steve Hawley

Crew	Loren J. Shriver, Charles F. Bolden Jr., Steven A. Hawley, Bruce McCandless II, Kathryn D. Sullivan

Mission: *Discovery*'s astronauts deployed the Hubble Space Telescope (HST) using the RMS on Flight Day 2. McCandless and Sullivan donned spacesuits to manually deploy a stuck Hubble solar array, but Mission Control commands cleared the jam, negating the need for a spacewalk. *Discovery* released HST at a record 386 miles (621 kilometers) to reduce atmospheric drag effects.

Charlie Bolden, Pilot:

We'll remember STS-31 for a long time because of the way it dramatically changed astronomy and astrophysics in textbooks. It is one of the most remarkable missions that any space agency ever flew. Of my four missions, it was the one that had to overcome more adversity than any other.

The average man and woman on the street had no idea of the difficulty we had with deployment of the HST. It turned out to be a nightmare because the RMS control laws that had been worked out by the experts for years and years turned out not to be exactly right, so Steve Hawley had to lift Hubble out of the payload bay joint by joint. What was supposed to take us maybe fifteen minutes took several hours.

We finally looked like we were back on track and then had a problem with the second solar array: It stopped extending about sixteen inches (forty centimeters) out. It took the rest of the day to solve that problem. If we didn't get in deploy attitude pretty dog-gone quickly, we might lose the telescope, because the battery was running out of power. So we ended up maneuvering contrary to the flight rules using the more powerful primary jets.

My most memorable experience was being scared to death. It had nothing to do with the flying portion of the mission. It was when we were told to go downstairs and get Bruce McCandless and Kathy Sullivan suited up in the airlock and ready to go outside. I was the intravehicular crewmember, and I had never done a real EVA before. All of a sudden it struck me that I was getting ready to put these guys in the airlock—their lives were going to depend on me. That was sobering. I was really surprised at how much it hit me as they put the gloves and helmets on, and we got a thumbs up. I floated out and locked the hatch. Then we started to depressurize. As the air whistled out of the airlock, I could feel how vulnerable my friends were outside the cabin, spacesuits or no. That was not a good time for me.

But some young engineer at Goddard realized that we didn't have a physical problem with Hubble at all; we had a software problem with something called the tension monitoring module. I think it took him a good portion of the day to convince the ground control teams that he was right. Bruce and Kathy had actually gotten down to an airlock pressure of five psi—they were within minutes of actually going out to manually deploy the solar array. The Goddard engineer saved us from having to put them out there unnecessarily.

Watching Hubble come off the arm and watching it as we moved away from it was really satisfying, but at the time we didn't know what the world would find out later: that there was a spherical aberration in the mirror. So in our minds we had just completed the perfect start to this incredible Hubble mission.

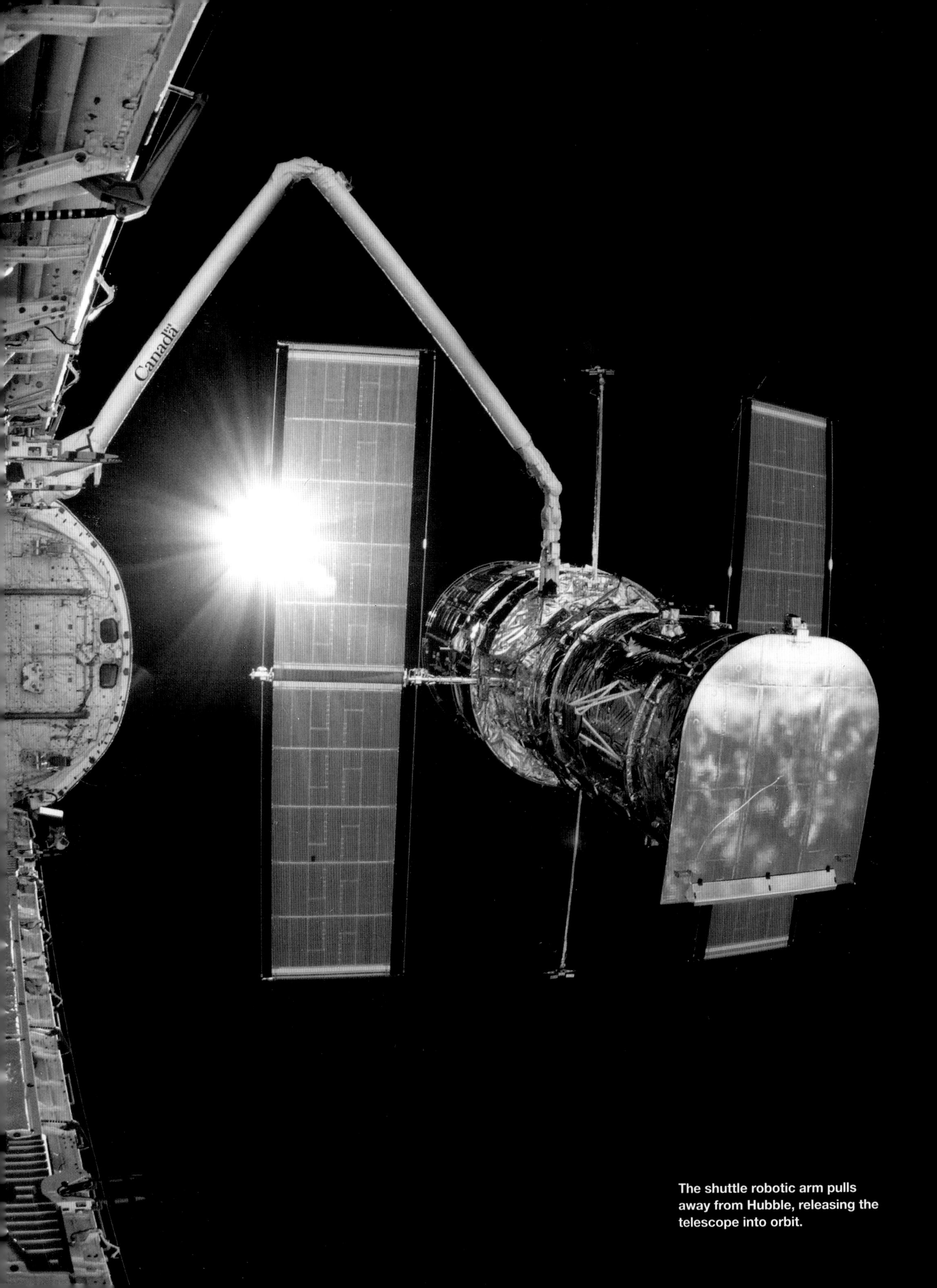

The shuttle robotic arm pulls away from Hubble, releasing the telescope into orbit.

Mission no. **36**

STS 41

Orbiter	Discovery
Launch	October 6, 1990
Landing	October 10, 1990
Duration	4 days, 2 hrs., 10 mins., 4 secs.

The STS-41 crew on the middeck: Tom Akers, Dick Richards, Bruce Melnick, Bob Cabana, and Bill Shepherd

Crew	Richard N. Richards, Robert D. Cabana, William M. Shepherd, Bruce E. Melnick, Thomas D. Akers

Mission: *Discovery*'s crew deployed the Ulysses probe atop its IUS six hours after launch, beginning its eighteen-year mission to observe the Sun's polar regions. Astronauts studied microgravity plant growth and fire behavior, measured the Sun's ultraviolet output, and tested the physiological response of lab rats to free fall.

Bruce Melnick, Mission Specialist:

No one's ever explored the polar regions of the Sun. When I talk to people, I say that it's as if we know everything about Earth's atmosphere around the equator, but nothing about what it's like at the north and south poles. That's the same gap in data we had with this mission.

We did get involved a little by making some public responses to flying a radioisotope thermoelectric generator (RTG) on Ulysses. People then were very concerned about launching a plutonium-powered RTG. We had to tell people why we felt safe flying it, that we were not going to ruin the world if something went wrong.

Flight Crew Operations pulled the crew together with Dick Richards as the commander. You could tell he was apprehensive about our relative inexperience; he said many times that this would be the shortest shelf life of any crew that's ever flown in space. Bill Shepherd was Mission Specialist 2, on his second flight, and then Bob Cabana, Tom Akers, and I were all rookies. So we trained and trained, and I think we even got some extra training. Dick knew he had a rookie crew with him, but he was proud of us when we got back.

Think of the astonishing jobs we got to train for. From starting out as a young charter fisherman, I became the flight doc on our mission! Then you'd become an expert photographer, and then you'd become an orbital scientist. It was pretty amazing to jam all of that into three years of training to go up there and have everything work perfectly.

When Ulysses deployed from its cradle, tilted up at its 58-degree angle from the payload bay, and cruised over our heads, it looked like a bus that was way too close to us. It left slowly, and seemed to just stay there, too close—it was big!

When we fired Ulysses off to go to Jupiter on its way to the Sun, it was not only the heaviest payload for the orbiter to date but was the fastest spacecraft at the time, going something like ten miles per second or 36,000 miles per hour (sixteen kilometers per second or 57,936 kilometers per hour). Everything went like clockwork and was over before we knew it.

We'd been in space for less than four days. When we landed, we came running down the ramp, walked around the orbiter, went and did our stuff with the flight medicine people, hopped in the jet, flew back to Houston, and did our post-landing meet-and-greet at Ellington Field. I hopped in the car with my family and drove us home! You don't do that after a seven-day flight or a ten-day flight. I just had never totally adapted to space.

I don't think anyone can plan to be an astronaut. What I tell school kids or anyone aspiring to that job is that everyone's got a chance—I'm proof that everyone's got a chance. No one in my family had ever been to college. I really feel in a lot of ways that I lived the American dream. But make sure that while you're working toward becoming an astronaut, you're doing something you really enjoy.

ABOVE: **Two shuttles on the launch pad: *Columbia* (foreground) on Pad 39A and *Discovery* on Pad 39B.**

BELOW: **Ulysses atop its IUS heads for Jupiter on its way to observe the Sun's poles.**

LEFT: **The Ulysses probe atop its three booster stages at the pad payload change-out room.**

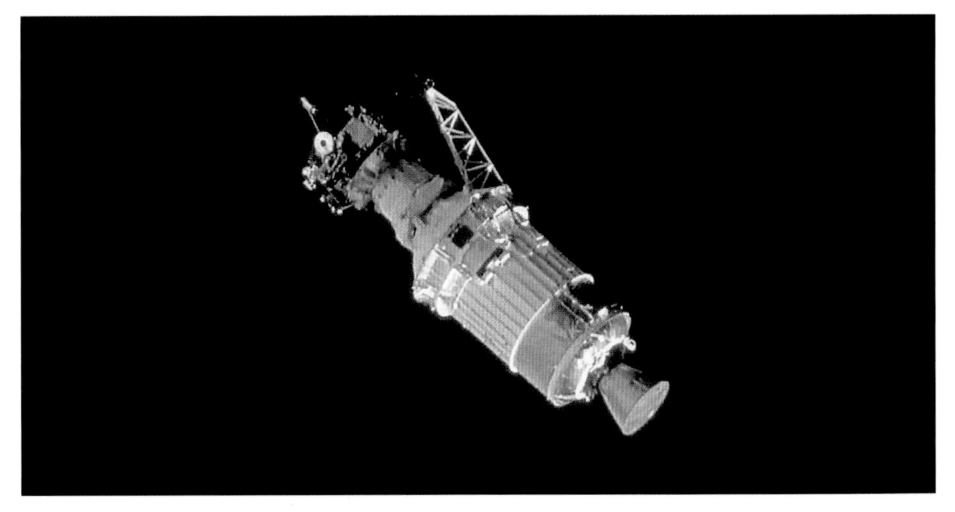

Mission no. **37**

STS

38

Orbiter	Atlantis
Launch	November 15, 1990
Landing	November 20, 1990
Duration	4 days, 21 hrs., 54 mins., 27 secs.

The STS-38 crew on the middeck: Bob Springer, Frank Culbertson, Dick Covey, Sam Gemar, and Carl Meade

Crew	Richard O. Covey, Frank L. Culbertson Jr., Robert C. Springer, Carl J. Meade, Charles D. "Sam" Gemar

Mission: On this seventh classified mission dedicated to DOD objectives, the *Atlantis* crew deployed a satellite known only as USA-67. This and other orbital activities of STS-38 remain classified.

Bob Springer, Mission Specialist:

STS-38 was a DOD flight with a unique customer within the classified world, the CIA. There's a building out in the back of the Johnson Space Center, not far from where the crew quarters were. That's where we did our meetings, three stories underground. Nobody knew about that room; it didn't exist.

During the flight we were at a very low altitude, in fact the lowest that any shuttle ever flew, between 105 to 120 miles (169 to 193 kilometers). The United States had gotten rid of most of its aerial overhead reconnaissance stuff, and the satellites were not good for area surveillance. So the Pentagon asked us to use our handheld cameras to take pictures of the Middle East. They had to get permission from the crew to do it because it came during our sleep period. Us five military guys said, "Absolutely!" We provided the maps and the basis for photo intelligence for Operation Desert Storm.

We had a good crew to work with. Dick Covey and I got along really well. We had roughly the same level of operational background, he in the Air Force and me in the Marine Corps. Frank Culbertson was the next most experienced. Carl Meade and Sam Gemar were less experienced, but we were all fascinated by the context of the mission, the kinds of things we were going to do, and the absolute need for the classification category.

The payload had a self-destruct mechanism, and we were not to land with that payload at any place other than Kennedy Space Center. If we had had to go to one of the emergency fields, they'd have pulled the pin. They really didn't want that payload exposed. The original reason for it was when the payload was on orbit, if there was any chance that somebody could make a grab for the satellite, they would blow the satellite up. They didn't really make a big deal about "You're not going to land with it," but we knew what that meant.

I hadn't done an EVA. But we had a contingency EVA for a critical antenna that had to deploy. If the antenna didn't deploy, you didn't deploy the spacecraft. With the challenges of landing with such a heavy payload, there was a real question of "What do you do now?"

We were going along the timeline, and Sam and I were the EVA guys, and the antenna didn't deploy. We were suited up, had done our pre-breathe, and were marching along to get out. We were just getting ready to start the depressurization for the EVA when—*sproing*! Absolutely broke my heart.

We had a great mission patch design. It shows one orbiter right side up and one upside down. But we struggled with it because you can't give away anything associated with the payload or the mission in general. You'll notice the orbiter underneath that's in the shadows is literally supporting the orbiter that's flying into orbit. It's our way of acknowledging the tens of thousands of people who worked on the program and never got the recognition.

ABOVE: ***Columbia*** **(left) rolls to the pad past** ***Atlantis*****, waiting to move inside the Vehicle Assembly Building for repairs.**

LEFT: **Sunlight glints off a serene ocean, as seen from** ***Atlantis*****.**

BELOW: ***Atlantis*** **receives attention from the orbiter safing convoy after landing at Kennedy Space Center Runway 33.**

Mission no. **38**

STS

35

Orbiter	Columbia
Launch	December 2, 1990
Landing	December 10, 1990
Duration	8 days, 23 hrs., 5 mins., 8 secs.

STS-35's crew: (clockwise from bottom left) Bob Parker, Ron Parise, Jeff Hoffman, Guy Gardner, Mike Lounge, Sam Durrance, and Vance Brand

Crew	Vance D. Brand, Guy S. Gardner, Jeffrey A. Hoffman, John M. "Mike" Lounge, Robert A. R. Parker, Samuel T. Durrance, Ronald A. Parise

Mission: A series of hydrogen propellant leaks delayed STS-35's launch by over six months. The crew operated the Astro-1 ultraviolet telescopes around the clock. Onboard computer failures forced controllers to point the telescopes from Earth, with astronauts making final corrections.

Jeff Hoffman, Mission Specialist:

The basic idea was that this was an ultraviolet observatory. Of course, you have to go above the atmosphere to see at those wavelengths. I think we had three different types of ultraviolet cameras going all the way to the far ultraviolet. And they all had different characteristics. One of them was a survey camera, one a spectroscope, and one a polarimeter. We were scheduled to fly in March 1986, which is when Halley's Comet was at its brightest. So there was a special Halley's Comet observing camera added. We were next in line after *Challenger*, and Halley was gone before return-to-flight. I don't need to go through that part of the story.

We had the two Spacelab computers, which did most of the coordination and pointing. Then we had a little handheld paddle, which is typical for operating a telescope, for the final correction, getting the object right down the boresight.

On the very first day of the mission, one of our Spacelab computers failed. It's not a pleasant feeling to smell smoke when you're in a shuttle, but we were still able to operate because we had another computer. These Spacelab computers had a whopping 64K of core memory. But they did have a color display—three colors!

Then on the fourth day, we smelled smoke again—the second Spacelab computer had just burned up. After the flight, they traced this back to the fact that while everything was sitting around in those months of delays between our first attempt and when we actually launched, they periodically would turn on the computers. But nobody had thought to clean the filters, so they were just totally clogged with dust. The computers were not getting proper cooling and they overheated and failed.

For a while, we thought, "We're out of business." But I have to say, controllers on the ground did a great job. They figured out how they could send from the ground most of the commands that we would normally put in from the computer. And then we would do the final bit of pointing using the onboard paddle. They actually did manage to get a few days of reasonable observations. The important thing is that they learned so much that when Astro-2 went up, it went off without a flaw.

Because four of us were astronomers, we arranged to do astronomy lessons from space. That's when I wore what was, if not the first space tie, certainly the first space tie in the American space program.

I was telling a group from the fashion design firm, Hermès, that on my next flight I was going to do this teacher-in-space lesson and that I was planning to wear a tie. The Hermès person I was talking with said, "Oh, Jeff, if you're going to wear the first tie in space, it has to be an Hermès tie." I said, "Well I'm really sorry, but Hermès ties are all made of silk and all of our clothing has to be cotton." Without missing a beat, he said, "Jeff, for you, we can make a cotton tie." And sure enough, they did, and I flew with it.

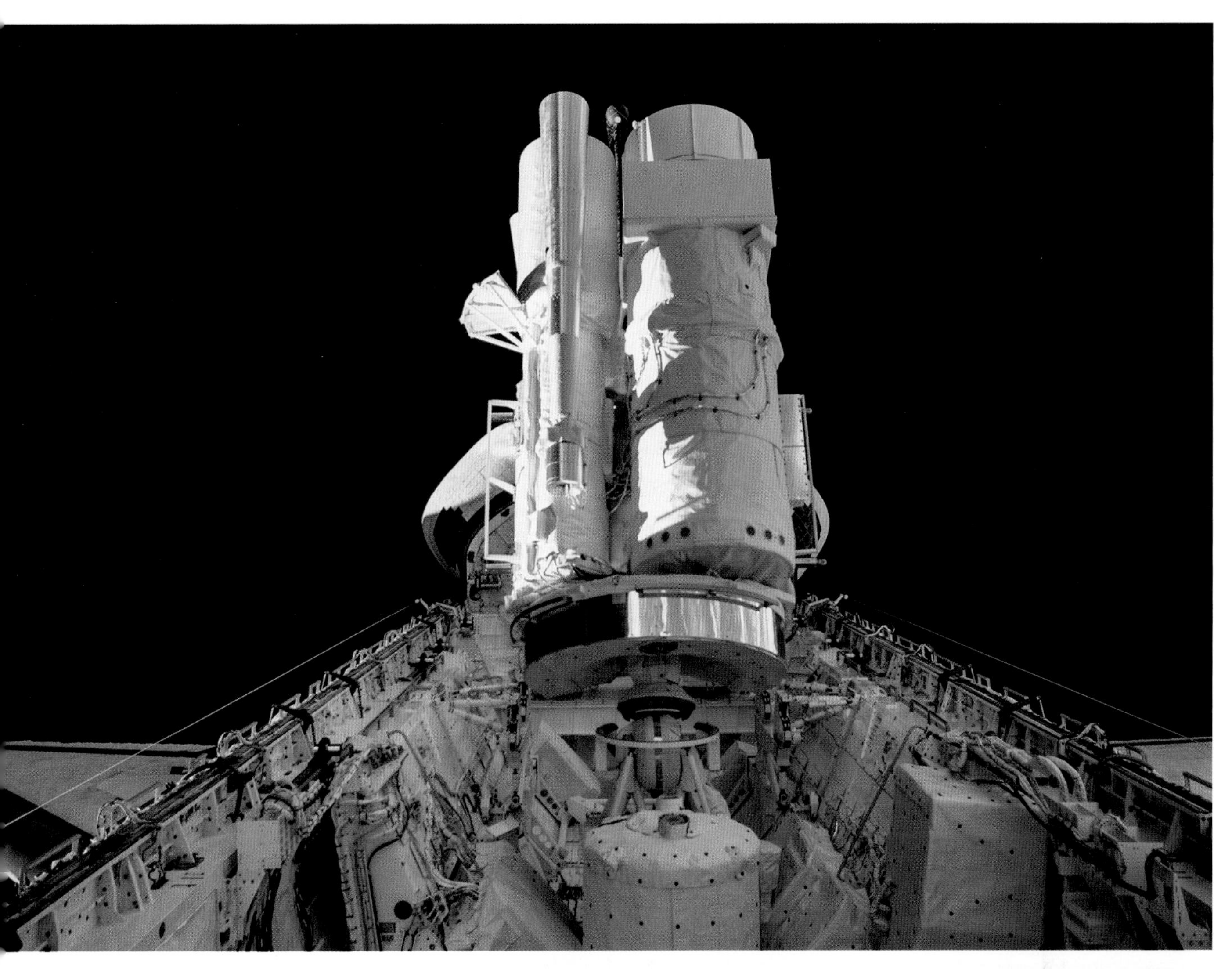

ABOVE: **Astro-1's cluster of ultraviolet telescopes and spectrometers tracks a stellar target from *Columbia*'s payload bay.**

RIGHT: **Ultraviolet image of the Cygnus supernova remnant, captured by Astro 1 observatory.**

BELOW: **Jeff Hoffman, sporting a Hermès necktie, joins Sam Durrance for a school astronomy lecture.**

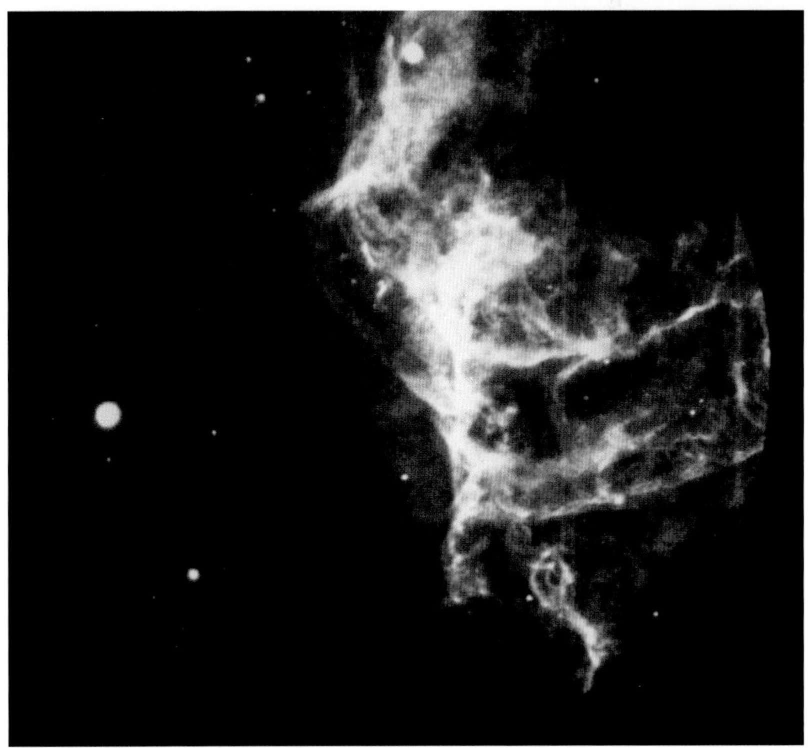

Mission no. **39**

STS

37

Orbiter	Atlantis
Launch	April 5, 1991
Landing	April 11, 1991
Duration	5 days, 23 hrs., 32 mins., 44 secs.

The STS-37 crew with *Atlantis* after landing: Ken Cameron, Steve Nagel, Linda Godwin, Jerry Ross, and Jay Apt

Crew	Steven R. Nagel, Kenneth D. Cameron, Jerry L. Ross, Jay Apt, Linda M. Godwin

Mission: STS-37 deployed the Compton Gamma Ray Observatory (GRO), one of NASA's Great Observatories. Before the RMS could release the GRO, Jerry Ross and Jay Apt had to perform an EVA to free GRO's stuck high-gain antenna. Their second EVA tested ISS mobility schemes. Compton completed its successful mission in June 2000.

Jerry Ross, Mission Specialist:

STS-37 carried the GRO into space. The discoveries it made helped astrophysicists rethink many things about the distribution and sources of gamma rays throughout the universe. I always say that anytime an engineer can help scientists rethink their theories, that's a pretty good day.

When the guys at the Goddard Space Flight Center commanded the GRO antenna boom to extend, it started to move but didn't fully deploy. We had talked about that eventuality and had come up with some alternate ways to deploy the antenna before doing a spacewalk. They included firing thrusters and moving the arm quickly and stopping it, all of which were very benign kinds of things. We had completed an oxygen pre-breathe before satellite deployment so that we were ready to go out at a moment's notice if things went wrong. As they started to get into those alternate strategies, I was on the flight deck already in my liquid cooling garment. I took off my wedding ring and said, "Steve, I'm going downstairs." He said, "I'll be right behind you."

Once I got outside, Jay Apt was bringing out some hardware—a foot restraint and other things. Linda Godwin would use the shuttle arm to bring the satellite down to where I could climb up onto it. There were huge hydrazine tanks right underneath where I was transiting, so I had to be very careful. There was some insulation around the antenna boom, and it looked to me like it was wedged in the boom's receptacle a little. I got up into a position where I could put some decent force onto the boom and not damage anything. I received concurrence from the ground to do that. So I started pushing on it with more and more vigor and after a while I could actually see it starting to move—I continued that process and finally it sprang free and opened up.

Jay had been the astronaut office person that had been following GRO, and he had worked on getting EVA features designed into the spacecraft so we could assist in deployment. That allowed us to save the satellite. Otherwise, it could have been a very big piece of space junk.

For the satellite release Jay and I were told to be in the airlock, but I can honestly say that only our toes were in there. We were hovering outside the airlock watching the satellite go, which was pretty cool.

On the planned EVA, Linda had me way up above the payload bay on the robotic arm. It was night, and during a break I turned off my helmet-mounted lights and just leaned back to suck in as much of the universe as I could. While doing that I had this sensation come over me that I was at unity with the universe. It gave me a sense of peace, that I was doing exactly what God had designed me to do, to use my hands and brain to go out and fix satellites and build space stations.

ABOVE: **Crewmates photographed Jerry Ross outside the aft flight deck after he and Jay Apt successfully freed the Compton observatory's high-gain antenna.**

RIGHT: **The RMS holds the Compton Gamma Ray Observatory ready for release into its observing orbit.**

BELOW: **Linda Godwin balances Ross on her fingertip in *Atlantis*'s middeck.**

Mission no. **40**

STS 39

Orbiter	Discovery
Launch	April 28, 1991
Landing	May 6, 1991
Duration	8 days, 7 hrs., 22 mins., 21 secs.

The STS-39 crew on the middeck: (front row) Don McMonagle, Mike Coats, Lacy Veach, and Greg Harbaugh; (back row) Guy Bluford, Blaine Hammond, and Rick Hieb

Crew	Michael L. Coats, L. Blaine Hammond Jr., Guion S. "Guy" Bluford Jr., Gregory J. Harbaugh, Richard J. Hieb, Donald R. McMonagle, Charles L. "Lacy" Veach

Mission: On this DOD flight, *Discovery*'s crew operated around the clock, using sensors to observe the atmosphere, the space environment, and the spectral signature of the maneuvering orbiter. They deployed five satellites and launched and retrieved the SPAS-II.

Rick Hieb, Mission Specialist:

This first unclassified DOD mission was paid for by the Strategic Defense Initiative Organization, informally known as Star Wars. It was unclassified, but the data taken were classified. One of the fundamental issues the organization was trying to solve had to do with the fact that rocket plumes are big, bright, hot, and easy to see with infrared cameras. However, rocket bodies are small, dark, cold, and hard to see. One important mission objective was to collect data to be able to write algorithms to go from seeing a rocket plume to figuring out where the actual missile is to make an intercept possible.

The reason we needed the shuttle to do this work was that there was no vacuum chamber on Earth big enough to characterize an active, maneuvering spacecraft. But we could deploy the Infrared Background Signature Survey (IBSS) spacecraft on SPAS, fly away from it, turn it back to point at us, and fire thrusters. The scientists would see the thruster plumes with the IBSS, and then be able to say, "Okay, we know exactly where the space shuttle is, so therefore we can start writing algorithms to take us from plume to hard body."

Years later I ran into Parney Albright, who was the chief scientist for the IBSS, and asked, "Parney, I know you can't tell us a lot of things, but did you get what you needed?" All he said was, "Yeah, we got what we needed."

About four months after Guy Bluford, Lacy Veach, and I were assigned, Mike Coats was assigned as Mission Commander, Blaine Hammond as Pilot, and Don McMonagle and Greg Harbaugh as additional mission specialists. In all we spent about two years developing the mission and training for it.

This was viewed then as the most complex mission we had ever tried to fly on shuttle. There were times when we needed six of the seven crewmembers to take data, even though we were nominally working two shifts. You'll see some pictures of us with both shifts on the flight deck in this really tightly choreographed activity. Everybody had a job to do.

Our mission was designed specifically to fly through an aurora to take data. Boom! There it is! Spectacular aurora, from nearly 250 miles (402 kilometers) up, all the way down to about sixty miles (ninety-seven kilometers). We were flying right through the curtain; it has all the colors that the cameras of those days couldn't capture.

Sometime during the mission, I was downstairs doing something and Lacy was floating horizontally with his feet toward the lockers, and all of a sudden, he says, "Oh my God! The floor is the wall!" I think, "Yeah, you're floating sideways, so of course." I didn't get it at the time but looking back it's really funny because Lacy had this dumbfounded look on his face.

It was later when I was doing something upside down, and suddenly the world was all new. I didn't know where anything was. Everything's wrong, and now the floor really is the ceiling, and the ceiling is the floor, and I was lost. I didn't understand until then the really cool experience you have when you make that transition and discover this new upside-down world that's so different.

ABOVE: **Gas flows from a canister as part of the Critical Ionization Velocity experiment aboard *Discovery*.**

RIGHT: **Lightning illuminates a turbulent storm system below during orbital night.**

BELOW: **Lacy Veach operates STS-39 payloads from the aft flight deck.**

Mission no. **41**

STS 40

Orbiter	Columbia
Launch	June 5, 1991
Landing	June 14, 1991
Duration	9 days, 2 hrs., 14 mins., 20 secs.

The STS-40 crew on the middeck: (front) Sid Gutierrez and Jim Bagian; (middle) Drew Gaffney and Rhea Seddon; (back) Bryan O'Connor, Tammy Jernigan, and Millie Hughes-Fulford

Crew	Bryan D. O'Connor, Sidney M. Gutierrez, James P. Bagian, Tamara E. "Tammy" Jernigan, M. Rhea Seddon, F. Drew Gaffney, Millie Hughes-Fulford

Mission: *Columbia*'s crew operated the Spacelab Life Sciences-1, with eighteen investigations focused on six body systems. To obtain the most detailed physiological data since the 1973–74 Skylab missions, astronauts served as research subjects along with thirty rodents and thousands of tiny jellyfish.

Sid Gutierrez, Pilot:

STS-40 was the first mission dedicated to the study of the human body in the microgravity environment of space. NASA had done a lot of experiments on the body here and there in the past, but this was the first time they dedicated an entire mission to it.

I was excited about getting to fly with Bryan O'Connor, Jim Bagian, and Rhea Seddon. And Tammy Jernigan was great to have on the flight. She was one class behind me, so we were both rookies. Later I got to work with our first-class Payload Specialists, Millie Hughes-Fulford and Drew Gaffney. I think the key to success was that everybody worked together as a team. Up until that time, there was a concept of the flight deck team and the payload team. Bryan made the decision that we could increase the study sample size from five to seven if the commander and pilot got involved in the experiments. So we agreed.

The mission was interesting but vastly oversubscribed. We wound up with too many experiments despite the simulations demonstrating they took too long to complete. We worked sixteen-hour days to get it all done.

Another issue we encountered was that the refrigerator up front had a strange odor coming from it. Rhea floated by at one point and said that to her it smelled like formaldehyde. Both Bryan and I were getting ill with headaches because of it, and Bryan started having issues with his eyes. We reported this over the private medical conference: "The two people who can land this vehicle are going downhill rapidly and we think it's caused by the odor coming from the refrigerator." Th ground reported there was no formaldehyde in the refrigerator but decided to shut it down.

It turned out that the motor armature was overheating, and when it did, it released formaldehyde. Rhea was correct after all. tell people: "If we had continued letting it run, the bad news was that it would have eventually kille us. But the good news was that when the recovery team arrived we would be well preserved."

I remember the night we had a close pass with Mir, passing just four to five miles (six to eight kilometers) from the Russian spac station. On the dark side of Earth, we would be looking for a bright star moving fast relative to us. We turned all the lights down and tried to gather around the window they told us to look out. You can imagine seven people all cramme around a window, trying to see. It was quiet since we had a little fre time, which was rare on a flight lik this. It was a good time to stop an reflect: I thought how fortunate I was to be one of the ten people o of billions able to fly in space on that day. We did see a fast-movin star that we assumed was Mir. W waved. I assume they waved bac

We succeeded because of teamwork, but also because of a lot of adjusting and improvising. I've spent a lot of time with Ken Cameron, a Marine, and one of th things that Marines say is that yo have to improvise. We did a lot o that on the flight.

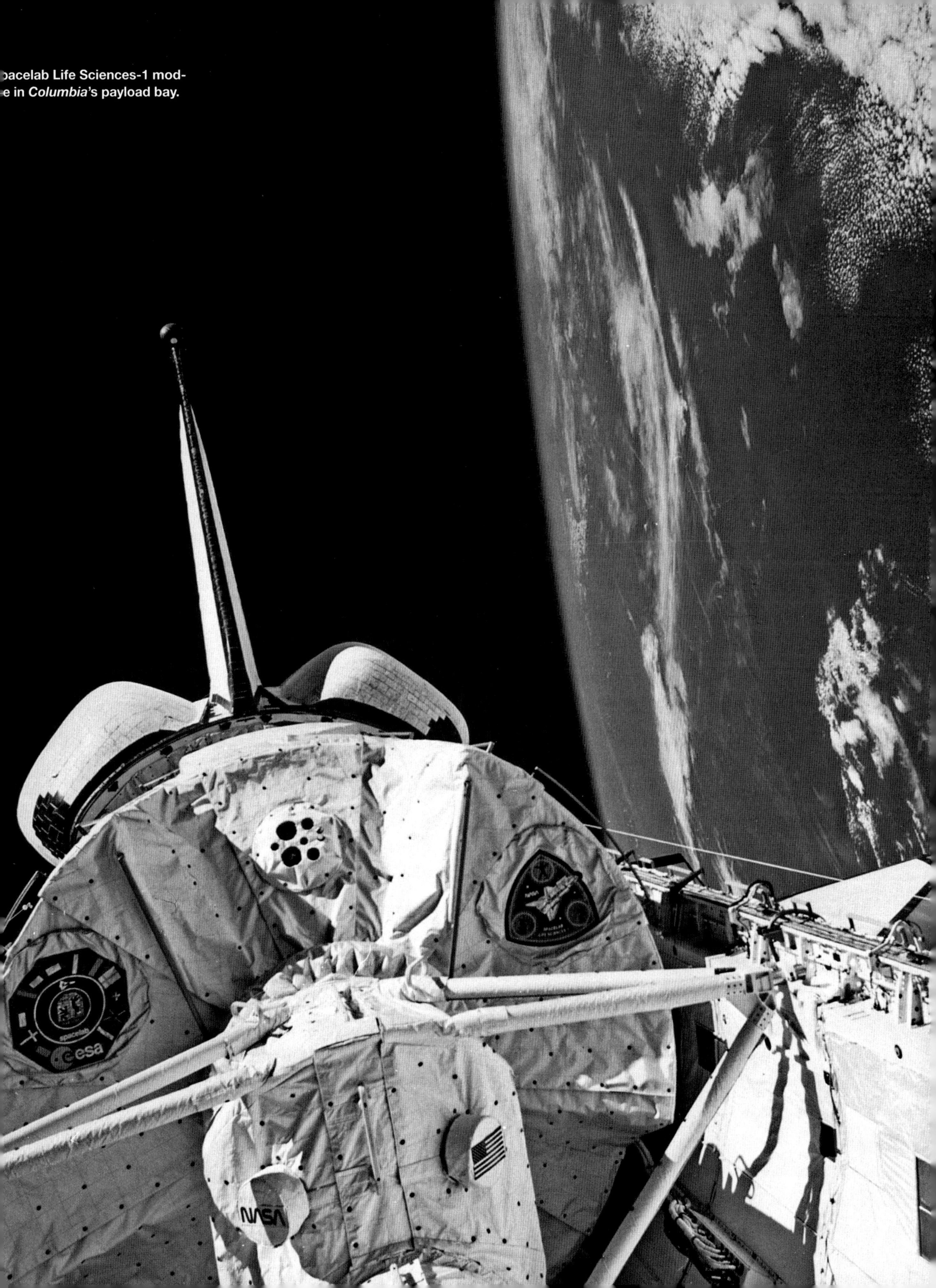

pacelab Life Sciences-1 mod-
e in *Columbia*'s payload bay.

Mission no. **42**

STS

43

Orbiter	Atlantis
Launch	August 2, 1991
Landing	August 11, 1991
Duration	8 days, 21 hrs., 21 mins., 25 secs.

The STS-43 crew on the middeck: G. David Low, Shannon Lucid, Jim Adamson, John Blaha, and Mike Baker

Crew	John E. Blaha, Michael E. Baker, Shannon W. Lucid, James C. Adamson, G. David Low

Mission: STS-43 launched the TDRS-E relay satellite six hours after liftoff, and its IUS boosted it to geosynchronous orbit. The crew sent the first email from space, via AppleLink: "Having a GREAT time, wish you were here,...send cryo and RCS! Hasta la vista, baby,...We'll be back!"

Shannon Lucid, Mission Specialist:

The TDRS was in the payload bay attached to an IUS. We were responsible for deploying this complex into low Earth orbit, and then the IUS ignited and launched the TDRS into geostationary orbit.

It was a great relief when it all went as smooth as could be, and I looked out the flight deck rear windows into an empty payload bay. I thought, "Wow, no matter what else happens during the rest of our flight, we have accomplished the main objective without a hitch." Now we could just relax and enjoy the next nine days in orbit.

John Blaha and I were the oldest crewmembers of STS-43, and Dave Low was the youngest. One day during training in the shuttle simulator, during a lull, Dave and I started talking and comparing our previous shuttle experiences. I was telling him about fun activities that I had participated in on previous flights. Then I started to tell him how much fun we were going to have on-orbit. During this conversation, he said to me, "No, we need to be serious." He then added, "Oh, I've never laughed on a shuttle mission."

Right then and there, I had a primary objective for STS-43. "My personal goal for our flight," I told Dave, "Is that you are going to laugh and smile." And you know what? That happened on the third day. Mission accomplished! That just made me feel good all over.

Because Dave Low was the youngest member of our crew, I felt a trifle maternal towards him. I told him, "You know what, Dave. Speaking from experience, in the grand scheme of life, what you'll remember most and what will be most meaningful to you is relationships, such as having a family, being a husband, a father—not shuttle flights." At that moment, he didn't understand where I was coming from.

We had a great mission and we landed. Dave flew once again, he got married, and moved away. Over the course of time, Dave became a father. I was so thrilled I sent him a little book for the baby, and he wrote back in a thank-you note: "Shannon, you know how hard it is for me to admit that I'm ever wrong. But I was wrong, and you were right. Having a child is much more rewarding than any shuttle flight!" I thought that spoke volumes about Dave.

But overall it was very satisfying to launch the TDRS. You feel that you have a great responsibility to all the many people who have put so much of their lives into building these payloads. All you had to do was throw one switch wrong and everything that they'd been looking forward to, and everything that they had planned for, would have been for naught.

ABOVE: **An orbital sunrise arrives, silhouetting distant thunderheads.**

RIGHT: **Springs propel the TDRS-E atop its IUS from its payload bay support carriage.**

BELOW: **Shannon Lucid activates a bioprocessing experiment on the middeck.**

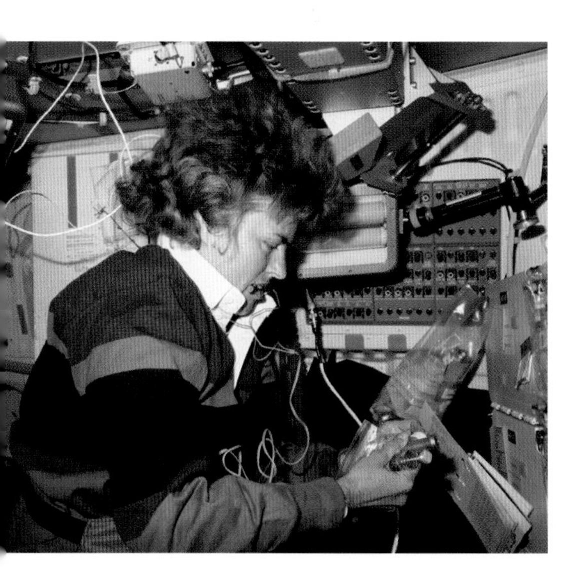

Mission no. **43**

STS

48

Orbiter	Discovery
Launch	September 12, 1991
Landing	September 18, 1991
Duration	5 days, 8 hrs., 27 mins., 38 secs.

The STS-48 crew on the middeck: (front) Ken Reightler and Jim Buchli; (middle) J. O. Creighton; (back) Mark Brown and Sam Gemar

Crew	John O. "J. O." Creighton, Kenneth S. Reightler Jr., Mark N. Brown, Charles D. "Sam" Gemar, James F. Buchli

Mission: *Discovery*'s crew released the UARS into its 313 nautical-mile-high (580 kilometer-high) working orbit, an atmospheric sciences mission that continued for fourteen years. Astronauts performed an array of biological and materials science experiments in the middeck.

Ken Reightler, Pilot:

This was the first attempt to use the space shuttle in NASA's Mission to Planet Earth. Our mission was to deliver a spacecraft that had been designed specifically to look at the upper atmosphere. We assume we know all about the atmosphere because we've lived on the planet for such a long time, but I was really amazed to learn what we didn't know.

One of those mysteries, probably the most significant one, had to do with the hole in the ozone layer that was forming over Antarctica. Why was that happening? There were theories, but we didn't really have the data. So one of the major experiments on the spacecraft was to measure that phenomenon and prove or disprove some of the theories. To help understand what it was that caused the hole in the ozone and then start to do something about it was incredibly satisfying.

In flight, once ascent is over you don't have too many rapid-action, immediate crisis situations—you can generally take your time, work through them, and talk to the ground. But there's one that is a quick-reaction issue—that's the "fuel cell reactant valve" procedure. You have to get on it, because your fuel cell is going to start building up pressure and explode if you don't shut it down within a few minutes.

The first night in orbit, I was just getting to sleep and had one of those internal eye flashes, caused by charged particles passing through my head. I cam flying out of my sleeping bag, trying to figure out what in the world that was. Of course, every one laughed at me because they knew exactly what it was. They were seeing them too, but they'd experienced them before. I settled down and had just gotten back to sleep when the Master Alarm went off. I came flying out of my sleeping bag again, went up to the flight deck looked at the computer screen, and saw that it was a "Fuel Cell Reac Valve" message.

Even in the early stages of sleep, I had an immediate reactio to the alarm and didn't have to think about where to go, how to get there. I was in my seat, I had my eyes on the screen, and I had my hand on the fuel cell shutdow valve switch, knowing that at that point if we shut the fuel cell dowr the mission was over. Later in the shuttle program, they approved restarting fuel cells, but at that time if you shut one down, it was an immediate mission abort. So, I'm thinking, "Man, my first night in space and I'm about ready to cause this mission to end—right now."

The next step was:

Me: "OK, Houston, what do yo see?"

Houston: "It's a faulty sensor—disregard."

Me: "Roger."

Houston: "Go back to sleep."

That was my "Welcome to space" moment.

LEFT: **An Antarctic iceberg, spanning 21.7 by 59 miles (35 by 95 kilometers), drifts in the Southern Ocean off South America.**

ABOVE: **Mark Brown and Jim Buchli assemble the MODE ISS structures experiment on the middeck.**

BELOW: **UARS satellite oriented by the shuttle arm for release into its science orbit.**

Mission no. **44**

STS

44

Orbiter	Atlantis
Launch	November 24, 1991
Landing	December 1, 1991
Duration	6 days, 22 hrs., 50 mins., 43 secs.

The STS-44 crew on the middeck: (clockwise from left) Jim Voss, Mario Runco, Tom Henricks, Story Musgrave, Fred Gregory, and Tom Hennen

Crew	Frederick D. Gregory, Terence T. "Tom" Henricks, Mario Runco Jr., James S. Voss, F. Story Musgrave, Thomas J. Hennen

Mission: The crew launched the DSP, an infrared missile launch detection satellite atop its IUS, carrying it to geosynchronous orbit. The failure of one of three inertial measurement units forced *Atlantis* to return home early. This was the shuttle's last dry lakebed landing.

Mario Runco, Mission Specialist:

STS-44 was the first unclassified DOD mission where the Defense Support Program (DSP) satellite was brought out from behind the curtain. The darn thing was so heavy that we barely had enough "oomph" to get into orbit. So they light-loaded us with cryogenic fuel cell reactants, mostly oxygen.

On ascent in the Shuttle Mission Simulator, we'd simulate Main Engine Cut Off, or MECO, then External Tank (ET) Separation. When ET Sep happens in the sim, it goes "poof!" I worked in the Shuttle Avionics Integration Laboratory a lot, and it had the same sound system: a pair of old six-inch-by-nine-inch (fifteen-by-twenty-two-centimeter) car speakers set in wooden boxes, stuck on the floor in the back. They probably got them out of a 1961 Studebaker that was sitting derelict in a junkyard, and they had the volume on them turned down to boot.

Now, when the ET separated on STS-44 for real, it was "Ka-Boom!" I jumped out of my seat and stretched those seatbelts because it startled the living daylights out of me. I remember thinking, "What was that?!" right after the explosion. "Oh, it's ET Sep." It was supposed to be "poof!"

On STS-44, I did the ET photography. I was Mission Specialist-3, on the middeck sitting with Tom Hennen next to me, and right after we arrived on orbit and jettisoned the tank, I was up out of the seat and floated upstairs. I went to the locker, pulled out the cameras, then proceeded to take pictures of the tank as it floated away. It was one of those moments that takes your breath away. I wasn't stunned in the sense of fear; it was more "Wow! That is such a gorgeous sight!" I paused and smelled the roses fo a few seconds, then went back to work taking pictures of the ET. Seeing our beautiful blue Earth at a glance out the window—the stars, galaxies, the universe all more vividly than from the confines of Earth—it would do the world good if everyone had the experience.

Later in the mission, we turned on Inertial Measurement Unit (IM 2, and lo and behold, it failed. We had to pack up to return home in the morning. The mission was shortened to almost seven days instead of the planned ten.

The good news, maybe not fro NASA's perspective, is that we g to land on the lakebed at Edward Air Force base on Runway 05. W were the only flight to have used that lakebed runway, and they were able to get our families out there on short notice to welcome us home. We got to spend the night with our families under som beautifully clear night skies at the nearby Silver Saddle Ranch, a rather idyllic desert resort.

After the mission, we found ou that they used the DSP satellite constellation during the 1991 Gu War to detect the Iraqi Scud mis sile launches and monitor their trajectories. The DSP system allowed US and coalition defend ing assets to know the Scuds were in the air and what their intended targets were long befo they arrived. That, in turn, enabled the defending Patriot missile batteries in Israel and Saudi Arabia to intercept them before any serious damage was done.

LEFT: ***Atlantis*'s liftoff showcases the blue shock cones in the shuttle main engine exhaust.**

ABOVE: **Jim Voss amid plumbing and water storage tanks in *Atlantis*'s lower equipment bay.**

BELOW: **The DSP infrared missile-warning satellite atop its IUS, ready for deployment.**

Mission no. **45**

STS

42

Orbiter	Discovery
Launch	January 22, 1992
Landing	January 30, 1992
Duration	8 days, 1 hr., 14 mins., 44 secs.

The STS-42 crew in Spacelab: (clockwise from bottom left) Roberta Bondar, Dave Hilmers, Ron Grabe, Ulf Merbold, Bill Readdy, Norm Thagard, and Steve Oswald

Crew	Ronald J. Grabe, Stephen S. "Oz" Oswald, Norman E. Thagard, David C. Hilmers, William F. Readdy, Roberta L. Bondar, Ulf D. Merbold

Mission: *Discovery*'s crew operated the IML-1 Spacelab to explore the effects of weightlessness on living organisms and materials processing. Complemented by middeck and payload bay experiments, the crew achieved all scientific objectives and guided *Discovery* to an Edwards Air Force Base landing.

Steve Oswald, Pilot:

STS-42 was the first iteration of the International Microgravity Laboratory (IML). This flight was a lot more interactive, partly because we had so many experiments in the lab that required the flight deck crew to participate.

This was Ron Grabe's third flight, his first command, and I was the right-seater and Bill Readdy was the flight engineer. Norm Thagard had the most flights: It was his fourth, and he was incredibly competent. It was Ulf Merbold's second flight, and he was the ESA representative. Then we had Bobbie Bondar as the Canadian payload specialist.

We had two of the smartest mission specialists—Thagard and Hilmers. It was wonderful. It was like flying with the checklist open all the time, but it was just embedded in their brains. The only bad thing about that was they were right 99.7 percent of the time. But the other 0.3 percent of the time when they weren't right, they were pretty damn sure they *were* right. You could end up going down the wrong rabbit hole.

On ascent in the simulator, you get to two minutes and four seconds, and there's this nice, soft, hydraulic thump and there's this salmon-colored glow on the windows that lasts about two seconds, and it goes away. In real life, it's a giant clang and orange and white and red flames on the windscreen that last about a week!

The Canadians had an experiment called Back Pain in Astronauts. And it wasn't all that elegant, actually, but they wanted to understand and measure how much folks were growing, as in stretching. So they set up this measurement system in the forward part of the lab, and abou every eight hours or so, you'd go stand up against it, and they could see how much you were growing. I grew about three and a half inches (nine centimeters) in the first two and a half days. And then, of course, you reenter and give it all back in about half a second when you stand up on Earth.

What highlights the fragility of Earth, especially to me, was realiz ing that the little band around the limb—the horizon—is all you've got, with fifty miles (eighty kilometers) of atmosphere separatin you from all that bad stuff out in space. I was a Boy Scout, and it was always "Leave the campground cleaner than you found it. I think everybody who has been to space comes back more of an environmentalist than they were than before they went.

The other thing: With my militar background, it was kind of odd working with the Russians at NASA, because I spent the first third of my working life trying to kill Russians more efficiently. Now we were working with them and it was way better. Because I'm old, I tend to think of maps in geopolitical terms like boundaries red countries, and blue countries, and green countries, and so on. Of course, you don't see any of that when you're on orbit. It's just oceans that turn into beaches, which turn into forests. I came back scratching my head about why it is that we continue nationa istically to try to make life miserable for each other.

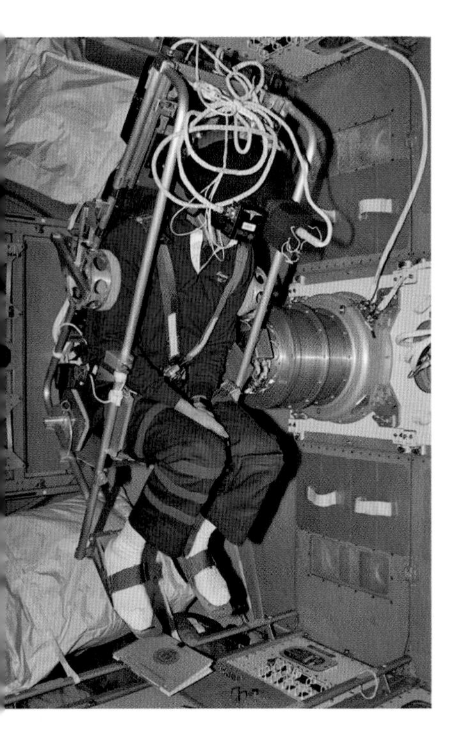
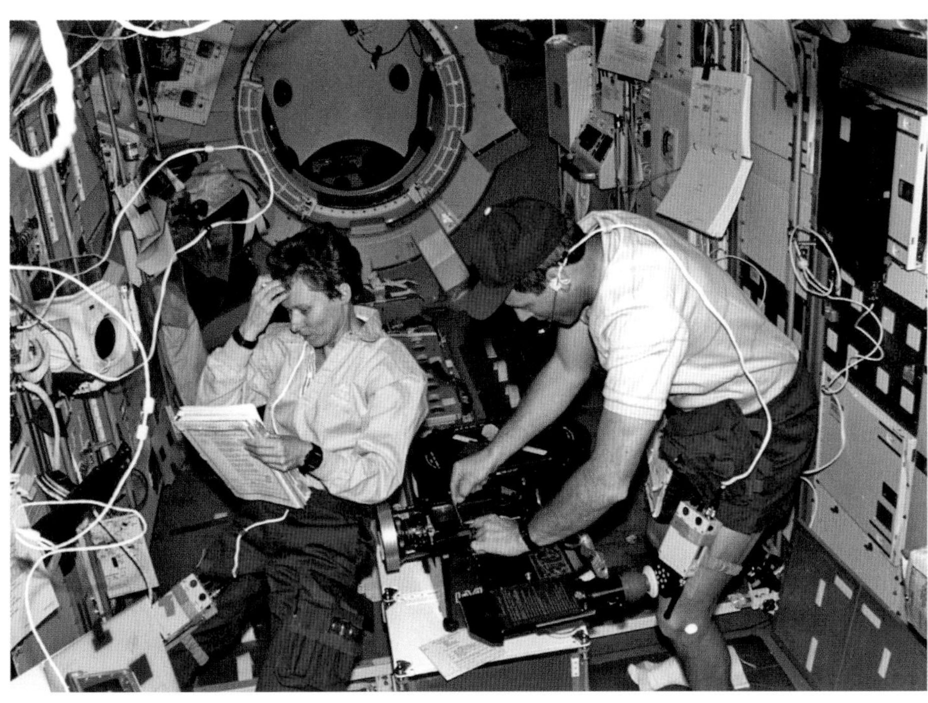

ABOVE LEFT: **Dave Hilmers rides the rotating chair in a Spacelab vestibular experiment.**

ABOVE RIGHT: **Roberta Bondar and Steve Oswald work inside IML-1, with the tunnel to the middeck at top.**

BELOW: **Ron Grabe and Oswald fly *Discovery* through reentry plasma generated by its hypersonic passage.**

Mission no. **46**

STS 45

Orbiter	Atlantis
Launch	March 24, 1992
Landing	April 2, 1992
Duration	8 days, 22 hrs., 9 min. 26 secs.

STS-45's crew on the flight deck: (front row) Kathy Sullivan and Charlie Bolden; (back row) Dave Leestma, Brian Duffy, Byron Lichtenberg, Dirk Frimout, and Mike Foale

Crew	Charles F. Bolden Jr., Brian Duffy, Kathryn D. Sullivan, David C. Leestma, C. Michael Foale, Byron K. Lichtenberg, Dirk D. Frimout

Mission: The astronauts worked with ground investigators to operate the Atmospheric Laboratory for Applications and Science (ATLAS-1), comprising twelve instruments from the US, France, Germany, Belgium, Switzerland, the Netherlands, and Japan. ATLAS-1 focused on atmospheric chemistry, solar radiation, plasma physics, and ultraviolet astronomy.

Dave Leestma, Mission Specialist:

The mission was a pallet Spacelab, with a dozen or so experiments all crammed together in the payload bay. We worked twenty-four-hour operations using two shifts: Charlie Bolden and Brian Duffy, the commander and pilot, were on the blue shift, and I was the head of the red shift. Kathy Sullivan, payload commander, was on Charlie's shift, and Mike Foale was on my shift, and we also had two payload specialists.

The theme really was atmospheric physics and plasma physics; we spent a lot of time looking at the Sun and its interaction with the atmosphere. For me, maneuvering the orbiter to point the payload sensors at specific targets was fun. We did three or four orbital maneuvers every orbit. We'd point various instruments at the Sun rising, then back down at the atmosphere, then over at the setting Sun. We did more than 300 maneuvers on that flight.

We were one of the first crews to use bright-light therapy before we launched to get our circadian rhythms adjusted. And so for my shift, right after we got on orbit and after post-insertion, we went to bed. That is not easy to do after you have about a quart of adrenaline in your body after launch. When we got up, we went right into the two-shift operation.

On about day three, they were first firing this plasma generator pointed at certain places on the Earth. It would build up a charge and they'd fire it, and it looked very much like a *Star Trek* starship firing a photon torpedo. These things came "POW!" out of the payload bay, and then you'd look way, way down at the Earth, where all of a sudden, this light streak—"PSHEW!"—would appear at the top of the atmosphere. And we were saying, "Wow! Is that neat?!"

Kathy, being the geographer, really enjoyed finding certain places on the ground to look for specific things that she may have studied or that interested her. I remember trying to find Mount Everest. We had some maps that showed a little lake that looked like a figure-eight. We called it "Dog Bone Lake." If you looked right down the axis of Dog Bone Lake just to the south, it pointed right at Mount Everest. And we had a couple of clear passes over the Himalayas where we could say, "Yep! There it is."

It was surprising how capable the space shuttle orbiter was as a platform for scientific instruments because you could operate twenty-four hours a day. The space shuttle is very maneuverable, whereas the Space Station is not. With the shuttle, you could mount these instruments with precise pointing requirements. Doing that for eight and a half days, we got a lot of good data that surprised the scientists.

ABOVE: **ATLAS experiments mounted on Spacelab pallets in the payload bay, flying over the western Sahara.**

BELOW LEFT: ***Atlantis*'s jettisoned external tank, blackened by main engine and separation motor firings.**

BELOW RIGHT: **ESA payload specialist Dirk Frimout controls an ATLAS spectrometer using an aft flight deck display.**

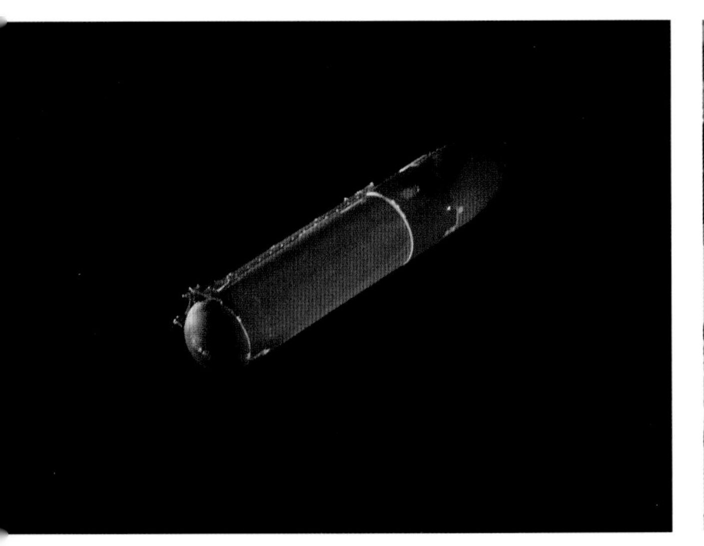

Mission no. **47**

STS

49

Orbiter	Endeavor
Launch	May 7, 1992
Landing	May 16, 1992
Duration	8 days, 21 hrs., 17 mins., 39 secs.

The STS-49 crew at Edwards Air Force Base, California, after landing: Rick Hieb, Kevin Chilton, Dan Brandenstein, Tom Akers, Pierre Thuot, Kathy Thornton, and Bruce Melnick

Crew	Daniel C. Brandenstein, Kevin P. Chilton, Pierre J. "Pepe" Thuot, Kathryn C. Thornton, Richard J. Hieb, Thomas D. Akers, Bruce E. Melnick

Mission: Astronauts on *Endeavour*'s first flight retrieved the crippled Intelsat VI satellite to give it a new perigee kick motor. During the only three-person EVA in space history, Pierre Thuot, Rick Hieb, and Tom Akers captured the satellite by hand, enabling its successful relaunch.

Kevin Chilton, Pilot:

I was the only rookie on the flight. I think the others enjoyed watching me learn, and I certainly enjoyed all the mentoring I got from folks who had flown before.

We had individual strengths that combined into common strengths. Bruce Melnick was unbelievable on the robotic arm, and our EVA crews—Tom, Rick, Pierre, and Kathy Thornton—they just knew their stuff. My job was to be the expert on the rendezvous procedures. Dan Brandenstein was a tremendous leader, with rendezvous experience from a previous flight. Our mission was the first flight of *Endeavour*, the orbiter built to replace *Challenger*. That was special from a historical perspective, but then the mission turned out to be a little more special than we'd anticipated.

As Dan flew the final approach to the first rendezvous and capture attempt, I remember looking over his shoulder out the window, and I made the comment: "This is just like the simulator, except it's in color!" The team thought we had a pretty good plan. But the EVA capture failed—twice—to grab Intelsat VI, so we had to come up with a better plan.

Dan, Rick, Bruce, and I were on the flight deck. I was looking out the back window, and as God is my witness, an Air Force class called "Problem Solving" came into my mind. I couldn't remember all the steps, but I remembered step one was to "Define the problem."

After a long discussion, we came to the conclusion that as Pierre was pushing the bar against the satellite, it was moving ever so slightly away, so when the bar's claws fired, it wasn't grabbing the satellite rim. To solve that problem we had to keep the satellite from moving away while Pierre applied pressure. Step one of problem solving.

Then step two was to "Inventory your assets." We looked outside and asked ourselves, "What do we have in the cargo bay that we don't ordinarily have on the shuttle?" We asked the same question about the inside of the shuttle. We had extra spacesuits. We went through all the different things we had at hand, including a tool inventory. It reminded me of the Apollo 13 movie scene where they come in with a box of stuff and dump it on the table and Gene Kranz says, "I suggest you gentlemen invent a way to put a square peg in a round hole. Rapidly."

Then we started the next step in problem solving: "Brainstorm." And we went around, and Bruce, our arm operator, said, "Why don't we put three people out?" What an out-of-the-box thought! The team dynamic was really magical.

All the things that went wrong, the team overcame. The various problems that had to be solved by both the crew on orbit and on the ground—I don't think there's another mission that ended successfully that compares.

ick Hieb, Tom Akers, nd Pierre Thuot install a andling fixture on Intelsat I after Dan Brandenstein ositioned *Endeavour* eneath the satellite.

Mission no. **48**

STS

50

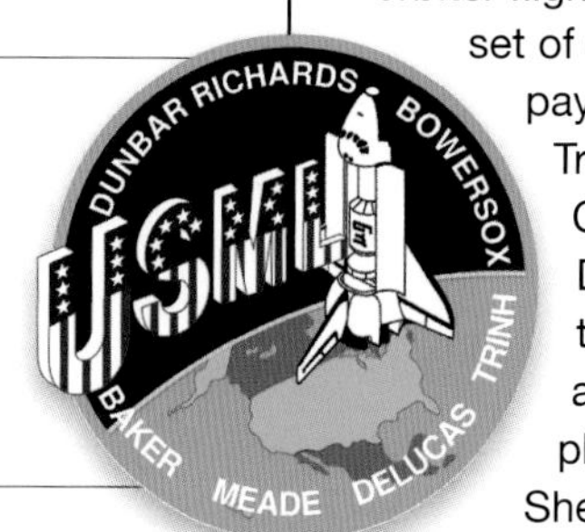

Orbiter	Columbia
Launch	June 25, 1992
Landing	July 9, 1992
Duration	13 days, 19 hrs., 30 mins., 4 secs.

The STS-50 crew inside Spacelab: (front row) Carl Meade, Ellen Baker, and Gene Trinh; (second row) Larry DeLucas, Dick Richards, and Bonnie Dunbar; (top) Ken Bowersox

Crew	Richard N. Richards, Kenneth D. "Sox" Bowersox, Bonnie J. Dunbar, Ellen S. Baker, Carl J. Meade, Lawrence J. DeLucas, Eugene H. Trinh

Mission: The crew ran the USML-1 with thirty-one experiments, laying the groundwork for future ISS science operations. Commander Dick Richards and Pilot Ken Bowersox showed that precise orbiter landings were possible even after extended free fall exposure.

Dick Richards, Commander:

For the first extended-duration orbiter flight, I had a really good set of crewmembers. The payload crew—Eugene Trinh, Larry DeLucas, Carl Meade, and Bonnie Dunbar—they did a fantastic job. Bonnie was all over the research plan from the get-go. She had the Spacelab activities organized, and the crew all had prepared and knew what they were going to do back there. If the orbiter behaved itself so that we could concentrate on the lab, I knew the mission was going to be a success.

Astronauts are almost all type-A personalities. My experience with them is that if you can remove the barriers and just give them general guidance, they'll all do the right thing. That's the way my first commander, Brewster Shaw, operated, and that was my philosophy.

In order to get the most benign microgravity environment, which is what our mission was all about, we ended up flying tail down, payload bay forward. It turned out that was not a good attitude to fly in because of micrometeoroid and orbital debris impacts. I detected several hits on the overhead windows during the flight. I could see little starburst patterns there, and of course I recorded what I saw, but I thought, "Boy, if that's what we're seeing on the windows, I wonder what the payload bay doors and radiators are getting." My recollection of the postflight report is that we took quite a few hits, and after our mission, I believe there were flight rules saying, "We're not gonna use that attitude anymore."

When Sox and I got off the flight, we were able to get out and walk around the orbiter. We got back by the right external tank door, and there was a huge, eight-by-ten-inch (twenty-by-twenty-five-centimeter) section of tile that had been hit by something. I think this was the second flight where a big piece of foam loosened and fell from the ET. That's what eventually undid *Columbia* on STS-107. On our mission, the debris missed the leading edge, but it created a big gouge. We never knew anything about it in orbit—we didn't have an arm for inspection. That was a close call for us.

There was a cyclone moving up from Mexico, just to the west of the United States, and it had spread a lot of clouds over the landing area at Edwards. It was the biggest storm I'd ever seen, and we were going over it in the dark. At the time, there must have been a hundred thunderstorms in that thing. All we could see was just cloud-to-cloud lightning everywhere.

Research had suggested that sometimes lightning travels up. All of a sudden, I got in the back of my mind, "Gosh, the next time we get over this hurricane, we're going to be a lot lower, and we're going to be streaming a fifty-mile-long river of plasma. I sure hope they are incorrect about lightning going up." Just then, Mission Control said, "Oh, we're going to go to Florida." That meant we were going to miss the thunderstorms over Edwards, and I just relaxed completely. I said, "I think this is really a good idea."

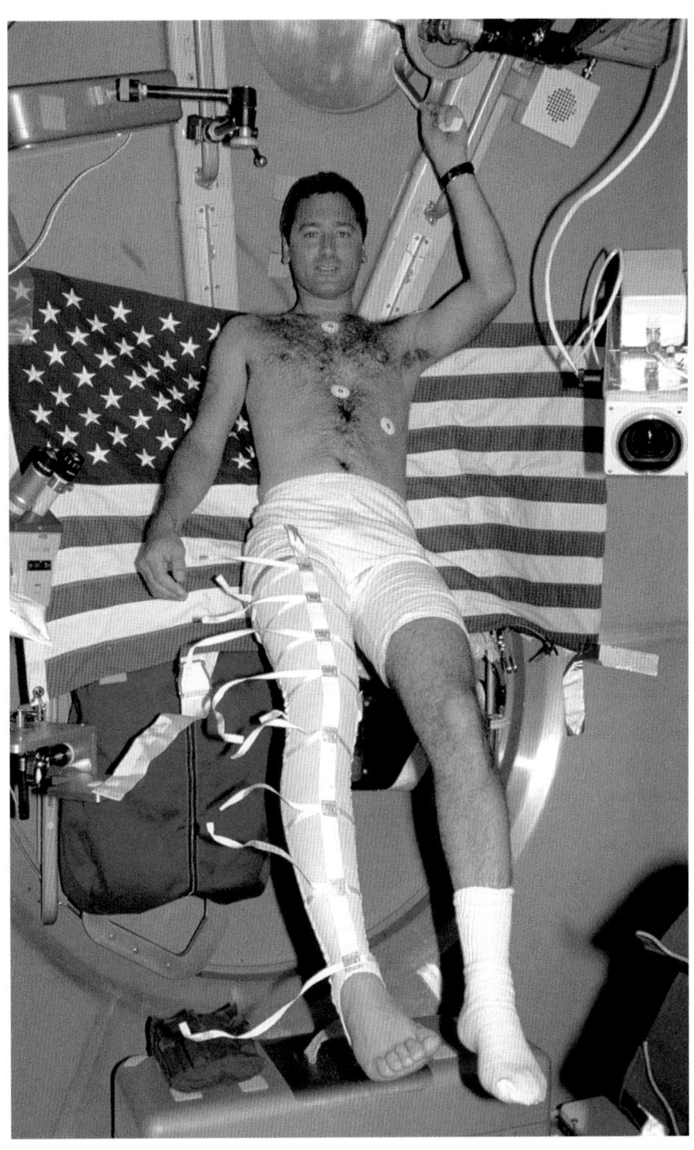

ABOVE LEFT: **Dick Richards (in red) and Bonnie Dunbar share a meal on *Columbia*'s middeck.**

ABOVE RIGHT: **Larry DeLucas measures his leg muscle volume inside Spacelab.**

LEFT: **After the 14-day mission, the drag chute slows *Columbia* during rollout on Kennedy Space Center Runway 33.**

Mission no. **49**

STS

46

Orbiter	Atlantis
Launch	July 31, 1992
Landing	August 8, 1992
Duration	7 days, 23 hrs., 15 mins., 2 secs.

The STS-46 crew on the middeck: (front) Jeff Hoffman, Marsha Ivins, and Franco Malerba; (back) Claude Nicollier, Loren Shriver, Andy Allen, and Franklin Chang-Diaz

Crew	Loren J. Shriver, Andrew M. Allen, Jeffrey A. Hoffman, Franklin R. Chang-Díaz, Claude Nicollier, Marsha S. Ivins, Franco Malerba

Mission: Astronauts deployed ESA's EURECA satellite for a microgravity research mission. The crew began unreeling the Tethered Satellite to explore the physics of Earth's magnetic field and the behavior of tethered, orbiting bodies. A tether reel jam stopped deployment at 850 feet (259 meters)—well short of the planned twelve miles (nineteen kilometers).

Andy Allen, Pilot:

The primary part of the mission was the Tethered Satellite. It's rather like Benjamin Franklin in space: We put this kite out, in the form of a conducting satellite on a string, which is twelve miles (nineteen kilometers) long, and we had electron guns and other things down in the payload bay conducting electrical experiments seeing how electricity and the Earth's magnetic field interact in space.

The European Retrievable Carrier (EURECA) consisted of fifteen passive experiments for long-duration exposure to space. We were to drop it off, and then another shuttle would come up later and pick it up. When we released EURECA, my job was to fly formation at a distance of 1,000 feet (305 meters) for about two hours while ground controllers tested it. We had the K_u-band radar working so you could get the range and the range rate on the satellite. In the dark, I would maintain the range using the radar, and in daylight I could fly visually and use the radar.

As luck would have it, we lost communication with the ground on one night pass during the checkout. Losing comm was pretty standard back in the early days. We had about thirty minutes with no connection to the ground, and I was up there on the flight deck by myself. I couldn't see the satellite or the reflectors, but I was watching the K_u-band radar, reading the range and range rate. Then EURECA's range started decreasing, and it was picking up speed as it was closing on us. The range rate started increasing pretty significantly.

I started firing jets, trying to keep my distance, but it was still closing in on us. Finally, it had come inside 600 feet (183 meters) with a high closure rate. It was probably within thirty seconds of impact if I hadn't taken action. I finally got aggressive, thinking that if the radar's right, then I had to start firing the bigger RCS thrusters—which may hurt EURECA's solar panels and damage the satellite.

I started firing. When you fire those big RCS jets, especially up and forward, it's like firing a little howitzer: "Ba-Boom, Ba-Boom, Ba-Boom!" Commander Loren Shriver hollered up from the middeck, "What are you doing up there?" My response was "I don't know."

The satellite kept coming, probably getting as close as 500 feet (152 meters). Then I finally sidestepped well enough with the thruster firings, and as I got out of the orbital plane we shared with EURECA, the radar lost its lock on the satellite. When that happened we had no idea where the satellite was. We didn't know if it was going to hit us or if it was damaged. But I got out of plane enough that I was pretty sure it wasn't going to impact.

Everyone got to a window and started looking outside. I was thinking that I had just destroyed the satellite on my first and what could be my only flight. Finally, Jeff Hoffman found the satellite, and the solar arrays were sticking out like they're supposed to, and it looked beautiful. It turns out that an errant command from EURECA's European control center had mistakenly sent the satellite thrusting toward us.

But hats off to the people who trained us. Sometimes we get unhappy with them, but they come up with every scenario they can think of for what possible failures might occur. The bottom line: When it's not right, evaluate it, and figure out what you're going to do

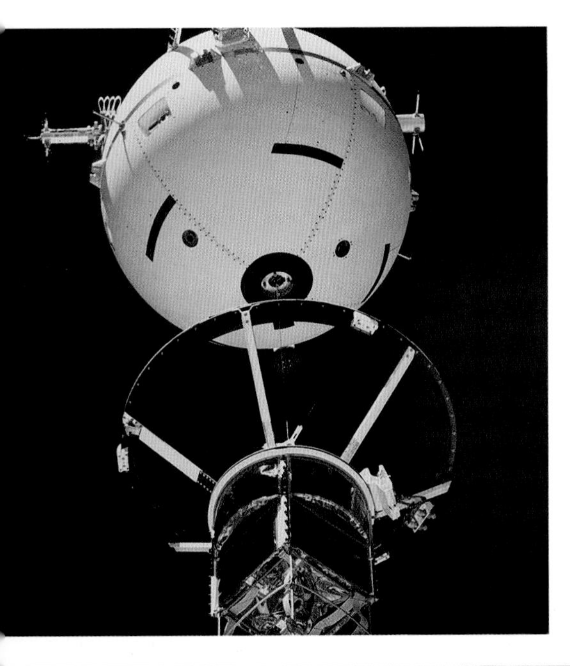

LEFT: **The Tethered Satellite reels out from its deployment mast on *Atlantis*.**

RIGHT: **Andy Allen prepares for reentry in *Atlantis*'s pilot's seat.**

BELOW: **EURECA orbits above Cape Canaveral after a long night pass.**

Mission no. **50**

STS

47

Orbiter	Endeavour
Launch	September 12, 1992
Landing	September 20, 1992
Duration	7 days, 22 hrs., 30 mins., 22 secs.

The STS-47 crew in Spacelab J: (front) Jan Davis, Mark Lee, and Mamoru Mohri; (back) Curt Brown, Jay Apt, Hoot Gibson, and Mae Jemison

Crew	Robert L. "Hoot" Gibson, Curtis L. Brown Jr., Mark C. Lee, N. Jan Davis, Jay Apt, Mae C. Jemison, Mamoru Mohri

Mission: The crew operated Spacelab J with its twenty-four materials science and twenty life sciences experiments. Mae Jemison, the first African American woman in space, joined the first Japanese shuttle astronaut, Mamoru Mohri. Jan Davis and Mark Lee were the first married couple to fly together in space.

Jay Apt, Mission Specialist:

Ours was the first cooperative mission between the United States and Japan. We called it Spacelab "J," for Japan, and it was one of the most successful pieces of space shuttle diplomacy. The Japanese Experiment Module (JEM) on the ISS is a direct result. All of the Japanese astronauts who flew on the shuttle and on the Station tapped into the experience of Mamoru Mohri on this mission.

Mamoru was the first professional Japanese astronaut to fly, and it was a big deal in Japan. I remember John Glenn's 1962 flight pretty well, when he became the first American to orbit Earth. In Japan, Mamoru's flight was even bigger than that. When we went to Japan after the flight, there were shopkeepers hanging out of the windows of the third and fourth floors of department stores as we passed by, shouting out his name.

The scientific cooperation started with the assignment of the payload crew. That was Jan Davis, Mark Lee, and Mae Jemison from our side of the house. Then there was Mamoru as the prime payload specialist. That gaggle of folks had been training over at Mitsubishi Heavy Industries and all of the other places in Japan for a good long time. Jan was my EVA partner, if we'd had to go outside. I went with her when she did her spacesuit checkout in the vacuum chamber preflight, and she was wonderful through all of that.

This was nearly a failed mission. On the second morning of the flight, there was a cooling problem that affected the furnaces and all the other chemistry-based experiments back in the Spacelab. Curt Brown and Mark, both being very mechanically inclined, figured out what the problem was. They did an inflight maintenance procedure with help from the folks on the ground. They really saved that flight.

We were flying at high enough inclination that we were able to see the aurora, mostly below us. It was principally a green aurora, with some red stuff. Seeing it dance was terrific.

Because the Spacelab was big, we were able to do some fun stuff. We spun Curt up in the tunnel and then stopped him. His eyeballs didn't stop; they kept rotating!

Mamoru was also able to bring up some Japanese food and chopsticks. He prepared three or four dinners. Anything that wasn't an MRE, or what the military calls a meal-ready-to-eat, was a welcome relief from our rations.

The most long-lived result of the flight is the tremendous—and not well-appreciated—Japanese-US cooperation, which has not had anywhere near the bumps in the road nor the publicity that the US-Russian cooperation has.

ABOVE: **Greens and reds of auroral curtains produced by excited oxygen atoms, photographed from *Endeavour.***

RIGHT: **Mae Jemison enjoys a free fall break from research in the Spacelab J module.**

BELOW: **Curt Brown tapes down a suit air circulation fan to add cooling capacity to a middeck experiment.**

Mission no. **51**

STS

52

Orbiter	Columbia
Launch	October 22, 1992
Landing	November 1, 1992
Duration	9 days, 20 hrs., 56 mins., 13 secs.

The STS-52 crew preflight portrait: (front row) Lacy Veach, Tammy Jernigan, and Bill Shepherd: (back row) Mike Baker, Jim Wetherbee, and Steve MacLean

Crew	James D. Wetherbee, Michael A. Baker, Charles L. "Lacy" Veach, William M. Shepherd, Tamara E. "Tammy" Jernigan, Steven A. MacLean

Mission: Astronauts and the ground team operated US Microgravity Payload-1 (USMP-1) and its seventeen experiments. The crew also deployed LAGEOS II, which rocketed into a 3,700-mile (5,957-kilometer) orbit, serving as a long-lived laser ranging benchmark.

Steve MacLean, Payload Specialist:

When the pad technicians close the side hatch and pull the swing-arm away, you're sitting there with your crew, who have become your brothers and sisters, and you feel a lot of emotions. You wonder how you are going to feel when you launch because it's still risky business. You're sitting on a veritable bomb. The pyrotechnics and the solid rockets have to light within milliseconds of each other. If they don't, it's a bad day—the unbalanced thrust would tear the shuttle apart. So, I wondered: Would my heart be beating faster? Would I be a little bit nervous?

The point where I felt the most reflective—I don't want to say afraid—was when I left my wife and my two small children the day before. That was the point where I felt irresponsible. I couldn't see my kids growing up by themselves. And I thought, "How can I do something like that?" But I tell you without hesitating for a second, I was so happy to be on the pad. After that long delay, that happiness displaced any sense of fear that I should have had.

And I really enjoyed the ride. Tammy Jernigan was on the middeck with me, and when we hit the fifty-mile mark, she gave me a big high five. "Welcome to space!"

The Laser Geodynamic Satellite (LAGEOS) was interesting to me because it was going to measure the motion of Earth's tectonic plates with a .4-inch (one-centimeter) resolution. That accuracy was brand new. It's a passive satellite, just a bunch of retroreflective mirrors, and it's the equipment on the ground that does all the work. The satellite contributed to the confirmation of Einstein's theory of frame dragging, a space-tim effect where a satellite's motion can distort the gravitational space time field around it. It will be eight million years or so before it falls back down to Earth, and it's easy for the ground to go on continuall measuring this forever. You just fe proud to be a part of that.

One of the medical experiments I performed was a leukem experiment, where we took up leukocyte cells with a population of normal cells and a population cancer cells. The concept was th you take stem cells—leukocyte cells from a child—separate them on orbit, bring them back, and then reinsert them into the body for stem cell transplants.

When we separated those cells on orbit, at first the cancerous cells and normal cells appeared homogeneous in the solution. The process of cell separation was like watching two wormlike entities growing and moving away from each other. It was like watching an impressionist paintin slowly evolve, and I became so involved in the artistic nature of this separation process that I act ally forgot for a few minutes that I was in space. These biomedical experiments were important—and also beautiful. Out of 135 shuttle flights, more than 100 had medic experiments related to the study of cancer and the immune system.

STS-52 advanced Canadian space cooperation with the United States. We flew medical experiments, materials science experiments, and upper atmospheric experiments, and we had an instrument out on the end of the robot arm that allowed us to measure how much atomic oxygen was present. We also tested the guidance system for th arm, which was a big part of the Space Vision System used in ISS assembly.

ABOVE: ***Columbia*'s microgravity payloads operate above a stormy Kalahari Desert in southwest Africa.**

RIGHT: **Steve MacLean, once a member of Canada's national gymnastics team, tumbles in free fall on the flight deck.**

BELOW: **Lacy Veach, raised in Honolulu, photographed his grandfather's stone adze with the Big Island below; this personal memento was quarried on Hawaii Island.**

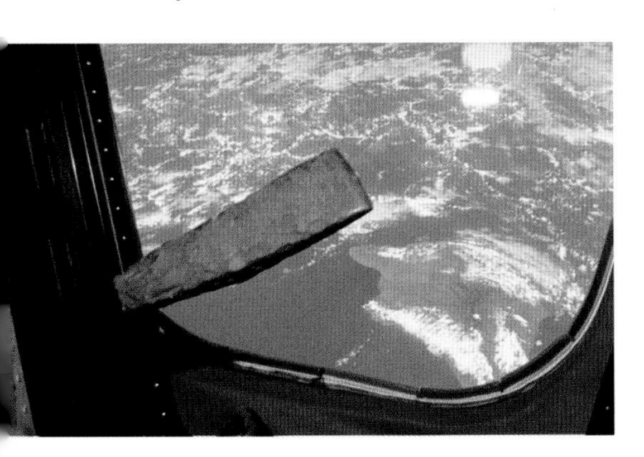

Mission no. **52**

STS 53

Orbiter	Discovery
Launch	December 2, 1992
Landing	December 9, 1992
Duration	7 days, 7 hrs., 19 mins., 47 secs.

The STS-53 crew before launch: Guy Bluford, Dave Walker, Bob Cabana, Jim Voss, and Rich Clifford

Crew	David M. Walker, Robert D. "Bob" Cabana, Guion S. "Guy" Bluford Jr., James S. Voss, Michael R. "Rich" Clifford

Mission: *Discovery* astronauts deployed the mission's primary classified payload, DOD-1, the last major spacecraft carried for the Department of Defense. Subsequent military payloads were launched on expendable rockets. The crew also operated eleven research and technology experiments in the crew cabin and cargo bay.

Jim Voss, Mission Specialist:

The payload we deployed on this last DOD flight is one that we still can't talk about today. This payload was pretty big, so it needed the volume of the payload bay. One of the requirements for the shuttle's design was a cargo bay big enough for DOD payloads.

The training team nicknamed us the Dog Crew. The way they put it, training astronauts was a lot like training dogs: There was a lot of repetition, rewards, and a lot of punishment. Dave Walker, the commander, with red hair and his Navy call sign "Red Flash," became Red Dog—a natural. Bob Cabana, because he was so good at everything he did in the cockpit, was Mighty Dog. Rich Clifford was the only rookie on the crew, so without his consent—or his liking—he became Puppy Dog. Guy Bluford joined us pretty late, and because he was often away doing public relations things, we named him Dog Gone. They didn't have a dog name for me, and when Guy came back, they asked him for ideas. "Sure! He's an infantry soldier, so 'Dog Face!'" That was it.

We had a truly wonderful, compatible crew. Dave was really good at fostering friendships and teamwork, he's the kind of guy who would do anything for a friend. Dave decided that we needed special transportation for the crew, so he bought this used station wagon. I think he paid $600 for it, and it was a piece of junk. It was spray-painted flat black, and Dave named it the "Dogmobile."

We launched into a 57-degree inclination orbit at a time of year where we ended up tracking along the terminator, the dividing line between day and night, for almost the entire flight. We had this partial light, partial darkness thing the whole time. We had told Rich Clifford about how beautiful the sunrises and sunsets were, and when we got up there and were ready to watch one with him, the Sun dipped down close to the horizon—and then it went back up again. I don't think we ever had a true, color-explosion sunrise or sunset the whole time.

Diverting into Edwards, Dave got distracted while he was turning toward final. You don't build up a lot of g's on reentry, just a tiny bit, but he had his sunglasses up on the overhead, stuck there with Velcro, and they fell during the turn. Dave decided to retrieve the sunglasses. He bent over, got off guidance, and Bob was saying, "Dave, Dave! We've got to get back!" Dave, instead of sitting up, forgetting the sunglasses, and getting us back on track, said, "Bob, take over." Bob got us back on, and then Dave sat up and took over again. It was kind of funny. If you hear Bob tell the story, he saved the day.

It was a very successful flight. We dealt with a lot of different people in the payload world—wonderful people. They knew they were doing important stuff for our nation by designing, developing, and running the operations for DOD-1.

Discovery and the Moon orbit above Earth's terminator on STS-53.

Mission no. **53**

STS 54

Orbiter	Endeavour
Launch	January 13, 1993
Landing	January 19, 1993
Duration	5 days 23 hrs. 38 mins. 17 secs.

The STS-54 crew in orbit: Greg Harbaugh, Don McMonagle, Susan Helms (top), John Casper, and Mario Runco

Crew	John H. Casper, Donald R. McMonagle, Mario Runco Jr., Gregory J. Harbaugh, Susan J. Helms

Mission: *Endeavour*'s astronauts deployed TDRS-F to support shuttle and scientific satellite communications and navigation. Its IUS propelled TDRS-F into geosynchronous orbit, where it is still in active service. Astronauts Mario Runco and Greg Harbaugh conducted a five-hour spacewalk to test ISS tools and assembly techniques.

John Casper, Commander:

The United States has a space-based communications system for all its low Earth orbit satellites, and a minimum of three operational satellites is required to provide continuous coverage. In 1993 only two of the four orbiting TDRS spacecraft were fully functional, and STS-54's mission to deploy an additional TDRS took on national significance.

Our educational activity, called "physics of toys," demonstrated toys in space to teach science and technology concepts to children in an exciting way. We downlinked real-time video with four of our hometown schools. The kids were amazed, and we laughed along with them while demonstrating toys like a wind-up flapping bird, which did cartwheels floating in the cabin, completely unlike its gliding behavior on Earth.

We also had a spacewalk added to the flight about two months before launch. With future EVAs to service the Hubble Space Telescope and assemble the ISS, the spacewalk was added to answer questions about difficult spacewalking tasks. For example, could one spacewalker easily move a large object down the length of the payload bay?

I thought, "Well, we've done EVAs before, we're trained to do them, and we know how to do them." It didn't really hit me until they actually stepped out of the airlock door, and I thought, "Oh my gosh! Those guys are out there in the vacuum of space and any one of a thousand things could go wrong and take them out. They're really hanging it out!" I was very happy when they came back inside the shuttle four and a half hours later.

Looking at Earth below, I was awestruck by the "oneness" of our humanity—every human being who has ever existed has lived on this one planet. Earth looks so beautiful against the blackness of space, with vibrant colors of blue oceans, green forests, and white clouds. Earth is the only planet we know of that has air we can breathe, water we can drink, plants growing, and animals running around. Seeing Earth from this perspective made me realize how much we need to take care of the "garden" we live in.

STS-54's reentry was visually spectacular, because the initial part was at night. As we descended into the atmosphere, I noticed a flickering of white light behind me, erratically flashing and lighting up the cockpit. I turned in my seat to look out the overhead windows and saw the charged plasma trail that our reentry was creating. It was snaking behind us, pulsing and flashing with white-hot flame. Out the front windows was the reddish-orange glow of reentry heat. We were literally inside a fireball speeding through the sky like a meteor.

What surprised me was the shuttle's tremendous versatility. It was a launch vehicle that took the huge TDRS and its booster rocket into orbit. It was a laboratory for our thirty-plus experiments outside in the payload bay and inside the crew cabin. We configured the middeck as a classroom to teach the physics of toys. The shuttle airlock enabled spacewalks, which Greg and Mario performed. The payload bay could return large, heavy satellites or equipment to Earth. Now that the shuttle is retired, I think it's going to be a long time before we fly another vehicle that can match all of its capabilities.

ABOVE: **Susan Helms helps Greg Harbaugh and Mario Runco suit up in the middeck.**

RIGHT: **Harbaugh hangs out below *Endeavour*'s starboard sill during the EVA.**

BELOW: **The TDRS-F satellite deploys from the payload bay atop its IUS.**

Mission no. **54**

STS

56

Orbiter	Discovery
Launch	April 8, 1993
Landing	April 17, 1993
Duration	9 days, 6 hrs., 8 mins., 19 secs.

The STS-56 crew on the flight deck: (front row) Ken Cameron, and Mike Foale; (back row) Ellen Ochoa, Steve Oswald, and Ken Cockrell

Crew	Kenneth D. Cameron, Stephen S. "Oz" Oswald, C. Michael Foale, Kenneth D. "Taco" Cockrell, Ellen Ochoa

Mission: *Discovery*'s crew operated the seven-instrument ATLAS-2 experiment suite in the cargo bay. It studied the Sun's energy output and its effects on Earth's middle atmosphere and ozone layer. The astronauts released SPARTAN 201 to study the solar wind and corona, later retrieving it with the RMS.

Ellen Ochoa, Mission Specialist:

It was the second of three flights of the ATLAS. The idea was to look at the upper atmosphere to try to understand the concentration of trace molecules, particularly the byproducts of chlorofluorocarbons, or CFCs, and really study them in detail. We also wanted to measure the total amount of solar radiation, by wavelength, trying to get the very detailed information that would help researchers tease out the differences between human activities and the natural variation of the Sun and how the Sun's activity contributed to ozone depletion. Two years after we flew, the scientists who had made the discoveries about what destroys ozone and the connection to chlorofluorocarbons got a Nobel Prize.

We were launching at night. Oz Oswald talked to Taco Cockrell and me on the flight deck: "Okay, when the SRBs separate, there's going to be this huge flash of light. I suggest you close one eye to keep it dark-adapted." I remember Taco saying, "Or maybe both eyes!"

After the solid boosters separate, you go from feeling three g's back to slightly over one g. I remember feeling that we'd just stopped. It's partly that the acceleration goes away, and partly that you had all this vibration from first stage, and that disappears. I remember thinking for just a few seconds, "This cannot be good!" Fortunately, I was on the flight deck and I could see the instruments and see that the engines were still ticking away. Not too long afterward, there was another performance call from Mission Control confirming our progress, and I thought, "Okay. It's all good It's all good."

I was the prime arm operator and Oz was my backup. We used it for deploying another spacecra —SPARTAN—into orbit, and then the satellite operated in free flight for two days. Then we rendezvoused with it so I could capture it.

When we actually deployed SPARTAN, it turned out we were over Greece. I remember Ken Cameron saying, "We just deployed SPARTAN over Sparta!" Then he turned to me and said, "And there's our 'Ellen of Troy.'"

Our 57-degree orbital inclinatio gave us amazing views of Earth. We also had to be on the flight deck for every sunrise and every sunset, because we had to start a video recorder for an instrument called the Atmospheric Trace Molecule Spectroscopy Experiment— ATMOS, for short. They wanted the video recording in addition to the spectrometer data they were collecting. So we got to watch every sunrise and every sunset up there.

I was almost a music major, but I made the right decision by choosing physics instead. I wanted to take my flute because it was important to me. I think Ken Cameron was the one who came up with the idea of using it for our "Liftoff to Learning" recording. As part of this video, we thought we could add a scene about how astronauts pursue their hobbies in space, just like kids do on Earth. That was the only way I got my flute onboard. I played the Marine Corps hymn, the Navy hymn, *God Save the Queen,* and then some Vivaldi and Mozart. From Vivaldi, I played a selection from the *Four Seasons,* and then the Mozart *First Flute Concerto in G Major.*

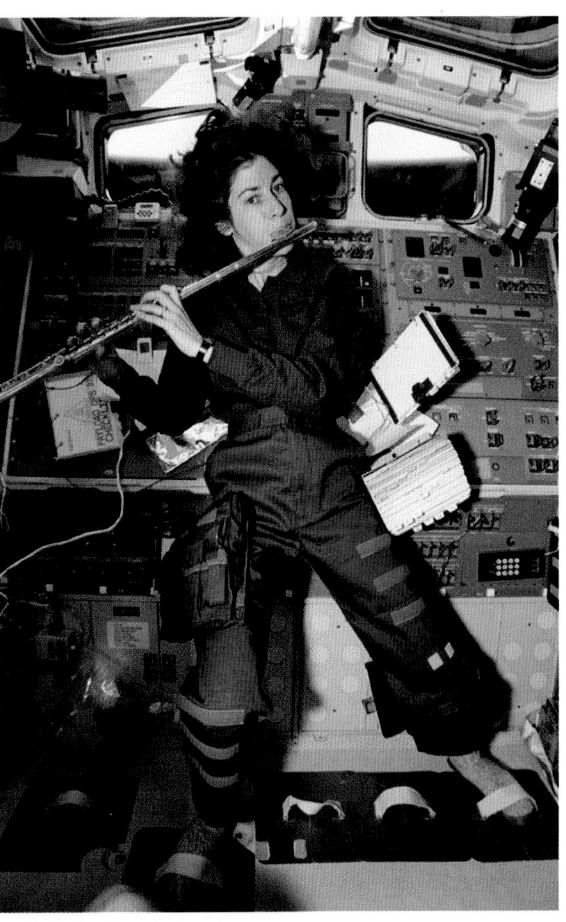

ABOVE: ***Discovery*'s STS-56 liftoff from the Kennedy Space Center's Pad 39B in Florida.**

RIGHT: **Ellen Ochoa plays the flute on *Discovery*'s flight deck.**

BELOW: **SPARTAN soars above Crete after its release from the shuttle's robotic arm.**

Mission no. **55**

STS

55

Orbiter	Columbia
Launch	April 26, 1993
Landing	May 6, 1993
Duration	9 days, 23 hrs., 39 mins., 59 secs.

The STS-55 crew in Spacelab: (front) Tom Henricks, Steve Nagel, Ulrich Walter, and Charlie Precourt; (back) Bernard Harris, Hans Schlegel, and Jerry Ross

Crew	Steven R. Nagel, Terence T. Henricks, Jerry L. Ross, Charles J. Precourt, Bernard A. Harris Jr., Ulrich Walter, Hans W. Schlegel

Mission: After an engine failure forced an abort three seconds before planned liftoff on March 22, *Columbia*'s astronauts returned in April to launch into a 190-mile (361-kilometer) high orbit. The crew worked alternating twelve-hour shifts to conduct experiment operations in the German-sponsored Spacelab D-2.

Bernard Harris, Mission Specialist:

Most notable about STS-55 was that we were only the third launch pad abort. After the main engines ignited and just as we were getting ready to lift off, the master alarm went off. The main engines got cut off after three and a half seconds. I was sitting in the right rear seat on the flight deck where I could look out the overhead windows. I had a wrist mirror, so I pointed the mirror out and looked down the side of the vehicle toward the flame trench. After the engines cut off and the master alarms were turned off, what I saw was a hydrogen fire rolling up the side of the vehicle. I thought, "Holy shit! We're gonna die!" Fortunately, the launch controllers quickly got everything safed.

Jerry Ross on the middeck said, "Are we moving?" My eyes shifted from the overhead window to the commander and pilot who were throwing switches. As I looked out Steve Nagel's window, I saw the gantry moving. I was thinking, "The launch pad's moving. No, it's not the pad. It's us!" We swayed six feet (two meters) one way, and then six feet (two meters) the other direction. Eight three-and-a-half-inch-diameter (nine-centimeter-diameter) bolts clamp the shuttle's boosters to the pad. Going back and forth, I was thinking, "I sure hope those damn bolts hold! If they don't, we're dead!" It took a good ten minutes or so for the swaying to damp out.

Then in April when we finally lifted off, a loud noise and vibration—what felt like an explosion—shook the vehicle six minutes into the flight. We were all thinking, "What the hell is that?" Steve Nagel calmly said to the crew that all the instruments looked nominal and we were going on to orbit. "We'll figure it out when we get up there," he said.

What we found was that one of the relief valves on the wastewater tank didn't open, causing an over-pressurization which ruptured the tank. That meant that we couldn't use the toilet until we put a workaround in place. Mission Control had us collect wastewater in some auxiliary bags, and every few days we'd have to discharge the liquid into space. Because the liquid immediately crystallized in a vacuum, this urine spray looked like it was snowing. It was beautiful.

Making a contribution to space medicine was very rewarding for me as a physician and flier. The Spacelab allowed us to conduct many different experiments: We had two aquariums, one with fish and the other with tadpoles. Microgravity was not good to them: Their swimming patterns in micro-g couldn't infuse their gills with enough oxygen. We monitored those animals and grew plants. We drew blood from one another. We did heart stress tests. One experiment involved evaluating the impact of fluid loading on the body. All this was to assess the overall cardiovascular condition of the crewmembers.

The Spacelab is a pretty big volume. Once when there was no crew around, I got on the end of the module and pushed off to live out my fantasy of being Superman. This was early in the flight before I learned how to navigate well, and I ended up pushing off too fast. It was great for about five or six seconds—until I slammed into the other end of the module. "Damn! That hurt!" I never did that again.

ABOVE: **Hans Schlegel (front) and Bernard Harris at work on Spacelab D-2 experiments.**

LEFT: ***Columbia* and Spacelab soar above lightning-lit storm clouds and the glow of city lights.**

BELOW: **Harris prepares a metal alloy sample at the Spacelab's Heater Facility.**

Mission no. **56**

STS

57

Orbiter	Endeavour
Launch	June 21, 1993
Landing	July 1, 1993
Duration	9 days, 23 hrs., 44 mins., 54 secs.

The STS-57 crew on the flight deck: (front row) G. David Low and Jeff Wisoff; (back row) Ron Grabe, Brian Duffy, Janice Voss, and Nancy Currie

Crew	Ronald J. Grabe, Brian Duffy, G. David Low, Nancy J. Currie, Peter J. K. "Jeff" Wisoff, Janice E. Voss

Mission: *Endeavour*'s crew grappled EURECA (deployed on STS-46) with the RMS, but its two antennas would not fully stow. Astronauts G. David Low and Jeff Wisoff completed their retraction on a spacewalk. Astronauts operated a variety of technology and scientific experiments in the new SPACEHAB's ample volume.

Nancy Currie, Mission Specialist:
All shuttle missions are good, but some are better than others. We had a rendezvous and retrieval with the arm, an EVA, and the first flight of SPACEHAB with all the experiments in the back. It was the best of everything.

The important thing about leadership on the flight that the public doesn't know is the ability to delegate responsibility. Ron Grabe thought long and hard about assigning me as an EVA person, but I didn't fit in the suit very well. I had flown with Ron and Brian Duffy in the shuttle training aircraft even before I was an astronaut. I figured that G. David Low would be the flight engineer, but he was the lead EVA guy. Ron said to me, "Well, you're the natural selection. I've flown with you, and I'm just used to you being in that center seat."

I used to say I always wanted to grow up to be an astronaut like G. David. The leadership that he provided, especially to three rookie mission specialists, was superb. He was a joy to fly with, and a joy to learn from—very patient and informative. To this day he epitomizes what I think of when I think of an astronaut.

I was in charge of opening the payload bay doors, so I was in the back with Brian, diligently timing all the microswitches, my hand at the ready in case I needed to stop anything. As the doors opened, Brian grabbed me by the neck and floated me up in front of the windows. The world was unveiled in front of me. It took my breath away. Then Brian pushed me back in front of the computer display, saying, "Back to work!"

During the EVA, I was flying the arm from up on the flight deck. All of a sudden, it felt like somebody kicked the orbiter—the entire vehicle shook. It literally shook the hand controllers. Ron said, "I think I know what that is." We did make sure that none of the EVA guys had kicked anything. Ron figured out that the jolt came from the SPACEHAB tunnel expanding in sunlight. As it expanded, the metal tunnel joints popped a little bit to relieve the stress.

My daughter Stephanie was only six years old, and I was a single mom. My best friends took care of her during the flight. At the time, the shuttle amateur radio experiment, or SAREX, was the only way to talk to your family. The ground operators trained Stephanie to always say, "Over," whenever she stopped talking. So the conversation went like this: "Hi, Mommy. Over." "I love you, Mommy. Over." Finally, I said, "Stephanie, you can say a little more than that." She said, "Okay. Over." Whoever trained her did a superb job!

I think people underestimate the challenge that landing the shuttle was. I had flown the shuttle training aircraft with these guys, so I had an idea of the concentration required. Ron did a spectacular job of putting the shuttle down with a minimal descent rate, pretty much directly on the centerline. At the postlanding briefing with the engineers, they'd say to the commander, "You landed 8.3 inches (twenty-one centimeters) off the centerline." And I'd think, "Seriously, you're measuring *that*?"

ABOVE: **Janice Voss carries video gear from the middeck through the tunnel to SPACEHAB.**

LEFT: **Jeff Wisoff (facing camera) anchors to G. David Low's spacesuit during EVA environmental testing.**

BELOW: **Nancy Currie grapples EURECA from orbit with the RMS robotic arm.**

Mission no. **57**

STS

51

Orbiter	Discovery
Launch	September 12, 1993
Landing	September 22, 1993
Duration	9 days, 20 hrs., 11 mins., 6 secs.

The STS-51 crew on the flight deck: Bill Readdy, Dan Bursch, Frank Culbertson, Carl Walz, and Jim Newman

Crew	Frank L. Culbertson Jr., William F. Readdy, James H. "Pluto" Newman, Daniel W. Bursch, Carl E. Walz

Mission: Three launch "scrubs"—including a pad abort—delayed the mission. Once in orbit, the astronauts deployed ACTS to its geosynchronous orbit and ORFEUS-SPAS on its six-day astronomy mission. Astronauts Carl Walz and Jim Newman completed a seven-hour, five-minute spacewalk, and the crew later retrieved ORFEUS-SPAS.

Dan Bursch, Mission Specialist:

It took us four tries to get off the launch pad. It was very unusual to have four full suit-ups before we ended up launching. The third attempt was the pad abort, where the engines shut down a little over two seconds before launch. The engines started quickly, and I think Bill Readdy said, "Three at a hundred," which meant that all three engines were at full power. You feel those shockwaves going through your back as the engines start up, and then all of a sudden, the sounds and vibrations of the engines go away, and then you hear a lot of bells and whistles with the emergency alarms.

Frank Culbertson was telling everyone, "Okay, slowly unstrap, get ready if we need to egress pretty quickly." Carl Walz was down on the middeck, getting the hatch ready to go. There was a lot of going through checklists and waiting, trying to determine whether we needed to get out in a hurry.

For STS-51, all three mission specialists—Jim, Carl Walz, and I—were first-time fliers, and Frank and Bill were flying their second missions but their first as commander and pilot. A lot of people assume astronauts come from the same mold, like the "Original Seven," all test pilots. But NASA hires people of different backgrounds. We had three Navy people, and Carl was an Air Force flight test engineer. Jim was a civilian, who had flown gliders. The chemistry worked. Frank was good at bringing the team together, inviting people over to his house all the time. But I think it just came down to anticipating "What is going to fail next?" Besides the launch scrubs and the abort, we had some failures in-flight, and we worked through those well.

It was my job to deploy and retrieve the SPAS free-flyer telescope that had flown several times before. On this mission, it carried an IMAX camera. After we released the free-flyer, it got some unique film footage looking back at the shuttle. The footage was featured in the IMAX movies *Destiny in Space* and *Space Station 3D*.

We had the ACTS, the first K_a-band communications satellite to fly. It worked well as a technology demonstrator. For further development the government will provide seed money, but let industry and the market mature the technology.

When we deployed ACTS, we saw some extra debris come out of the payload bay. We were scratching our heads: "That doesn't seem normal, but maybe it is for this." The release mechanism used something called the Super*Zip. Think of a tube, oval in cross-section, girdling the satellite. Within that tube are two explosive cords, only one of which is supposed to ignite. The detonation expands the oval into a circular cross-section, pushing apart some flanges around the satellite cradle. That releases the satellite without any roll, pitch, or yaw errors.

After landing, we found out that both cords of explosive ignited. The tube could not take that stress and exploded. We looked in the payload bay and could see all of this jagged metal. Later we learned that shrapnel penetrated the inside of the payload bay, with one of the pieces coming really close to a shuttle hydraulic line. We literally dodged a bullet on that one.

ABOVE: **The Canadarm grasps ORFEUS-SPAS just before releasing it on its astronomy mission.**

LEFT: **A sleeping Bill Readdy and other STS-51 crewmembers in middeck sleep restraints.**

BELOW: **Jim Newman at work on *Discovery*'s starboard payload bay sill.**

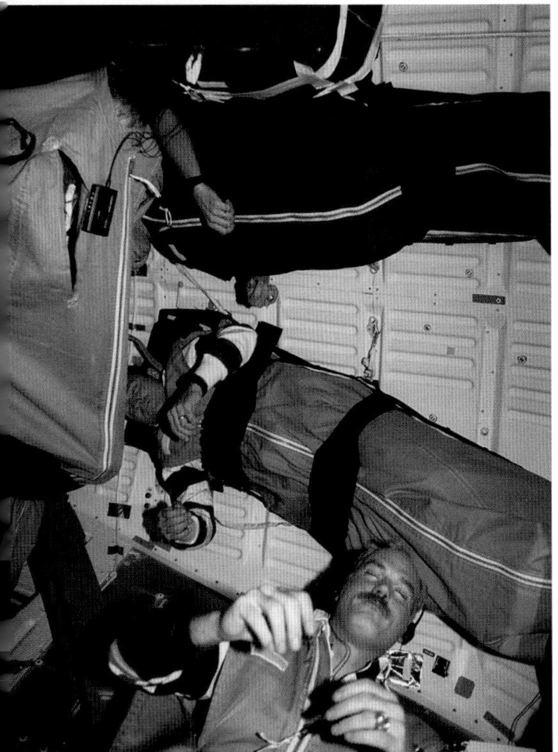

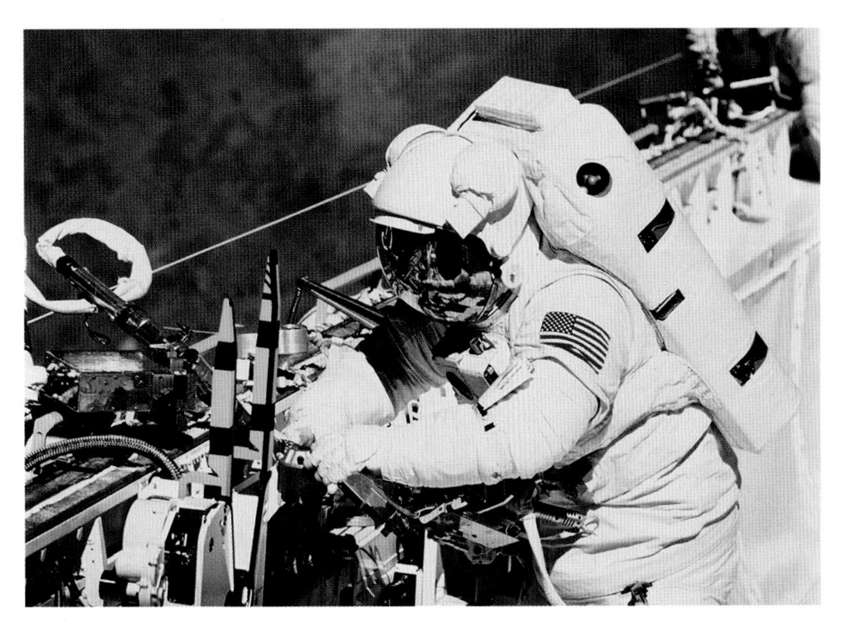

The ORFEUS-SPAS satellite captured this IMAX image of *Discovery* high above the Gulf of Venezuela.

Mission no. **58**

STS

58

Orbiter	Columbia
Launch	October 18, 1993
Landing	November 1, 1993
Duration	14 days, 0 hrs., 12 mins., 32 secs.

The STS-58 crew in the Spacelab: (clockwise from top left) Rhea Seddon, Shannon Lucid, Bill McArthur, Marty Fettman, Dave Wolf, Rick Searfoss, and John Blaha

Crew	John E. Blaha, Richard A. Searfoss, M. Rhea Seddon, William S. McArthur Jr., David A. Wolf, Shannon W. Lucid, Martin J. Fettman

Mission: *Columbia*'s astronauts served as subjects for SLS-2's eight experiments focused on the human body. Six investigations focused on forty-eight rats living in twenty-four habitats. SLS-2 results, combined with those of SLS-1, yielded the most detailed orbital life sciences measurements since Skylab.

Rhea Seddon, Payload Commander:

This was Spacelab Life Sciences 2 (SLS), the most extensive human physiology research mission to date, and probably the most extensive ever. It lasted fourteen days—the longest shuttle mission at that time.

Our bodies belonged to the scientists, and we took that very seriously. We stuck to the rules, even the one that said we couldn't have any caffeine. Even in preflight and postflight testing, we couldn't have a cup of coffee because it increased our heart rates and metabolism.

We had to keep ourselves healthy, logging our exercise and what we ate and drank, and collecting all our urine. Can you imagine? Two weeks preflight, all the days in flight, two weeks postflight, all in a jug or a bottle. The things we do for science.

We had learned a lot from the failures we had on SLS-1—how to write malfunction procedures and how to practice them. But we had a gas analyzer that sucked air through a small porthole and analyzed the respiratory gases—how much oxygen, how much CO_2—and it quit working. It was a major part of our musculoskeletal study, and we couldn't figure out what was wrong. Somebody bumped into the gas analyzer, and it started working again. We suddenly realized that with gas going through a tiny orifice, it can clog if there are dust particles in there. After that, the malfunction procedure was to just hit it really hard.

Bill McArthur made sure we had the ham radio on our flight. My eleven-year-old son Paul had been kind of at a distance from missions—we'd go down and watch launches, and he saw video of Mommy floating around—but his school requested a talk with the STS-58 crew in orbit, so now he'd be a part of it.

They all had to have a question, and he would come home and practice his. He seemed very uncomfortable. I asked, "Are you a little anxious about doing this?" He said, "No, Mom." "Well, what's wrong? You do well when you practice it." He said, "I'm just afraid when I ask the question, you'll say something mushy, and I'll get embarrassed, and my school friends will laugh at me." And I said, "I promise I won't."

When we came over the horizon the kids started asking questions. Paul asked his, and I answered it succinctly, just like the other ones. We were going out of contact, and communications were fading. I was looking out the window at Earth going by, and I heard a little voice out of the void say, "I love you, Mom. Have a safe trip home." So he said something mushy on his own accord. It was a memorable moment in the midst of all the other frenzied stuff that you did in space.

It was an amazing mission. We were there to set the stage for longer missions: See what changes happened, what changes needed to be reversed, and what we would need to adapt to. Could we stay in space longer as we looked ahead to the space station? I think we proved that we could. On those two missions, we had four males and four females, and we could compare data on whether one or the other sex adapted better or just the same. We proved that in space, there is no "weaker sex."

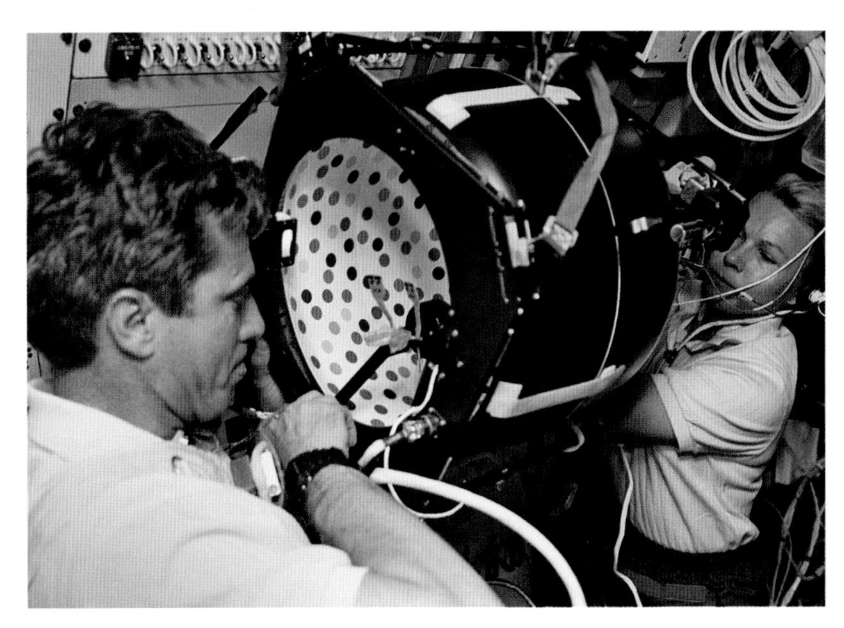

LEFT: **In 1993, *Columbia* roars from Pad 39B on its thirteenth mission, carrying Spacelab SLS-2.**

ABOVE: **John Blaha and Rhea Seddon conduct a neurological experiment with a rotating visual field.**

BELOW: ***Columbia* at rest on Edwards Air Force Base Runway 22, with the NASA crew transport vehicle attached at left.**

Mission no. **59**

STS

61

Orbiter	Endeavour
Launch	December 2, 1993
Landing	December 13, 1993
Duration	10 days, 19 hrs., 58 mins., 33 secs.

The STS-61 crew backdropped by Hubble: Story Musgrave, Claude Nicollier (front), Dick Covey, Jeff Hoffman, Ken Bowersox, Kathy Thornton, and Tom Akers

Crew	Richard O. Covey, Kenneth D. "Sox" Bowersox, F. Story Musgrave, Kathryn C. Thornton, Claude Nicollier, Jeffrey A. Hoffman, Thomas D. Akers

Mission: *Endeavour*'s crew grappled the HST, enabling two astronaut teams to perform a record five spacewalks totaling thirty-five hours and twenty-eight minutes. During the spacewalks, they installed corrective optics, replaced warped solar arrays, and upgraded computers and gyroscopes, restoring Hubble to design performance.

Kathy Thornton, Mission Specialist:

STS-61 was the first flight to service Hubble, and was one of several planned servicing missions. We put in corrective optics which took care of the spherical aberration that had been discovered shortly after the 1990 deployment. It was billed through most of our training period as "make or break" for NASA and the space station. If we screwed this up, that would have been the end of the space station.

What was so different from STS-49 was that everything went right. We planned for so many contingencies—none of which ever materialized. We had plans so that any of us could have done any task, and we always had the tools at our fingertips, whether in the cargo bay or the depressurized airlock, to do anything we could have imagined. We tried to have enormous flexibility to change the plan on the fly if things weren't going the way they were supposed to go.

On the second EVA, Tom Akers and I removed a warped Hubble solar array that wouldn't retract properly. The array was a giant nothing. It didn't weigh very much. Tom was the one who was detaching it from the telescope, and I was just there on the robot arm with a hand hold on it. Even though I was out there for six hours, I didn't get a lot of time for Earth- or stargazing. Before I let the array go, I was pretty much just looking at the panel.

Once we did let it go, I didn't shove it. I just let go and left it floating there. They pulled me back on the arm, and it was just amazing as Sox fired the jets to separate from it. The exhaust plume hit that solar array and it bent over almost double, then just sprang out and kept oscillating while it was cruising over the Middle East. The array was like this giant pterodactyl cruising over the deserts. That's how we left it. Mesmerizing.

It was very satisfying at the end of every day to say, "We did it. All those things for today, they're done. They worked. The ground sent the commands to the Wide Field and Planetary Camera 2 (WFPC) and the Corrective Optics Space Telescope Axial Replacement, or COSTAR, which supplied the correction for the other instruments. We would bask in that feeling for about ten minutes and then say, "Okay, tomorrow. What's happening tomorrow?" We'd roll into the next thing. It was an amazing rhythm we got going. We probably could have sustained it for longer, as we had two crews alternating EVAs.

During the fourth EVA, the COSTAR one, I was being moved from one place to another, paying no attention to what was going on around me, other than just doing what I was supposed to be doing. Unlike training in the water tank, when you're being moved around on the arm out there, you have no sensation of motion.

The guys said, "You've got to stop what you're doing and *look*." We were over the Gulf of Mexico, and we could see this beautiful aurora up over Canada and all of North America. I remember looking at the East Coast and the West Coast at the same time, another moment that will stay with me.

The other thing that's special to me about that mission of getting Hubble up to its full capability is that when we flew it, Carol, my oldest, was eleven years old. She went on to get her PhD in astrophysics using Hubble data. I hope Hubble lasts forever.

ubble, with corrective optics,
pgraded instruments, and new
olar arrays, ready for release.

Mission no. **60**

STS

60

Orbiter	Discovery
Launch	February 3, 1994
Landing	February 11, 1994
Duration	8 days, 7 hrs., 9 mins., 22 secs.

The STS-60 crew in SPACEHAB: (clockwise from bottom left) Franklin Chang-Diaz, Sergei Krikalev, Ken Reightler, Charlie Bolden, Ron Sega, and Jan Davis

Crew	Charles F. Bolden Jr., Kenneth S. Reightler Jr., N. Jan Davis, Ronald M. Sega, Franklin R. Chang-Díaz, Sergei K. Krikalev

Mission: STS-60's crew operated materials science, life science, and space dust experiments in the SPACEHAB-2 module. Malfunctions prevented release of the Wake Shield 1 satellite, but its semiconductor growth experiment was exercised. Sergei Krikalev was the first cosmonaut to fly aboard the shuttle.

Franklin Chang-Diaz, Mission Specialist and Payload Commander:

STS-60 was the first flight of the Shuttle-Mir program, the mission that really broke the ice between the United States and the former Soviet Union. We contributed to the success of the relationship between Russia and the United States and helped further the internationalization of space that continues today.

When we got to Moscow to be introduced for the first time, it was cold. And it was not just cold in terms of the weather, but the environment was very antagonistic. We didn't speak Russian, and of course hardly any of them spoke English. They had assigned two cosmonauts—Sergei Krikalev and Vladimir Titov—to work with us on the first shuttle flight with a Russian cosmonaut.

The Russians didn't think they could learn anything from us, and we didn't think we could learn anything from them. And we were both wrong. Both Sergei and Volodya—Vladimir—were as good ambassadors as they were astronauts. Sergei had a real ease and familiarity with space. On this, my fourth flight, I was pretty familiar and comfortable with the environment. But Sergei impressed me: how he handled the food, for example, and how he saved his various eating utensils and put them in different places, so that he would have them available and not lose them—little tricks like that. I was eager to learn, and I would share with him little tricks of my own, such as how to sleep so you wouldn't get a backache. The Russians had a lot to say about spaceflight that was ver wise and true.

To keep the center of mass of the shuttle within operational limits, the SPACEHAB module sat a fair distance behind the crew cabin, and a long and rather narrow pressurized tunnel connected the two. Despite all the equipment operating inside the module, it was cozy, quiet, and warm, and I had decided to sleep there. That plan all changed the first evening as I was making my overnight nest. I heard this tremendous clunking noise through the ship. My immediate thought: "The tunnel is losin structural integrity, and I am on th wrong side of the hatch!" It scare the heck out of me, and I thought, "I'm getting out of here!" The clunk was attributed to thermal stress during the day-night cycle, and I believe it repeated multiple times. It was an effective reminde that we were actually flying in the vacuum of space.

Charlie and I—he, a Marine pilo and I, a physicist—became really good at drawing blood samples, though it was painful for the flight controllers who volunteered to let us stick needles into them during training. In space, Charlie and I di quite a few phlebotomy sticks on each other. It was an odd feeling, sticking each other with needles, and saying, "Okay, you do me now!" and, "I do you next!" Yet there we were, orbiting Earth and commanding one of America's premier science missions.

ABOVE: **The Wake Shield on the RMS, silhouetted against an auroral curtain's glow.**

RIGHT: **Franklin Chang-Diaz organizes flight-plan updates on *Discovery*'s flight deck.**

BELOW: **Space-grown crystals, left, compared to samples grown on Earth, right.**

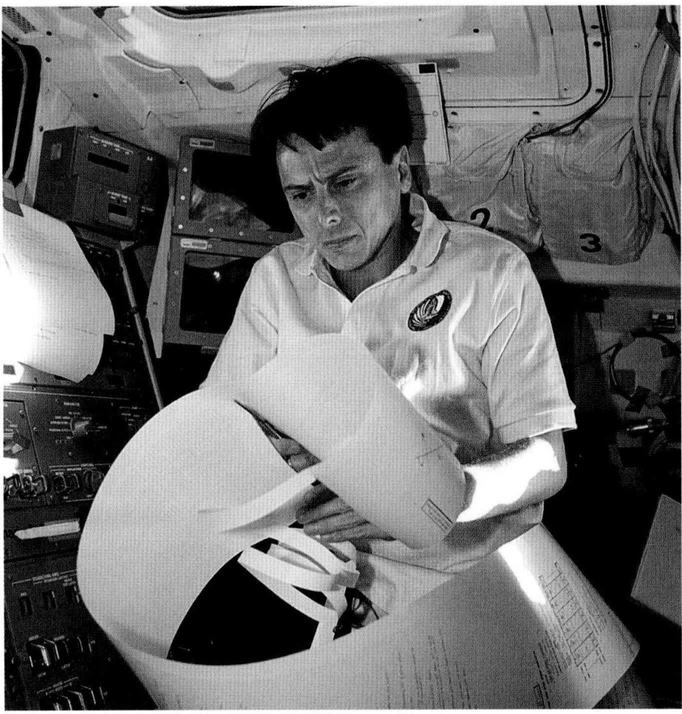

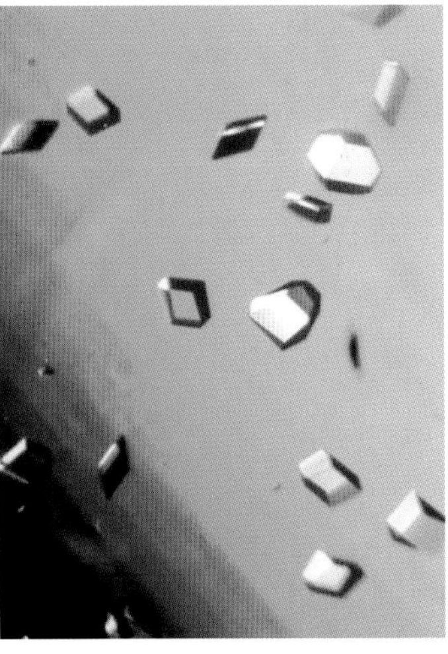

Mission no. **61**

STS

62

Orbiter	Columbia
Launch	March 4, 1994
Landing	March 18, 1994
Duration	13 days, 23 hrs., 16 mins., 41 secs.

The STS-62 crew on the flight deck: (front row) John Casper and Pierre Thuot; (back row) Andy Allen, Marsha Ivins, and Sam Gemar

Crew	John H. Casper, Andrew M. Allen, Pierre J. "Pepe" Thuot, Charles D. "Sam" Gemar, Marsha S. Ivins

Mission: *Columbia*'s astronauts operated the USMP-2, a suite of five materials processing and crystal growth experiments, and the OAST-2 payload with six spaceflight technology experiments. The long-duration mission provided these payload bay and cabin experiments with a nearly two-week run in quiescent free fall.

Pierre Thuot, Mission Specialist:

STS-62 was somewhat of a precursor to space station science. The USMP-2 and the OAST-2 were both turned on by the crew, then operated from the ground. We had approximately sixty-five experiments or demonstrations, and there were only five of us on a single shift. We were so busy that in our flight plan there were five columns, one for each person. There were a couple of days when we had lighter schedules, but most of the time each of us was booked solid. I give credit to the flight planning people who developed the plans—we worked with them a lot.

We had the Dexterous End Effector, which was about two feet (half a meter) long, about the same diameter as the end effector on the RMS, or the shuttle arm. After grappling it with the arm, we were to use the Dexterous End Effector to insert truncated cones mounted on a demonstration panel into holes on a target plate. The holes were divided into three sets with progressively tighter clearances around the cones.

Of the three of us, Marsha Ivins was the only one who was able to insert the three pins in the smallest holes. She did it once. Sam and I were unable to do it at all. We attribute that to the TV monitor crosshairs being too wide—they were too big for what we were trying to do. It was more of a display problem than an actual arm problem. We managed to do some pretty precise things out there at the end of this fifty-foot (fifteen-meter) arm.

One night I was sleeping on the middeck overhead. Sam and John were on the wall, and Andy was sleeping in the pilot's seat. The master alarm went off, and unfortunately the speaker was right behind my head. I was dead asleep and it just scared the crap out of me. The first thing Andy did was to start turning on the displays instead of punching the master alarm off so it would stop making noise. And I was just flailing around in my sleep restraint. I think it was an OMS pod getting cold, and the ground had to turn a heater on, but we went out of communications before they could do it. They were apologetic, but that was pretty scary.

We got really low, down to about 105 nautical miles (194 kilometers) altitude, doing the Experimental Investigation of Spacecraft Glow experiment. We started at 160 nautical miles (296 kilometers) and then got into an elliptical orbit with a low of 105 nautical miles (194 kilometers). We were closer to the clouds—they were only 105 nautical miles (194 kilometers) away—and I got this sense of speed. I almost felt like I wanted to grab on, because it sure looked like we were going faster. That was a cool sensation after having been at 160 nautical miles (296 kilometers) for ten or eleven days.

After fourteen days up there of concentrating—you're focused, you don't want to make any mistakes—we got mentally worn out. I think this was unique to a five-person crew. We told them that four hours off one day and four hours off another day is not enough. You really need a mental break somewhere if you're going to be flying that long. You think fourteen days isn't that long, until you've been there. On Flight Day 12 or 13, you think, "I'm ready to go home."

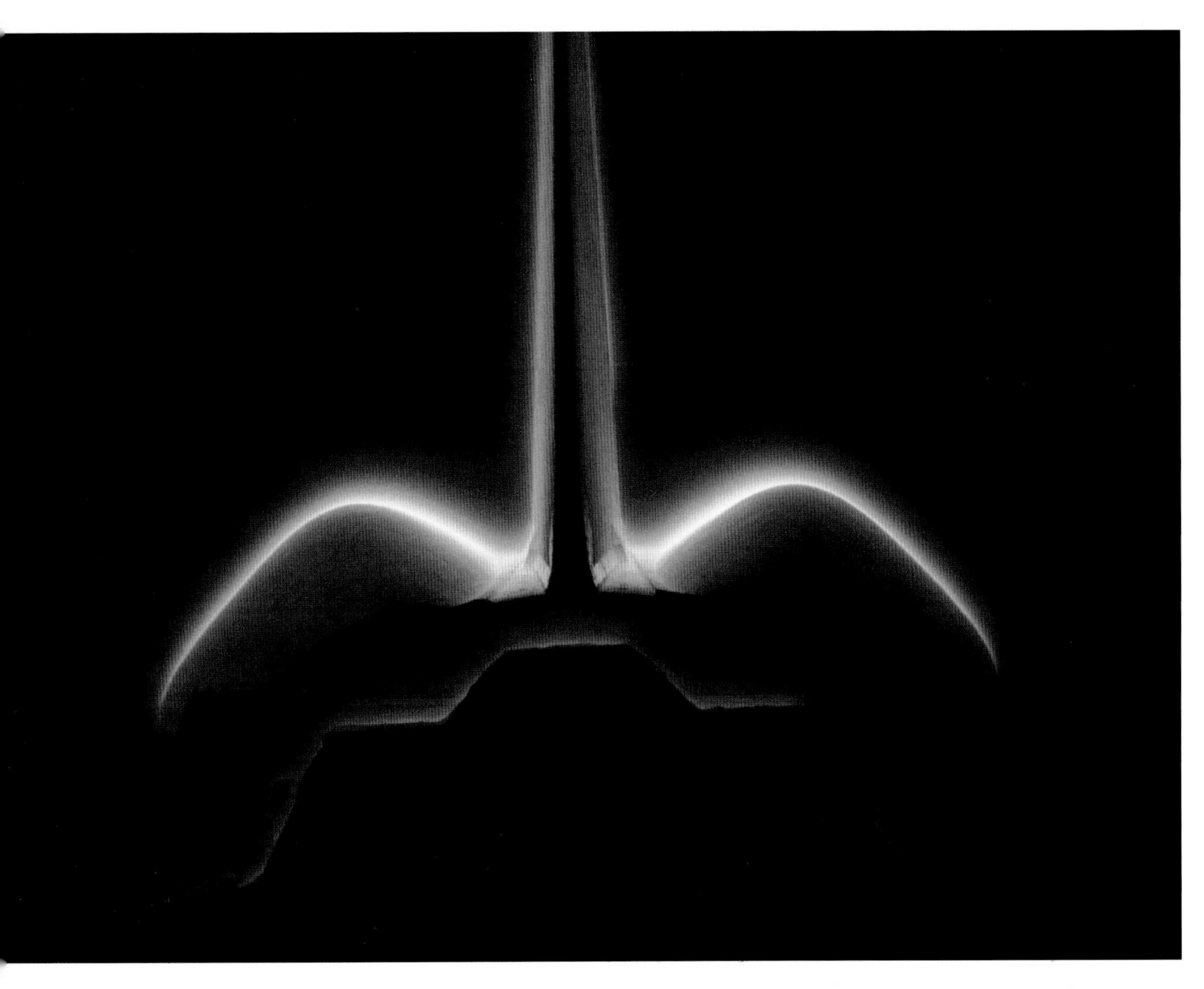

ABOVE: ***Columbia*'s low altitude flight through the tenuous upper atmosphere wrapped the tail in the glow from excited oxygen atoms.**

BELOW LEFT: **Marsha Ivins and Pierre Thuot operate a thermal imaging experiment on *Columbia*'s flight deck.**

BELOW RIGHT: **Pilot Andy Allen monitors *Columbia*'s reentry performance on the darkened flight deck.**

Mission no. **62**

STS

59

Orbiter	Endeavour
Launch	April 9, 1994
Landing	April 20, 1994
Duration	11 days, 5 hrs., 49 mins., 30 secs.

The STS-59 crew on *Endeavour*'s middeck: (front row) Jay Apt, Sid Gutierrez, and Tom Jones; (back row) Kevin Chilton, Linda Godwin, and Rich Clifford

Crew	Sidney M. Gutierrez, Kevin P. Chilton, Linda M. Godwin, Jay Apt, Michael R. "Rich" Clifford, Thomas D. Jones

Mission: *Endeavour*'s astronauts and the ground science team operated the Space Radar Laboratory-1 around-the-clock, studying natural and human-caused change on Earth's surface. SRL-1 scanned more than 400 science targets, mapped about 20 percent of the home planet, and tracked carbon monoxide pollution.

Tom Jones, Mission Specialist:

Space Radar Lab-1 was a NASA "Mission to Planet Earth," applying planetary science observation techniques to our own world. *Endeavour* would beam radar energy at Earth and turn the resulting echoes into detailed images, exploring the ability of space-based radar to map and monitor Earth's dynamic surface. Those of us in orbit were one part of a far-flung science team, while on the ground were forty-nine principal investigators and more than one hundred scientists studying every aspect of Earth science.

An hour after reaching orbit, I stared out the side hatch window at a sliver of light in the blackness outside. Palms cupping my eyes, I pressed close to the window and peered at inky darkness where a planet should be. The faint glow of starlight barely lit the cloud tops below. As *Endeavour* streaked toward sunrise, the robin's-egg-blue of our thin atmosphere was backlit against the shrouded horizon. That first dawn, shifting rapidly from blue to gold to dazzling white sunlight, triggered a flood of emotions and memories of thirty years of dreams and ambitions about flying and working in space. It may just have been that intense sunlight, but I had to fight back unexpected tears. With a silent prayer of thanks, I reluctantly turned back to work.

To point the radars at Earth, *Endeavour* flew upside down, the cargo bay angled nearly straight down, 26 degrees from the vertical. The ground team aimed the radars out to the left or right side of track, imaging a swath of the planet nine to fifty-six miles (fifteen to ninety kilometers) wide. The three radars swept Earth for thousands of miles along our orbital path.

To document our radar science targets, we shot more than 11,000 film images with four 70mm Hasselblads and two large-format Linhof mapping cameras. The view from our cabin windows was stunning. It was like staring up through the roof of a greenhouse at a planet filling half the sky. With our cameras pressed against the triple-paned glass, we were like a crowd of tourists mobbing the bus windows at a Grand Canyon overlook. A glance "up" at the scenery rolling by just 138 miles (222 kilometers) distant always made me draw a sharp breath. One could never grow bored with this view.

South of New Zealand, we sometimes flew right through the long, thin streamers of the aurora projecting straight up through the atmosphere as a yellow-green curtain. We could see long, spectral streamers rising above us, and often flew right over the thin auroral arcs below. As we gazed down at the invisible Southern Ocean, the aurora rippled in fantastic ribbons of atomic glow.

Through the American Radio Relay League, each of us before launch had coordinated a talk with home using *Endeavour*'s small amateur radio set and a window-mounted antenna. The orbiter was in darkness over the central Pacific when I transmitted our ham radio call letters into the ether. A strong signal brought the voice of a Hawaiian ham operator into my headphones. He immediately dialed my wife Liz, and we had about five minutes of precious conversation, Hawaii to Houston, before the islands disappeared astern, and I lost the wonderful sound of her voice. I wished I could touch her, keenly aware of the daunting challenge still ahead: getting six humans and our spaceship safely home from 138 miles (193 kilometers) up, circling the Earth at Mach 25.

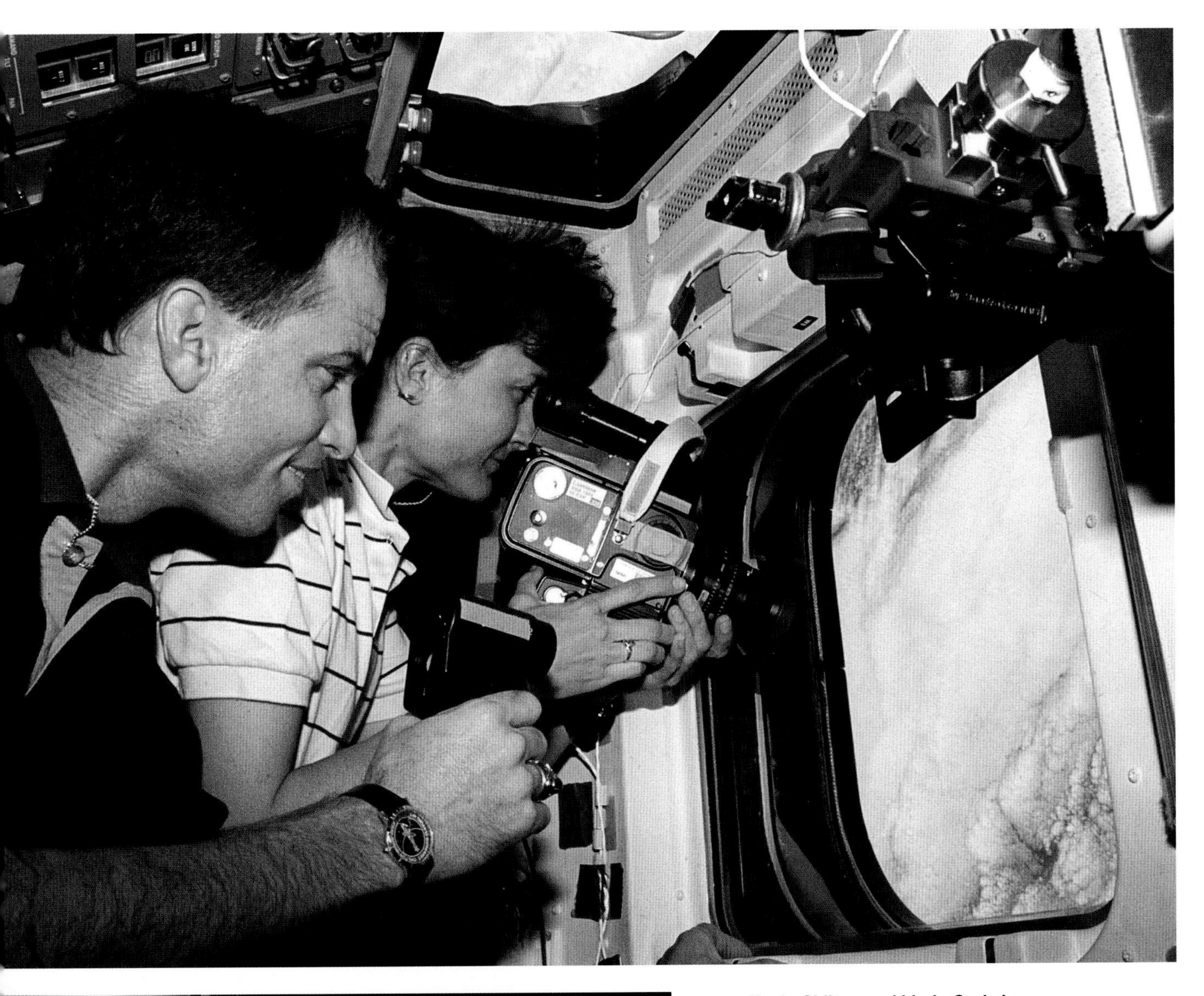

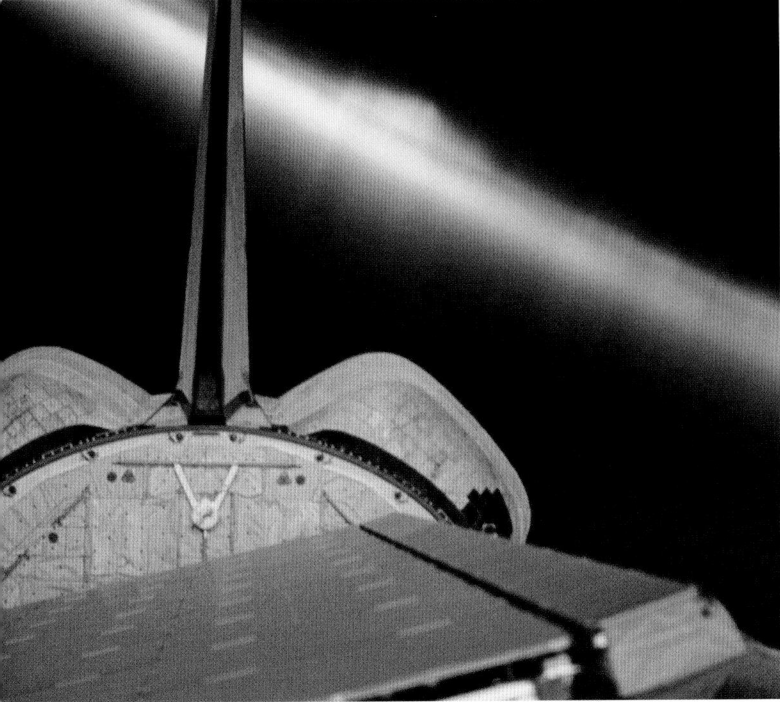

ABOVE: **Kevin Chilton and Linda Godwin photograph science targets for Space Radar Lab-1.**

LEFT: **The aurora australis glows above the SIR-C/X-SAR radar antennas.**

BELOW: **Sid Gutierrez and Chilton fly *Endeavour* to a smooth touchdown at Edwards Air Force Base, California.**

Mission no. **63**

STS 65

Orbiter	Columbia
Launch	July 8, 1994
Landing	July 23, 1994
Duration	14 days, 17 hrs., 55 mins., 00 secs.

The STS-65 crew in Spacelab: (front row) Carl Walz and Don Thomas; (back row) Rick Hieb, Chiaki Mukai, Bob Cabana, Leroy Chiao, and Jim Halsell

Crew	Robert D. Cabana, James D. Halsell Jr., Richard J. Hieb, Carl E. Walz, Leroy Chiao, Donald A. Thomas, Chiaki Naito-Mukai

Mission: The STS-65 crew operated the IML-2 housed in the Spacelab module. IML-2 had twice the number of experiments flown on IML-1 on STS-42, spread across life sciences and materials processing, with many ground-based investigators participating in real-time via telescience links.

Leroy Chiao, Mission Specialist:

We flew the second IML with scientists from all around the world: Japan, the countries of the ESA, and Canada. We had Chiaki Mukai, the first Japanese woman to fly into space. We travelled around quite a bit, constantly jetlagged, going two weeks to Europe, two weeks back in Houston, two weeks to Japan, two weeks back in Houston. We did that for the better part of a year and a half, and were pretty wiped out by the time we actually went to space. But it was a great experience travelling overseas and training. Each country gave me a deeper understanding and appreciation of the different cultures participating.

The mission was an emotional one for me, as it was my first, and I had dreamed about flying in space since I was an eight-year-old. I had worked hard, gotten through all the hoops and gotten a couple of lucky breaks, made it through the interview, and finally I was actually up there in space.

One of the first things I had to do was to shoot video of the ET with the camcorder. When I got to weightlessness, I got the full-headedness, the dizziness, and my internal gyros tumbled. But I had a plan: I took my helmet and gloves off and stuffed them into a helmet bag right next to me. Then I unbuckled, and I was floating.

Seeing that first view of Earth took me aback. It was so much brighter and more colorful than shown by photos and IMAX. At the Earth's limb, I could see the sunlight going through the atmosphere, creating different shades of fluorescent blue—a fantastic sight! I had to remind myself to get back to work, put the camcorder together, and videotape the tank as we flew away from it. The beauty of Earth and the sense of accomplishment was more emotional than I expected.

We had a scare during Flight Day 2. It was the first flight of the Extended Duration Orbiter (EDO) toilet. You've got to follow the checklist, because it's got a solid waste compactor—you want to make sure you don't try to rotate the compactor away with the plunger still extended. We messed up the steps, and the thing came up with an error message. It stopped. We said, "Oh my God—if we lose the toilet on Day 2..." Fortunately, Jim Halsell, the pilot, was able to fix it, so that was okay.

The Spacelab shifts had us on duty for twelve hours a day. You've got time in there for meals, workouts, and breaks. But on our shift, Don Thomas and I were getting stressed out. Because we kept finishing things early, we called down to Huntsville to tell them we were finished—and then they'd just load us up with more work. That had us going along at such a clip that they were calling us when we were going to the bathroom. I told Don: "For the rest of the mission, we're just going to stick to the timeline, and if you finish early, just take those few minutes to go look out the window, take a picture, get a snack, go to the bathroom." So we did that, and then the stress shifted from us to the ground. But that helped us get through. At the end, we still ended up getting a lot more accomplished than had been planned. We were very proud of that.

ABOVE: **Wetlands near Pad 39A reflect *Columbia*'s incandescent booster plume at liftoff.**

RIGHT: **Don Thomas, Leroy Chiao, Rick Hieb, and Chiaki Mukai work with IML experiments in Spacelab.**

BELOW: ***Columbia*'s sunlit payload bay cradles the IML-1 Spacelab.**

Mission no. **64**

STS

64

Orbiter	Discovery
Launch	September 9, 1994
Landing	September 20, 1994
Duration	10 days, 22 hrs., 49 mins., 57 secs.

The STS-64 crew on flight deck: Jerry Linenger, Mark Lee, Blaine Hammond, Susan Helms, Carl Meade, and Dick Richards

Crew	Richard N. Richards, L. Blaine Hammond Jr., Jerry M. Linenger, Susan J. Helms, Carl J. Meade, Mark C. Lee

Mission: *Discovery*'s astronauts operated the LITE laser-ranging experiment to probe Earth's atmosphere. They launched and retrieved SPARTAN 201 for two days of solar observations. Spacewalkers Mark Lee and Carl Meade test-flew the SAFER, a cold-gas-powered rescue jetpack for future ISS use.

Mark Lee, Mission Specialist:

We had three really important experiments on STS-64. First, the SPIFEX instrument was the thirty-five-foot-long (ten-meters-long) extension to the end of the arm. Its name—Space Shuttle Plume Impingement Flight Experiment—is almost as long as the instrument. It was an impressive object to be waving around up in orbit. SPIFEX helped prepare for Mir dockings by understanding how the shuttle's RCS thruster firings affect other vehicles in space. Next was the LITE, or LIDAR In-Space Technology Experiment, which collected laser data regarding dust, water vapor, and cloud reflectivity to compare with airborne and ground-based data over about twenty different countries, with that ground truth validating the space data.

On a mission dedicated to studying changes in the atmosphere because of dust storms, we observed a volcanic eruption at Rabaul in Papua New Guinea. Although we didn't do any laser operations specifically over the eruption, from orbit we were able to watch this plume form. This was in the middle of the Pacific, and there was not a cloud around except for this plume. It was cool to see how the cloud of volcanic dust from this eruption took form, with different layers in the atmosphere—dark, heavy particles down toward the Earth, and a thin layer of dust particles on top of the cloud.

The most significant part of the mission for me was testing the Simplified Aid for EVA Rescue (SAFER). We were given an idea, a requirement, and we had to come up with a way to make it work. The jetpack had to interface with the spacesuit backpack and meet certain requirements to get a spacewalker back to the station, overcoming a specified separation velocity. SAFER opened up a whole new arena for training: virtual reality.

The most controversial aspect about SAFER was that it was untethered. Then-NASA administrator Dan Goldin commented, "I don't want any space cowboy up there flying around with a jetpack." But we demonstrated that the tether reel forces would overcome the jet thrust, so for an accurate test, you either had to lock out the tether, or use the SAFER untethered. In the end, th flight director gave us approval to do it. We then worked with th engineering team to come up wi the scenarios.

We did about 200 maneuver sequences to get used to SAFER four-degrees-of-freedom hand controller. Each one of us flew fo different flight segments, coming up with all the data that the engineers wanted, not only to prove that SAFER would work but also that the training methods were realistic.

The first maneuver I flew was to go straight up about fifteen to twenty feet (4.6 to 6 meters) abov the shuttle. At the time I wasn't paying attention to the shuttle's orientation, so fifteen feet (4.6 meters) "up" was actually fifteen feet (4.6 meters) sideways in orbi All of a sudden, I looked down below my feet, and I was looking at Earth. At the same time, I was coming up on a sunset, and the hair stood up on the back of my neck. The view and the situation nearly overwhelmed me—I was flying a jet pack in space!

Flying SAFER was exactly like being in the virtual reality lab. We thought we wouldn't hear the jets firing, but it turned out that we di simply because of vibration conducted through the suit. In trainir we had no sound, no feedback, yet in the suit in space, we did hear the jets fire.

'ark Lee tests the SAFER
·tpack during untethered flight
ɔove *Discovery*'s payload bay.

Mission no. **65**

STS

68

Orbiter	Endeavour
Launch	September 30, 1994
Landing	October 11, 1994
Duration	11 days, 5 hrs., 46 mins., 8 secs.

The STS-68 crew on the middeck: (front row) Dan Bursch, Mike Baker, and Tom Jones; (back row) Terry Wilcutt, Steve Smith, and Jeff Wisoff

Crew	Michael A. Baker, Terrence W. "Fencer" Wilcutt, Thomas D. Jones, Steven L. Smith, Daniel W. Bursch, Peter J. K. "Jeff" Wisoff

Mission: *Endeavour* experienced a main engine failure just under two seconds before liftoff. Six weeks later, the astronauts worked with SRL-2 ground scientists to map Earth and create elevation and crustal motion maps over regions of volcanic, tectonic, and glacial activity.

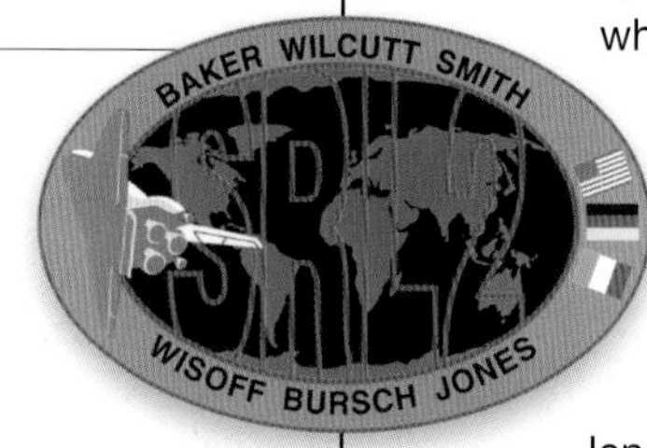

Jeff Wisoff, Mission Specialist:

As part of being good stewards of Earth, we have to understand what's happening to our planet over the long term. STS-68 was focused on testing a radar facility capable of imaging Earth in three different wavelengths of radar energy. We wanted to assess its effectiveness at being able to observe from space on a large scale the processes and resources of Earth.

On the pad abort, we counted down and the main engines lit, and then everything shut down, so we got unstrapped. I went over to the hatch to be ready to open it if we had to get out as we had been trained. Unlike in training, we had the "twang" going on the vehicle. The shuttle rocked forward under the initial main engine thrust, swung back to vertical, then oscillated. When I looked outside, the White Room vestibule at the near end of the crew access arm was swaying back and forth, and I thought, "I really don't want to get out and try to jump across to that swing arm."

The control team, as usual, was very professional. They communicated quickly that they wanted us to sit tight and wait it out while they were checking for noxious gases and fires. I thought: "One: I'm not going to go today. And two: I'm going to have to go do remedial training for six weeks." All of this after I felt I had already achieved peak readiness.

In the press the next day, they ran with their idea of how close a call the abort had been, so close to booster ignition. But in reality, within a few minutes Tom Jones and I were on the middeck eating our sandwiches. It played out exactly the way it was supposed to play out.

If there was a bright side to the abort—other than that nobody was hurt, and the vehicle was fine—it was that we got out of quarantine and could go visit with our friends and family that had come to see the launch. Down in Cocoa Beach, I told my dad, "I always knew there was a chance I could end up in the water today but I didn't think I'd be body surfing!"

Once we did get to orbit, our whole crew was challenged by the loss of a data recorder on Flight Day 10. The ground team delayed the shift handover so that our two teams could work together to get that recorder swapped out. It took us about two hours to do it, with minimal loss of data because the other two recorders were still working.

Because STS-68 was an Earth-observing mission, we got more window time than you normally do. Lightning storms would start up in Central America then ignite a cascade of lightning down over South America. As we passed the rocky, craggy Kerguelen Islands near Antarctica, I remember how happy I was that I was up in the shuttle and not living down on those pieces of rock.

We got an extra day on orbit. What I liked to do before sleep, when everything was done, was put on the headphones, listen to jazz and watch Earth turn. Seeing the weather patterns and the lightning around the planet, you can almost imagine the music being in sync with Earth. You think about how everybody and everything that matters to you is sitting on that little blue ball, and you are in this unique, lucky place, able to observe it from this vantage point.

ABOVE: **The rising Sun illuminates the MAPS experiment (foreground) and the flat SRL-2 radar antennas.**

BELOW: **Late in the mission, Jeff Wisoff and Steve Smith replace a failed SRL digital data recorder.**

RIGHT: **Shortly after *Endeavour*'s landing at Edwards Air Force Base, a 747 shuttle carrier aircraft carried *Columbia* toward Florida and its next mission.**

Mission no. **66**

STS

66

Orbiter	Atlantis
Launch	November 3, 1994
Landing	November 14, 1994
Duration	10 days, 22 hrs., 34 mins., 2 secs.

The STS-66 crew on the flight deck: (bottom row) Joe Tanner, Don McMonagle, Scott Parazynski, and Curt Brown; (top row) Ellen Ochoa and Jean-Francois Çlervoy

Crew	Donald R. McMonagle, Curtis L. Brown Jr., Ellen Ochoa, Scott E. Parazynski, Joseph R. Tanner, Jean-Francois Çlervoy

Mission: *Atlantis*'s crew operated the ATLAS-3 experiments to study the Sun's energy output and chemistry of Earth's middle atmosphere. The astronauts deployed and retrieved CRISTA-SPAS for observations of trace gases and ozone-destroying compounds in the middle atmosphere.

Joe Tanner, Mission Specialist:

This ATLAS-3 studied the ozone hole and performed other scientific investigations of our atmosphere. We also flew the Cryogenic Infrared Spectrometers and Telescopes for the Atmosphere Shuttle Pallet Satellite, or CRISTA-SPAS, from Germany, which used spectroscopy to identify trace chemicals and pollutants in the lower and middle atmosphere at high northern and southern latitudes. CRISTA-SPAS drove us to a 57-degree orbital inclination—really cool, because you get to fly over much more of Earth's land mass.

At the end of ascent, the engines cut off and everything just started floating. I told the guys I had to unstrap for a second and look outside. I leaned back to look out the overhead windows. We were flying upside down, so the first thing I saw was Earth. My first impressions: "It's round, it's absolutely beautiful, space is blacker than anything I could imagine, and God is good."

We carried an experiment in the middeck containing ten pregnant rats. Taking care of them was my responsibility; the crew referred to me as the "rats' nanny." Tending the rats was a lot of fun. I think we—the rats and I—all enjoyed our first trip to space. The animals lived in enclosure modules, so all I really needed to do was make sure they had fresh water every day. I couldn't handle them, but I did photograph them regularly until the viewing plate into their enclosure got so dirty that I couldn't see them well. Before exiting the orbiter after landing, I checked on them one last time: They didn't look very happy to be back on Earth.

I really enjoyed being Mission Specialist 2, the flight engineer. It's a lot of work but being an integral part of the flight deck team was really rewarding. I was also the rendezvous checklist and compu er expert, heavily involved in our SPAS retrieval work.

We had to maneuver four time a day for one of our science objectives, alternately looking at the Sun and Earth to compare the Sun's energy output to the energy reflected from the atmosphere. That told the scientists how much energy the atmosphe was absorbing. Don McMonagle and Curt Brown even let me fly *Atlantis*. They had me fly a coupl of orbiter maneuvers actually firir the thrusters through the control stick—a big deal for this astrona

When we retrieved CRISTA-SPAS after its free flight, we wer the first to use an R-bar rendezvous approach, testing it for future Mir docking missions. Th R-bar approach, where we fly u from beneath the target, is more fuel-efficient and provides natur braking, adding a margin of safe against collisions with Mir. I'm sure the Russians appreciated that.

I had more STA approaches than Don and Curt combined. I'd trained both of them when I was an STA instructor pilot before becoming an astronaut. During pre for reentry and landing, I said, "You guys don't look so good. A you sure you feel alright? You su you don't want me to do this lan ing?" Curt, the pilot, replied with something typically "Curt-esque, like, "Sit down, strap in, and shu up." We all had a good chuckle.

Ellen Ochoa and I were on the same shift and would often eat dinner together up on the flight deck. I said to her once, "Ellen, I finally figured out what I want to do when I grow up." (I was forty-four at the time.) Surprised, she looked at me and asked, "What, Joe? What do you want t do?" I said, "This."

ABOVE: **Joe Tanner works with the Crystal Observation System on *Atlantis*'s forward middeck.**

RIGHT: **RMS holds the CRISTA-SPAS research satellite after retrieval.**

BELOW: **Atlantic Hurricane Florence spins below *Atlantis*'s extended robot arm.**

Mission no. **67**

STS

63

Orbiter	Discovery
Launch	February 3, 1995
Landing	February 11, 1995
Duration	8 days, 6 hrs., 28 mins., 15 secs.

The STS-63 crew on the flight deck: (front) Bernard Harris and Mike Foale; (middle) Janice Voss and Jim Wetherbee; (back) Vladimir Titov and Eileen Collins

Crew	James D. Wetherbee, Eileen M. Collins, C. Michael Foale, Janice E. Voss, Bernard A. Harris Jr., Vladimir G. Titov

Mission: This "near-Mir" mission took *Discovery* within thirty-six feet (eleven meters) of the Russian station to rehearse docking methods. Astronauts operated SPACEHAB experiments and deployed SPARTAN for astronomical research. Eileen Collins became the first woman shuttle pilot, and Bernard Harris the first African American to spacewalk.

Mike Foale, Mission Specialist:

Bernard Harris did the first EVA by an African American, and it was a big deal. Bernard was cool about it, but from a historical and US—even a world—societal perspective, it was important. The spacewalk occupied my concentration for most of the preparation, training, and the flight. We went to Daryl Schuck, from the mission operations EVA branch: "Hey, Daryl, what can we do to get ourselves a spacewalk?" He knew about Story Musgrave's frostbite incident, where he damaged his fingertips during a cold vacuum chamber test—and came back to us: "Why don't we just propose doing a suit test to see how cold you get."

We went into the flight knowing that we were going to be suspended above the cargo bay for thirty minutes, day and night, to see how chilled we would get. We asked if we could do other things, and they added handling the mass of the SPARTAN satellite, which was about one and a half tons (1.3 metric tons). After we came out, we attached the foot restraint to the robot arm. I was in the foot restraint, and with Bernard tethered to me, Janice Voss hoisted us up to about thirty feet (nine meters) above the payload bay.

It was absolutely gorgeous. I had nothing to do except look at the view—or Bernard. After four hours in the airlock where we were trying *not* to look at each other, I preferred the space view.

In the last twenty minutes of the cold attitude, I attached this handling aid, like a large, aluminum steering wheel, to SPARTAN. We'd gone into darkness, and I was starting to feel pretty cold. The metal steering wheel was very cold, and I tried not to hold it too tightly. I was on the RMS with Bernard on the starboard sill, and I lifted SPARTAN up above the bay and moved it about, but not very far. The goal was to move it higher, but I just pulled it out of the guides and put back in. I was getting very cold.

Now it was Bernard's turn. He lifted it up, and I saw he was just very gently touching the wheel. He was letting go of it, and it wasn't tethered. I pulled my elbows all the way back in the suit arms; my hands were not in the gloves at all. When I put them back in, there was frost inside. The suit engineers had instrumented the gloves, and the finger temperature was 21 degrees Fahrenheit (-6 degrees Celsius).

I said, "Bernard, we're done." Immediately, they rolled the shuttle toward sunlight. It was just *wonderful*. Cleaning up equipment at the forward, starboard sill of the payload bay, I was in the focus of the radiator. I was being barbecued by it—and I loved it. Later, NASA added glove heaters and the cooling water bypass to the spacesuit.

Then we did the Mir rendezvous. Victor Blagov, the lead flight director in Russia, was really nervous about the effect of shuttle thruster plumes on Mir. They first wanted Jim Wetherbee to come in only about 330 feet (101 meters), but Jim pressed, and after Victor better understood the shuttle's maneuvering capability, he said, "Okay, you can come in to thirty feet (nine meters)," knowing he was going to have to let the shuttle dock someday.

As we came in, Vladimir Titov was talking rapidly in Russian to Mir, and Elena Kondokova was in the Mir's big, Base Block window looking at us, answering. None of us could understand her. Later, Titov told us she said, "Hey, come and have tea."

Discovery's crew photographed cosmonaut Valeri Polyakov framed in the Mir Base Block window.

Mission no. **68**

STS

67

Orbiter	Endeavour
Launch	March 2, 1995
Landing	March 18, 1995
Duration	16 days, 15 hrs., 8 mins., 48 secs.

The STS-67 crew on the flight deck: (front row) Tammy Jernigan, Steve Oswald, and Bill Gregory; (back row) Wendy Lawrence, Ron Parise, Sam Durrance, and John Grunsfeld

Crew	Stephen S. "Oz" Oswald, William G. Gregory, Tamara E. "Tammy" Jernigan, John M. Grunsfeld, Wendy B. Lawrence, Ronald A. Parise, Samuel T. Durrance

Mission: *Endeavour*'s crew split into alternating shifts to manage Astro-2 astronomy observations using three UV telescopes mounted in the cargo bay. Observations focused on intergalactic helium, planetary atmospheres, faint galaxies, and super-hot stars, all viewed from above Earth's obscuring atmosphere.

Wendy Lawrence, Mission Specialist:

We were doing astronomy in the UV, farther into the UV spectrum than Hubble could go. Somebody could spend an entire career analyzing what we would gather during fourteen days of telescope operation on Astro-2.

Up on the flight deck after launch, the engines cut off, and we got through all the immediate things that we needed to do. Oz Oswald had cleared us to get up out of our seats, and we were flailing around a bit. I floated over to the commander's window and got that classic view where you're out over the ocean with Sun glint. I will never forget that view, and the emotion: "Oh my gosh, I finally made it. Twenty-five years. If this is the one view I get during the mission, it's more than enough." Then Oz grabbed my helmet ring and said, "Okay, enough looking. Time to go back to work."

Once we were on orbit, Oz looked at me and said, "Borneo (Bill Gregory) and I are on the red shift. You're on the blue shift, you're the on-orbit commander, and you're in charge. You get to run the ship when I'm asleep." Having that level of responsibility was really a great way to start off my space career. I sat in the commander's seat during my shift and put in all of the maneuvers to point the telescope. For this lowly "rotor-head" helicopter driver, it was an amazing opportunity to truly be a pilot on my first flight.

We spent most of the flight with the telescopes and overhead windows pointed to deep space. You could literally feel the cold of space seeping through those overhead windows, and it really surprised me how cold it was up on the flight deck. We had to live in our sweaters when we were up there.

While sleeping in space, the sensation of floating in your sleeping bag was incredible. It felt so incredibly stable and yet perfectly relaxed. It was cold in the sleep stations, and once while I was asleep, I got so cold that I crawled down to the other end of my sleeping bag, curled up in a ball. When I woke up, it was completely dark in the bunk, and I had lost all sense of orientation. I thought, "Okay, I know I'm in my sleeping bag, but I'm not anywhere near the opening." I had to go hand over hand to get back up to the top.

I'd have to go around cleaning up after Oz—he'd lose blobs of coffee all over the flight deck, up on the circuit breaker panels. One time when he was coming up with his coffee, I kicked him off the flight deck: "Get off my flight deck. You see that blob of coffee? Get off." Between maneuvers, I'd go around and clean everything up, because I knew I had to turn the orbiter over to Marsha Ivins at the end of the flight. She was our lead astronaut at Kennedy, and she'd put the fear of God into me. I was not going to turn over a dirty orbiter to Marsha. Besides keeping *Endeavour* ship-shape, we achieved a lot on STS-67. I was really honored to contribute to a science mission so valuable in the eyes of the astronomers.

ABOVE: ***Endeavour*'s crawler-transporter ascends the incline to Pad 39A for its STS-67 launch.**

BELOW LEFT: **The ASTRO-2 telescopes observe science targets from above Australia's outback.**

BELOW RIGHT: **Sunset's waning glow as seen from *Endeavour*'s flight deck.**

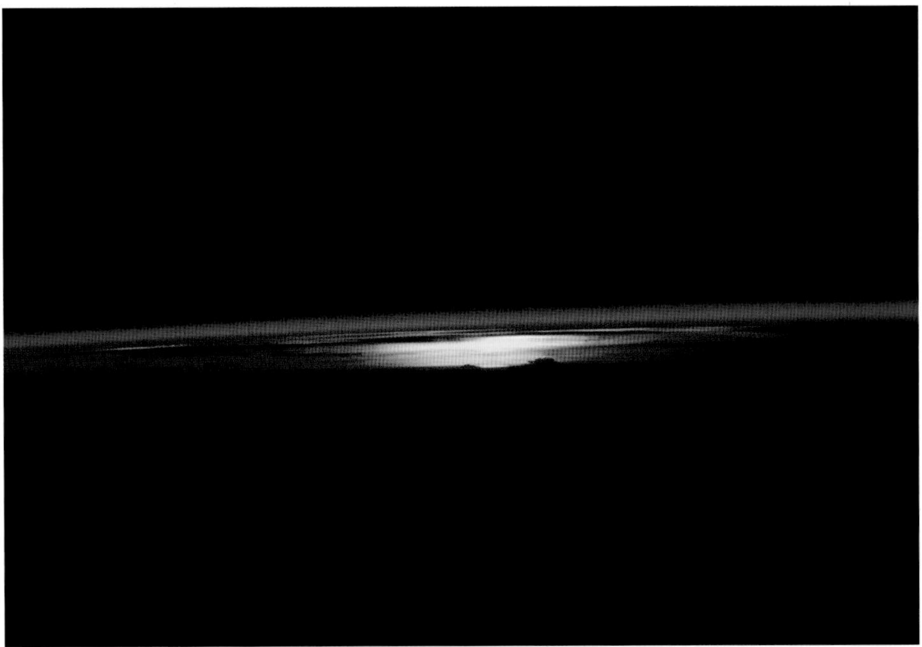

Mission no. **69**

STS 71

Orbiter	Atlantis
Launch	June 27, 1995
Landing	July 7, 1995
Duration	9 days, 19 hrs., 22 mins., 15 secs.

The STS-71 and Mir crews in Spacelab: (clockwise from bottom left) Charlie Precourt, Nikolai Budarin, Ellen Baker, Bonnie Dunbar, Norm Thagard, Gennady Strekalov, Vladimir Dezhurov, Anatoly Solovyev, Greg Harbaugh, and Hoot Gibson

Crew	Robert L. "Hoot" Gibson, Charles J. Precourt, Ellen S. Baker, Bonnie J. Dunbar, Gregory J. Harbaugh Up: Anatoly Solovyev, Nikolai Budarin Down: Norman E. Thagard, Vladimir Dezhurov, Gennady Strekalov

Mission: STS-71 accomplished the first US shuttle-Mir docking, creating the largest orbiting spacecraft to date. Spacelab's fifteen biomedical and scientific experiments focused on long-duration astronaut health. This was the 100th US human space launch.

Greg Harbaugh, Mission Specialist:

STS-71 was the first rendezvous and docking of an American spacecraft with a Russian one since the 1975 Apollo-Soyuz mission. It was a landmark flight; that's why I wanted to be part of it. The cultural bridge that STS-71 formed was a continuation of the process of fulfilling the international partnership between the United States and Russia. A pretty big deal.

I was the handheld laser rangefinder guy for the final approach to Mir. Unfortunately, the morning of the rendezvous I unpacked the laser to discover that we had no batteries. So I was given the real-time opportunity to hotwire the handheld laser through an electrical circuit "break out" box powered from the aft flight deck. I finished the reconfiguration just as we began the final approach. From 2,000 feet (608 meters) in, I tracked Mir and read off the range and range rate to Charlie Precourt, who entered them into his laptop approach corridor display.

I heard a rumor that the Russians were still sore at Deke Slayton for the way he rammed the Soyuz with his Apollo spacecraft back in 1975. They wanted to make sure that we were very gentle with this final approach. Hoot Gibson showed his skills as a pilot and did an absolutely stellar job. We nailed the docking parameters with less than a degree of offset in attitude, spot on time. The docking closure rate was supposed to be 0.1 feet (.03 meters) per second, and we absolutely nailed it.

It was my responsibility to operate the Russian-built docking mechanism. When we made initial contact with Mir, Hoot backed away from the controls and said, "Okay, Greg, it's all yours." A couple of switch throws, and we were fully mated, and it was a non-event—exactly how I was hoping it would go. Hoot flew the first 250 miles (402 kilometers) or so, and I had the last two feet (half a meter).

The Russians invited us over to their place for dinner the first night. We gathered around their table in the Base Block; it had a very homey feel to it. The cosmonauts passed around a container of some unidentifiable dark brown liquid, which I passed on, but it looked like maybe something that they used to help while away long winter nights. Sharing that experience with everybody was really neat, like being in a high-flying Russian dacha.

The day before our undocking, the Russians decided that they wanted to fly around and stand off in the Soyuz to document our departure. When Nikolai Budarin and Anatoly Solovyev undocked the Soyuz from Mir, they looked like they were right outside our window, like you could almost reach up and touch them. We saw them back off, and then we undocked and began our fly-around. Evidently Mir's main computer shut down, and the cosmonauts had to return to dock fairly quickly because Mir was in free drift, with no attitude control. That's a leadership lesson: Sometimes as the leader, you just have to say no. I think the Soyuz undocking was not a great idea.

Norm Thagard did a superb job. After we docked, Norm came aboard and was in the middeck, eating and chatting for a long time. He'd been with two Russian colleagues for the last three months, and probably spoke Russian 99.9 percent of the time. We were hustling by, doing stuff, and he was just hanging out and eating and chatting with everybody that went by. It just was really sweet, and pretty clear to me that part of him was happy to be on his way home.

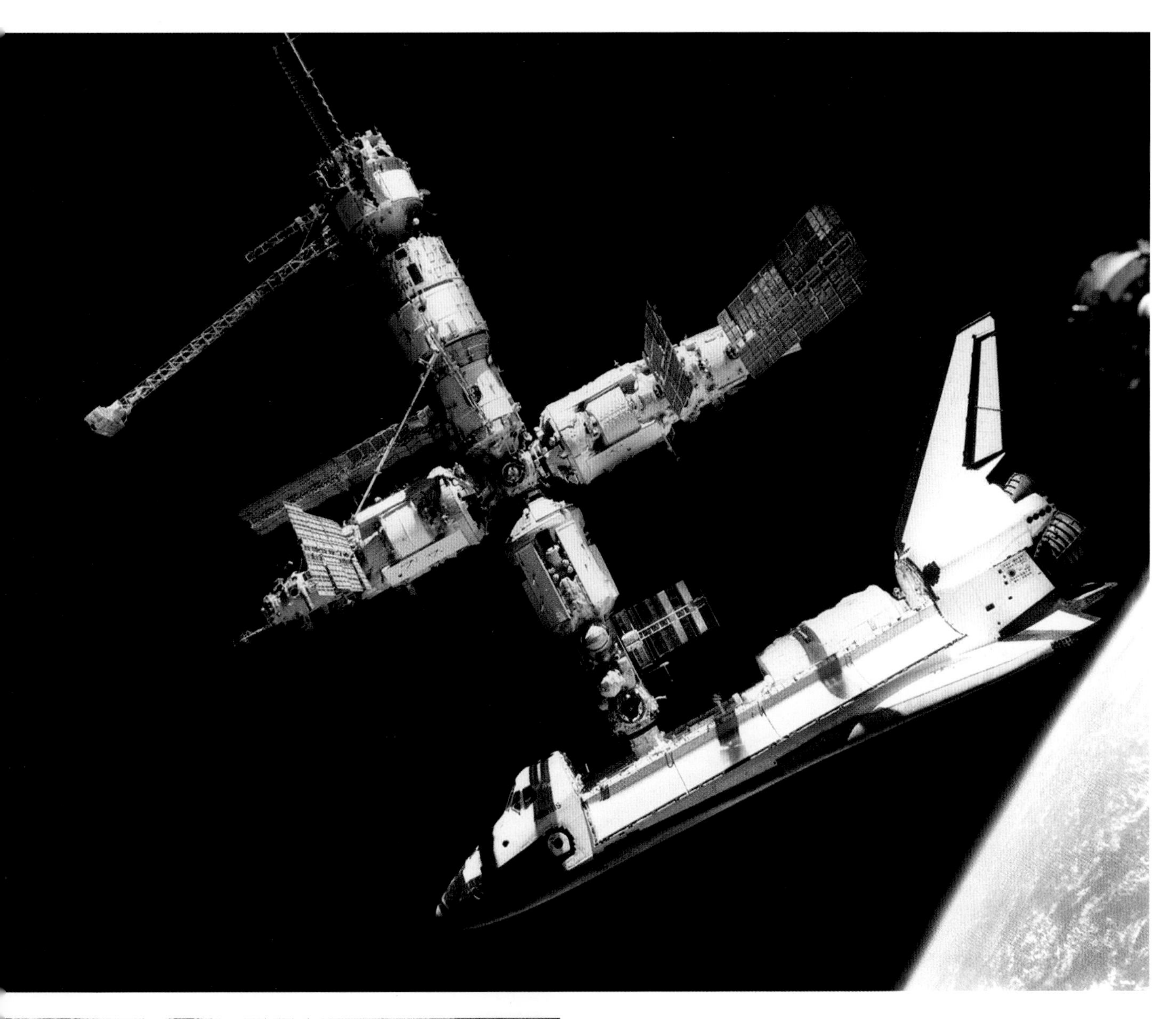

ABOVE: ***Atlantis* as photographed by a Soyuz crew while docked to Mir's Kvant module.**

LEFT: **A wide-angle view of *Atlantis*'s nose as shot from Mir.**

BELOW: **Vladimir Dezhurov and Hoot Gibson meet at the docking hatch into Mir.**

Mission no. **70**

STS

70

Orbiter	Discovery
Launch	July 13, 1995
Landing	July 22, 1995
Duration	8 days, 22 hrs., 20 mins., 5 secs.

The STS-70 crew in front of the Ohio flag on *Discovery*'s middeck: Don Thomas, Nancy Currie, Tom Henricks, Mary Ellen Weber, and Kevin Kregel. All but Kregel were Ohio natives.

Crew	Terence T. Henricks, Kevin R. Kregel, Nancy J. Currie, Donald A. Thomas, Mary Ellen Weber

Mission: Two months before the STS-70 launch, northern flicker woodpeckers bored 181 holes in the foam insulation of *Discovery*'s external tank, causing a six-week launch delay. With the tank insulation repaired, the STS-70 crew deployed TDRS-G into geosynchronous orbit, where it remains active today.

Mary Ellen Weber, Mission Specialist:

On STS-70, we were taking up a communications satellite used to support all the shuttle missions and other communications between spacecraft. Simply having the honor of contributing one brick to the infrastructure in order to create a spacefaring civilization was one of the things that made STS-70 so important.

Launch was delayed because a woodpecker was trying to make a nest in the foam insulation coating our external tank. We were actually at the fixed base simulator, doing a run right before we headed into quarantine, seven days prior to launch. Management called during the sim, saying, "You may not be going into quarantine." They were at first hoping they could repair the foam on the launchpad to minimize the delay, but NASA finally had to roll the shuttle back into the VAB for a fix. We became the woodpecker flight, and it struck a chord with a lot of folks.

After launch we had to capture video and still camera footage of the ET to assess how well those foam repairs held up. I floated up from the middeck, taking off the window sunshades, and as I got in position to take these ET photos, I saw Earth for the first time as this…planet.

I watched Neil Armstrong walk on the Moon; I was at my grandmother's house watching a little grainy TV. Women weren't allowed to be astronauts then. When I graduated from high school in 1980, no American woman had flown in space. Even then, it wasn't clear that women could ever be astronauts. It wasn't until I started skydiving while in graduate school that my eyes were opened to this whole world of aviation. That's when I thought, "I'm going to be an astronaut."

I was very goal-oriented. What stands out in my memory is not the length of time that I wanted to be an astronaut, but how important it was to me once I decided it's what I wanted to do. It was an all-consuming passion for years. Even when I got my astronaut interview, I still didn't think it would become a reality. But just having that goal out there was so meaningful to me.

I really did not expect Earth at night to be as spectacular as it was. Just seeing the airglow, this glowing blanket around Earth, was memorable. The Delta Aquariids meteor shower was happening during our mission, and while watching Earth at night for a full forty-five minutes, I saw sixteen meteors going into that ionosphere below. All five of us were crowded at the back windows of the shuttle, looking at Earth, and then somebody said, "Oh look! A meteor!" and both Kevin Kregel and I said, "Where?" Our first instinct was to look up, but it was below us.

As a scientist, I really wanted the rotating Bioreactor Development System experiment to be successful, so I spent a great deal of time before flight working with their team. In taking samples from the bioreactor, I saw colon cancer cells growing in three dimensions. Looking at these complex tissues forming, I was struck by how fortunate we all were to see the culmination of this experiment. People had spent years waiting for this moment, so I made a call down to the team and told them how proud of them I was and how I so wished that they could all be there with me.

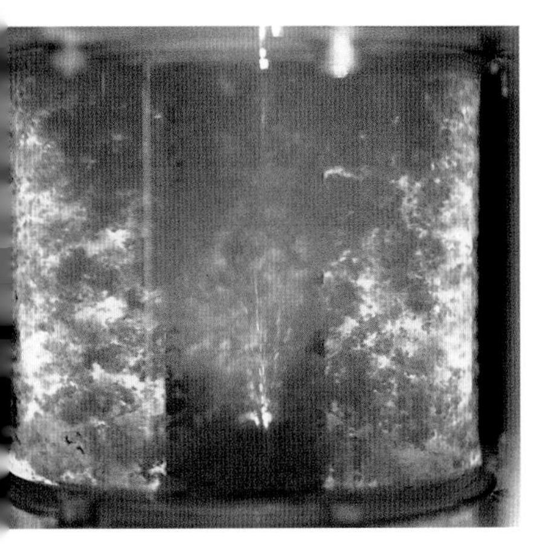

ABOVE: **Human colon cancer cells grow in three-dimensional structures in the Bioreactor experiment.**

RIGHT: **The *Discovery* crew fires a primary reaction control system thruster as part of an exhaust interaction study.**

BELOW: **TDRS-G's IUS booster nozzle is visible as the satellite leaves *Discovery*, escorted by water ice fragments.**

Mission no. **71**

STS

69

Orbiter	Endeavour
Launch	September 7, 1995
Landing	September 18, 1995
Duration	10 days, 20 hrs., 28 mins., 55 secs.

The STS-69 crew on the middeck: Jim Voss, Ken Cockrell, Mike Gernhardt, Dave Walker, and Jim Newman

Crew	David M. Walker, Kenneth D. "Taco" Cockrell, James S. Voss, James H. "Pluto" Newman, Michael L. Gernhardt

Mission: STS-69 astronauts deployed SPARTAN 201 to study the solar wind and atmosphere. The crew later launched the Wake Shield Facility 2, demonstrating manufacture of semiconductor films using its clean, high-vacuum wake. Spacewalkers Jim Voss and Mike Gernhardt tested ISS construction tools and evaluated spacesuit thermal performance.

Mike Gernhardt, Mission Specialist:

STS-69 ranked high as a complex shuttle mission because we deployed and recovered the Wake Shield and deployed and recovered the SPARTAN. And we pulled off a pivotal EVA that evaluated techniques and hardware to be used to assemble and maintain the space station.

We deployed the SPARTAN, and then two days later, with the observatory in what was supposed to be a stable attitude, we rendezvoused with it. But SPARTAN had lost attitude control. We were trained to grapple it with 0.3 degrees per second of rotation in one axis. From 10,000 feet (3,048 meters), we could tell it wasn't pointing in the right direction—and it was slowly spinning in all three axes. We said, "This will be fun."

The gold foil on SPARTAN was shining in the Sun, so Jim was holding a checklist up against the window, blocking the Sun so we could look out and see what was going on. Dave Walker was flying the orbiter around, trying to match SPARTAN's rotation rates. Finally, the ground said, "Okay, we're running low on propellant, so Underdog (me), you've got to get it."

I went in with the arm to snare SPARTAN, and I hit the reach limit right as I grappled it. If SPARTAN had been another two inches (five centimeters) away, the arm would have frozen, and I couldn't have gotten it. I grappled it, but the fact that the retrieval was successful didn't sink in until Dave—Red Dog—looked over with this huge smile and said, "What a fantastic team!" We were really calm talking to Mission Control, but on board, it was: "Jeez, that was frickin' tough!"

Getting that satellite was a huge team effort. Red Dog used to say there's a fine line between "shit" and "shit hot." Had I bumped the satellite and set it spinning, you know the newspaper headlines would have read: "Astronauts screw up the satellite, and it was Mike Gernhardt's fault."

The day before quarantine, I was running through a parking lot in this torrential downpour, and I tripped, fell, and dislocated my shoulder. I was in a sling, and I was rehabbing really hard every day. I was even rehabbing on orbit. But we went with our plan. The day before the EVA I was playing with the shoulder to see if I could get in the suit, and I did, but it was crazy painful. But we did the EVA.

During the EVA, we were evaluating the suit's sublimator cooling system bypass and the glove heaters. My job for twenty minutes was to ride the arm above the payload bay with all the lights out to see if I would get cold. I don't know if there had been any EVAs with no lights anywhere. I could see Jupiter and its four moons with the naked eye, and the Milky Way a thousand times brighter than on a sailboat in the middle of the ocean.

I looked down and saw a fine line of white light off the shuttle's port wing, then a crescent of blue as the Sun rose. For about twenty seconds, I was just hanging there between day and night. The blue light from the ocean propagated out into space and just took my breath away. We emerged from the terminator above Hurricane Marilyn, close to St. Thomas, where I used to teach scuba diving. There I was, just a diver above the clouds looking down on a hurricane.

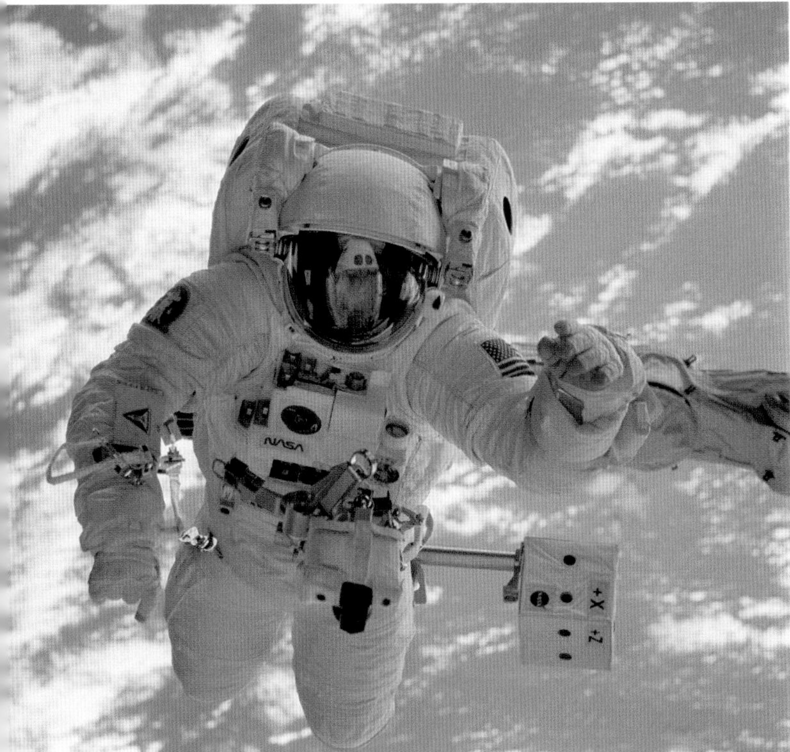

ABOVE: **Mike Gernhardt suspended on the robotic arm above Hurricane Marilyn.**

LEFT: **Gernhardt on an RMS-mounted foot restraint; a temperature sensor package is attached to the boot plate.**

BELOW: **Wake Shield 2 begins its free-flying materials science mission.**

Mission no. **72**

STS 73

Orbiter	Columbia
Launch	October 20, 1995
Landing	November 5, 1995
Duration	15 days, 21 hrs., 52 mins., 21 secs.

The STS-73 crew inside Spacelab: (front row) Kathy Thornton, Michael Lopez-Alegria, and Ken Bowersox; (middle) Cady Coleman and Fred Leslie; (back row) Al Sacco and Kent Rominger

Crew	Kenneth D. "Sox" Bowersox, Kent V. "Rommel" Rominger, Kathryn C. Thornton, Catherine G. "Cady" Coleman, Michael E. Lopez-Alegria, Fred W. Leslie, Albert Sacco Jr.

Mission: STS-73 tied the record for six launch countdown delays before the successful liftoff on October 20. Two shifts of astronauts operated the USML-2 Spacelab around the clock, conducting experiments across five scientific disciplines in the payload bay's Spacelab module. The nearly trouble-free mission proved laboratory operation concepts set for the ISS.

Cady Coleman, Mission Specialist:

STS-73 USML-2 was the pathfinder mission for science on the ISS. The goal was to figure out how we would simultaneously conduct a number of science experiments across different disciplines. We needed to find a way for the scientists on the ground to work with us as we performed their experiments in space. And we needed to develop a way to transmit their data to them in near-real-time so they could adjust their experimental approach.

We were fortunate to have scientists as payload specialists on the mission. Fred Leslie, a meteorologist and specialist in fluid dynamics, and chemical engineer Al Sacco trained with the mission specialists in Huntsville, Alabama, for a good eight months before the flight deck crew was assigned.

I came to grips with the fact that I was in space when I was pedaling the exercise bike on the flight deck, listening to music. Everyone left me alone while I was exercising, and, alone with my thoughts, I felt, "I live here now." I was on the ergometer, looking out the window, thinking of how privileged and happy I was to be doing my work.

In the Spacelab we worked hard. I had a relationship with the scientists on the ground that helped me understand what they wanted, what type of results were most important to them. For example, if there were no communications possible, I would have to decide whether the scientists wanted fewer runs that were perfect, or more runs that showed how wide the range of results could be. Automated experiments—ones with no planned crew intervention—were both the best and the worst. They were the best because they didn't depend on the astronauts' time. But they were the worst because if the experiment broke, we wouldn't know that, and it would be an enormous disappointment to scientists who had spent so much time and effort on something that returned no data. The experiments that meant the most to us were the ones where we felt as though our eyes and hands and senses made a difference in the data. But they were also the ones—because of all the intervention needed—that were the least practical to do on mission.

One of the experiments we conducted on that mission produced a truly magical science moment. We had a crystallization experiment with small test tubes containing dissolved spheres of polymethyl acrylate polymers, like those used in contact lenses. The goal was to find out how crystal shapes or spheres of the same size organize themselves without gravity. Every day we'd take a picture of the test tubes, and the ground would ask, "What do you see? How big are the crystals?" And I answered, "About two millimeters." The ground said "Not two. You mean less than one, right?" "No," I answered, "they're two millimeters."

Those crystals grew much faster and became much larger in microgravity than they would have in an Earth-bound laboratory. Down on Earth, they would have resembled little snowflakes. But in space, the snowflakes just kept growing until they reached the limit of the test tube walls, and the snowflake "arms" would fall off. What we were seeing was actually collections of these arms. The crystals assembled differently, formed a different structure, and grew quite large before they ran out of room. It was a shocking, joyful, astonishing moment.

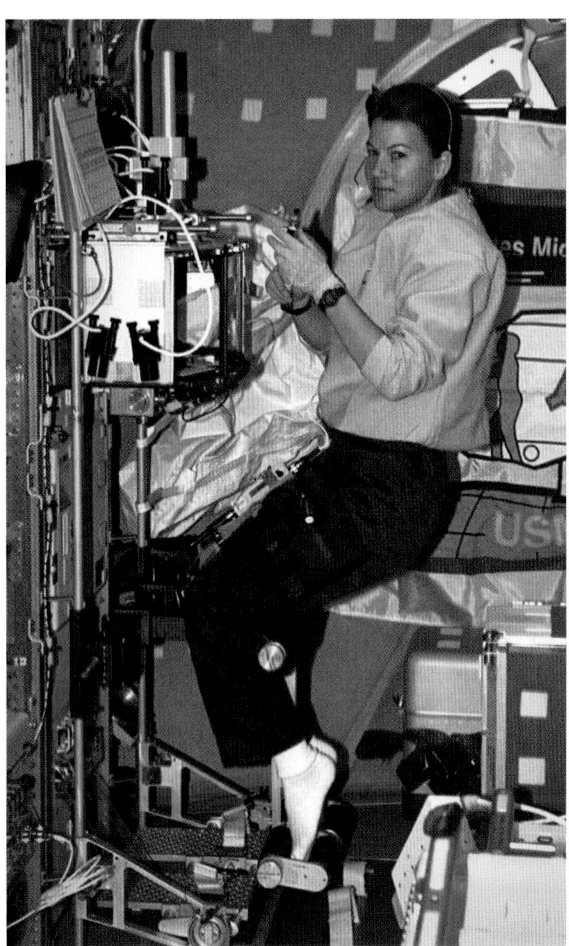

ABOVE: **US Microgravity Laboratory-2 Spacelab in *Columbia*'s payload bay above the Chilean coastline of South America.**

RIGHT: **Cady Coleman working on the USML-2 glovebox in Spacelab.**

BELOW: **Ken Bowersox guides *Columbia* to touchdown on Kennedy Space Center's Runway 33 after a 16-day research mission.**

Mission no. **73**

STS 74

Orbiter	Atlantis
Launch	November 12, 1995
Landing	November 20, 1995
Duration	8 days, 4 hrs., 30 mins., 44 secs.

The STS-74 crew in Mir with Mir EO-20 crew: (in blue) Sergei Avdeyev, Thomas Reiter, and Yuri Gidzenko; (in stripes, clockwise from left) Ken Cameron, Jim Halsell, Chris Hadfield, Jerry Ross, and Bill McArthur

Crew	Kenneth D. Cameron, James D. Halsell Jr., Jerry L. Ross, William S. McArthur Jr., Chris A. Hadfield

Mission: STS-74's crew delivered the Russian-built Docking Module to Mir, facilitating future shuttle dockings. After a precise link-up, the crew left the Docking Module at Mir as a permanent shuttle "front door." Astronauts transferred nearly half a ton (.4 metric tons) of water and supplies to Mir.

Bill McArthur, Mission Specialist:

STS-74 was the second shuttle docking with Mir. That Docking Module really served two purposes: first, it provided a port to which we and all subsequent shuttle missions to Mir were able to dock; second, it explored the technique that STS-88 later used to install the Unity node to the Russian Functional Cargo Block (FGB) module, creating the ISS.

It was so satisfying to work with the Russian crew. I felt we were really transcending national goals. We've always felt that human spaceflight, even when individual countries are pursuing it themselves, is certainly a worthwhile goal for humanity. But I think you get a greater sense of that value when you're working with an international crew. For STS-74, our counterparts were the Russians on Mir and Thomas Reiter and Christer Fuglesang of ESA.

The Mir crew had been there three months already, for a planned four-month mission, but they had been extended to six months. We saw how much it meant to them to have other people visit, with new faces and new people to talk to.

Jim Halsell, Pilot:

After docking and the hatch-opening ceremony, the flight plan directed each crew to give their counterparts a safety "walkdown" of their vessel, after which we were free to roam at will. That's exactly what we did—we exchanged space ships. Excited to explore the world's reigning long-duration space station, my crew cruised the maz of confusing passages connectin Mir's numerous modules, made much narrower by years of accumulated equipment bungee to the walls. At the same time, al three cosmonauts rushed over to the American shuttle without waiting for escort. Because they were at the halfway point in their six-month mission and suffered from more than a little cabin feve they savored the chance to change surroundings and view Earth through our expansive fligh deck windows.

Bill McArthur:

One of my tasks while docked was to position the plume impingement contamination sensors on the RMS end effector. I was to reposition the robot arm so that th Mir thruster exhaust would strike the sensors. I got the arm in the starting position, and from there I would bring the end effector over the payload bay, but not over the orbiter nose. I was driving the arm and looking at the cameras. When I looked overhead, that view about scared me to death. It looked like I was about to drive the arm right in the Mir solar arrays. I later asked myself why we would perform procedures like that with only one cre member, where we could easily have a bad outcome. Why not hav one person focused on the arm, and someone looking outside at th same time? On later missions we used a prime and a backup operator on critical operations like that.

IMAX view of Mir looming above the payload bay and the new, brown docking module just before contact.

Mission no. **74**

STS 72

Orbiter	Endeavour
Launch	January 11, 1996
Landing	January 20, 1996
Duration	8 days, 22 hrs., 0 mins., 40 secs.

The STS-72 crew on *Endeavour*'s flight deck: (front row) Dan Barry, Brian Duffy, and Leroy Chiao; (back row) Koichi Wakata, Brent Jett, and Winston Scott

Crew	Brian Duffy, Brent W. Jett Jr., Leroy Chiao, Daniel T. Barry, Winston E. Scott, Koichi Wakata

Mission: The STS-72 crew first retrieved the orbiting Japanese SFU spacecraft. Using the RMS, the astronauts deployed and retrieved the OAST-Flyer, a reusable SPARTAN spacecraft. During two EVAs, Leroy Chiao, Dan Barry, and Winston Scott tested ISS tools and the spacesuit's ability to withstand extreme cold.

Dan Barry, Mission Specialist:

When we went to capture the Japanese Space Flyer Unit (SFU), the solar panels didn't retract properly, and we immediately went into a contingency procedure. Because of the combined, international simulations we had done with the Japanese team, we quickly reached the conclusion that the arrays had to be jettisoned. That was a big step, because once the solar arrays were jettisoned—pretty far away from shuttle—we had about an hour to get the SFU berthed so we could power the heaters on the fuel lines. If the fuel lines froze, they could rupture, and we would have to throw away the satellite, because if the lines thawed during reentry, we could end up with fuel in the payload bay.

We brought it in, and we got the SFU lowered all the way down in the guides to where the microswitches were starting to show good contact. We had to get three out of four to latch the spacecraft down, but we could only get two.

It was really something to see how we came together as a crew. The burden was nearly all on Koichi Wakata's shoulders, because he was flying the arm and it was his job to get the SFU down. He tried over and over again, and we could get ready-to-latch indicators one and three to light up (indicating "contact"). But those would turn off when two and four lit up, and we were really at a loss about what we were going to do. The ground couldn't help; they were as confused as we were.

I will never forget watching Koichi during this episode. He was the first Japanese astronaut to train as part of a NASA astronau class, with tremendous pressure on him to get this satellite put away. But that guy was cool as a cucumber. He was trying everything he knew. He wore this look of concentration—not sweating, not yelling—just methodically try ing one thing after another. Soon he was trying things that were nc in the book anymore. Finally, we all said, "You know, we play a lot of basketball together. And one thing we never did was slam dur But we're going to have to do th right now." Good old Koichi raise that satellite up about twelve inches (thirty centimeters) and ju slammed it down with the arm in those latch contacts and boom, boom, boom, boom—they lit up! We flipped the switches and clar clang, clang, clang—they locked in! It was a beautiful thing to see My comrade and friend, a rookie astronaut with all that pressure o him, got it done.

On the EVA, we were doing an evaluation of ISS tools, with me in the foot plate on the robot arm and Koichi driving. The original plan was to do the work just above the payload bay. But we decided, "The hell with that." He and I had some code word like, "Ready to roll," which meant "Drive the arm as high as you ca possibly get it." Koichi started driving the arm up to the very top of its reach. I looked down and saw Leroy way down there in the payload bay. The beauty of the shuttle and Earth below just blew me away. Sitting up there, I said myself, "This is the thing I worke toward for forty years, and I'm here. I'm doing it." It's an indelib memory.

LEFT: **Dan Barry smiles in at flight deck crewmates during the EVA.**

ABOVE: **Koichi Wakata flying the RMS positions the Space Flyer Unit above the payload bay for berthing.**

BELOW: **Brian Duffy guides *Endeavour* to touchdown on Kennedy Space Center's Runway 15.**

Mission no. **75**

STS

75

Orbiter	Columbia
Launch	February 22, 1996
Landing	March 9, 1996
Duration	15 days, 17 hrs., 40 mins., 21 secs.

The STS-75 crew on *Columbia*'s flight deck: (front row) Franklin Chang-Diaz, Andy Allen, and Jeff Hoffman; (back row) Maurizio Cheli, Claude Nicollier, Scott Horowitz, and Umberto Guidoni

Crew	Andrew M. Allen, Scott J. Horowitz, Franklin R. Chang-Díaz, Maurizio Cheli, Jeffrey A. Hoffman, Claude Nicollier, Umberto Guidoni

Mission: *Columbia*'s crew unreeled the Tethered Satellite out to almost twelve miles (nineteen kilometers) before the tether insulation failed, and high-voltage plasma burned through the line. The TSS continued to transmit scientific data until it burned up in the atmosphere. The crew operated USMP-3 experiments throughout the mission.

Scott Horowitz, Pilot:

The main engines lit, and the three main engine indicators came up, and they normally go right to 100 percent. But on our launch, I had the right and center engines at 100 percent, and the left one was at 36 percent. I looked over at Andy Allen and said, "Left engine's at 36 percent!" He looked at me, and then the solids lit!

I looked back and said, "Where are we going?" And he said, "To space. Didn't they tell you?" "Okay, great," I said, "But the left engine's at 36 percent." And Andy said, "Well, I guess we're going to abort." I brought up the backup display, and I told Andy the engine looked normal. He called the roll program to Houston, adding, "We show the left engine at 36 percent." Of course, the Booster controller in Mission Control fell out of his chair. It was quiet for a minute, then they came back with, "We think it's okay." The data was good, and the indicator was bad. Everyone around me was saying, "Wheee! We're going to space!" and I was thinking, "Oh, no!" That distracted me for most of the ascent.

Maurizio Cheli, Umberto Guidoni, and I were on one shift, and we were there for the beginning of the Tethered Satellite deployment. Everybody was on deck for that; everyone had a job. When we released the satellite, it had jets on it firing for the first few feet to get some tension on the line. But once the satellite and orbiter were apart a certain distance, they traveled at different velocities, and that created tether tension all by itself. We got it up and everything was looking good, so our shift went to bed.

I woke up in the morning, and Franklin Chang-Diaz, the first guy I saw, had this frown on his face. Franklin never frowns—he's the happiest guy I know. I asked him what was wrong. He said, "We lost the satellite."

I looked out the window and there was the tether trailing up. I couldn't believe it. Then he said, "We're tired. We're going to bed." I asked him, "What am I supposed to do?" He said, "Whatever the ground tells you to do." It was my first day on orbit with a payload and they left the rookie crew up there with this broken wire.

There were about twelve miles (nineteen kilometers) of tether, all balled up into a big tangle. Controllers on the ground were discussing what to do. One thought was, "Okay, we'll go rendezvous with it, then we'll send a couple of guys out in spacesuits." And I said, "Isn't this wire carrying thousands of volts across it?" They answered, "Yeah, that's probably not a good idea." We reeled the broken tether in and brought the deployment tower down, and that was it. But the satellite was still conducting experiments, using its instruments to measure currents and fields. There's a lot of stuff discovered from things you never planned.

Andy Allen, Commander:

Johnson Space Center director George Abbey wanted us to do a live link with the Houston Livestock Show and Rodeo, and Jeff Hoffman and I were the only two crew available. We did the one-way video downlink to the Astrodome. Jeff and I answered a bunch of standard questions, and then Constable Bill Bailey said, "Andy, last question. What is your favorite rodeo event?" I had only ever been to one rodeo and without moving my lips I asked Jeff Hoffman if he knew the names of any rodeo events, and he whispered back "No." So without thinking I said, "I like watching the cowgirls." Bailey gave a nice but conservative laugh and we signed off. Afterwards, I was sure I would get in trouble, but those Texans liked it!

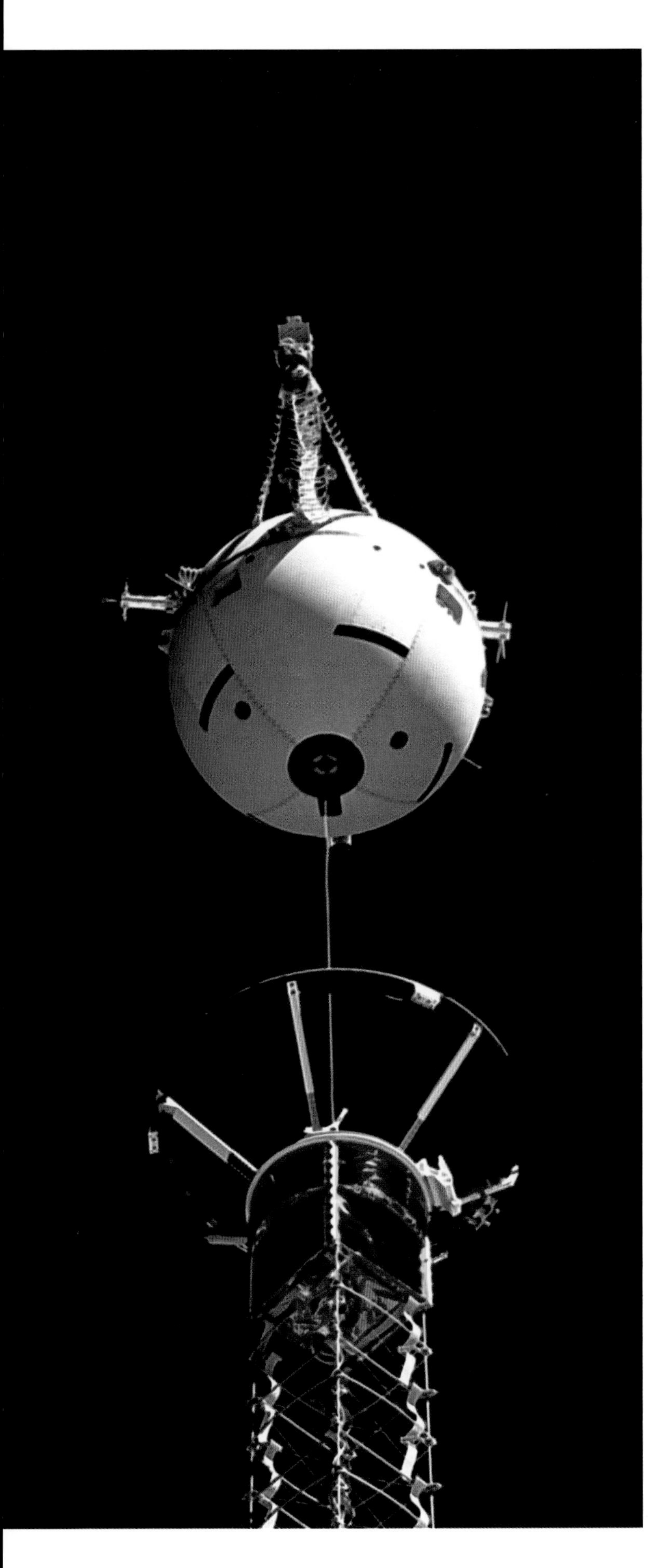

LEFT: **The Tethered Satellite unreels from its deployment mast on its way to a 12.2-mile (19.7-km) separation from *Columbia*.**

ABOVE: ***Columbia* vaults from Pad 39B carrying the Tethered Satellite System and the USMP-3.**

BELOW: **The burned-through tether hangs from the deployment tower truss.**

BOTTOM: **Scott Horowitz with a combustion glovebox and tray of shuttle maintenance tools.**

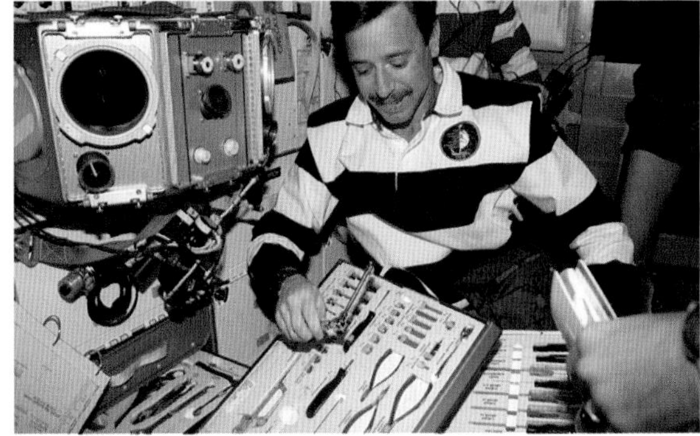

Mission no. **76**

STS 76

Orbiter	Atlantis
Launch	March 22, 1996
Landing	March 31, 1996
Duration	9 days, 5 hrs., 15 mins., 53 secs.

The STS-76 crew on *Atlantis*'s middeck: Linda Godwin, Rich Clifford, Kevin Chilton, Shannon Lucid, Rick Searfoss, and Ron Sega

Crew	Kevin P. Chilton, Richard A. Searfoss, Linda M. Godwin, Michael R. "Rich" Clifford, Ronald M. Sega Up: Shannon W. Lucid

Mission: *Atlantis* docked with Mir to begin Shannon Lucid's four-and-a-half-month stay there. Astronauts transferred water and scientific gear to Mir while operating eleven European Biorack experiments in SPACEHAB. Spacewalkers Linda Godwin and Rich Clifford installed four experiment packages on Mir's docking module.

Linda Godwin, Mission Specialist:

This mission showcased many shuttle capabilities: a rendezvous with Mir, experiments in a SPACEHAB laboratory, a spacewalk, and transporting Shannon Lucid up to Mir.

As we closed in on Mir on rendezvous day, I was struck by the amazing sight of another structure in orbit. Mir was unique; it had many modules all locked together. To think that we could launch to and rendezvous with that complex and knock on the door, and somebody was in there, and we were going to go in! We had met Yuri and Yuri in Russia—Yuri Onufriyenko and Yuri Usachov—and here they were in space! What we learned during the Shuttle-Mir missions benefited both the United States and Russia as we built the ISS.

What struck me right out of the gate was that Shannon was built for space. This was her element. That's why she was such a good choice as our second long-duration crewmember at Mir. On orbit, she had such an easy transition. I have to credit her determination even before launch. Part of it was natural, but she was also mentally ready to be part of the Mir crew—totally immersed—as soon as we docked.

STS-76 was the first and only time I enjoyed an extension to the crew module, with the long tunnel going back to SPACEHAB. Floating through that very long tunnel into another shirt-sleeved environment was a whole new experience, like going to the basement of your house.

SPACEHAB was deluxe. We had worked with locker-housed experiments in the cabin before, but SPACEHAB was like a real lab, with a glove box—a transparent enclosure with a contained atmosphere, and ports through which we could reach our gloved hands to handle materials. I was glad to be assigned to this part of the mission because it used my science background. The principal investigators on the ground are the real experts, but it was still rewarding to carry out some nine or ten of their experiments in space, collecting data on subjects ranging from nematode worms to bone cells, leukemia cells, seeds, and yeast. After the mission, the science team told us that they obtained almost 100 percent of their expected data return.

All astronauts hope to get to do an EVA, and this one was especially interesting because we spacewalked across two docked vehicles. The Russians made it clear to us that we were not to spacewalk on Mir beyond the docking module—and we didn't. When I opened the hatch and got out there, it didn't look like an alien view. All those hours in the altitude chamber, the water tank, and in the suits prepared me for this. I remember that moving down the sill on the port side was so much easier than moving through the water, but it was harder to stop. Because I had all that nice momentum, I overshot my first stop to pick up some equipment from that side.

I enjoyed going outside with Rich, who was just an excellent spacewalking partner. We worked well together, and we made a lot of trips back and forth from the payload bay to the docking module. When Rich and I finally got both those MEEPs—Mir Environmental Effects Payloads—installed and opened, up on those Mir handrails, that felt good.

Saying our goodbyes to Shannon and the Mir crew, I remembered what my husband Steve Nagel, another astronaut, had told me—we were married the December prior to launch. "Make sure you're on the correct side of the hatch when they close it."

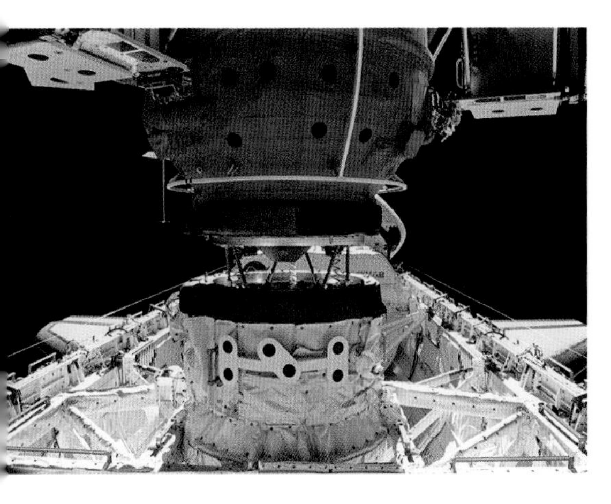

ABOVE: ***Atlantis*** **mated with Mir's brown docking module. Once mated, the orbiter's docking ring will retract to create an airtight tunnel.**

RIGHT: ***Atlantis*** **approaches Mir with SPACE-HAB in the aft payload bay.**

BELOW: **Linda Godwin transports a materials exposure experiment from the payload bay to Mir.**

Mission no. **77**

STS 77

Orbiter	Endeavour
Launch	May 19, 1996
Landing	May 29, 1996
Duration	10 days, 0 hrs., 39 mins., 20 secs.

The STS-77 crew at Pad 39B for preflight safety briefings: Dan Bursch, John Casper, Curt Brown, Mario Runco, Andy Thomas, and Marc Garneau

Crew	John H. Casper, Curtis L. Brown Jr., Daniel W. Bursch, Mario Runco Jr., Marc Garneau, Andrew S. W. Thomas

Mission: STS-77's astronauts released SPARTAN with its Inflatable Antenna Experiment, a tennis-court-size antenna dish, while John Casper and Curt Brown flew formation. The crew operated SPACEHAB and payload bay space technology experiments throughout the mission, and flew formation with the drag-stabilized PAMS satellite.

Dan Bursch, Mission Specialist:

STS-77 had many payloads related to SPACEHAB, and I think we set a record by completing four separate rendezvous during the mission. As the flight engineer, I spent a lot of time together with John and Curt, not only in launch and landing sims, but probably double the normal number of rendezvous sims too.

The SPARTAN inflatable antenna experiment deployed a large-aperture dish, important for applications like communications or imaging. The larger the aperture, the more imaging detail you can get, or better signal gain on a communications antenna. The Inflatable Antenna Experiment (IAE) was quite ambitious, with the antenna itself forty-six feet (fourteen meters) in diameter and more than ninety-two feet (twenty-eight meters) long. The premise was that if you can build an inflatable antenna, you would get the size you need with a cost an order of magnitude lower than a rigid antenna.

John Casper, Commander:

Mario Runco deployed the SPARTAN with the robot arm, and shortly after, the IAE began inflating with nitrogen gas. The reflector dish and long, hollow struts were made of thin Mylar, similar to helium party balloons. As we filmed the five-minute inflation process, our hearts nearly stopped when the struts and dish unexpectedly wrapped around the SPARTAN, tumbling the entire structure. It looked like a huge spider tangled in its own legs. But we were relieved to see the struts and dish slowly begin to swell, untangle, and take shape as intended. Lasers on the SPARTAN recorded the surface dimensions of the parabolic dish for later evaluation. During the next ninety-minute orbit I manually maintained position, a intense process. I traded off with Curt Brown periodically to take a break.

Dan Bursch:

As we were flying formation wi the antenna, it slowly started to tumble. A small leak somewhere must have propelled it into a rotation. We stayed close to get vide of the antenna and document its shape. After the antenna separated, we flew away from it, and the the first rendezvous was to go back and get the SPARTAN.

We flew a Coca-Cola dispense in the middeck. When you drink a carbonated beverage on Earth, there are certain things you're used to. You pop it open, and it makes a sound. Then you see some bubbles come out of solution. Drinking "Space Coke" was different. It tasted fine. You taste the fizziness, the carbonation in your mouth. But it wasn't like opening a beverage can on Earth The Coke tasted like it was put together from syrup, water, and carbonation from CO_2.

This was Andy Thomas's first mission, and he was designated the payload commander. That jus showed how much trust the astro naut office had in him. He ended up flying later than a lot of other people, but he did well as payloa commander. He had his hands full with not only the SPACEHAB, but all the other experiments that were on the mission. He is a very detail-oriented guy, and he met the challenges.

I think the public tends to think that all astronauts are from the same mold. But we're all unique in how our journey brought us to space and to the shuttle program Bringing these unique personalities together is challenging. People assume that it's all going to work, yet it's not without its challenges and bumps in the roa along the way.

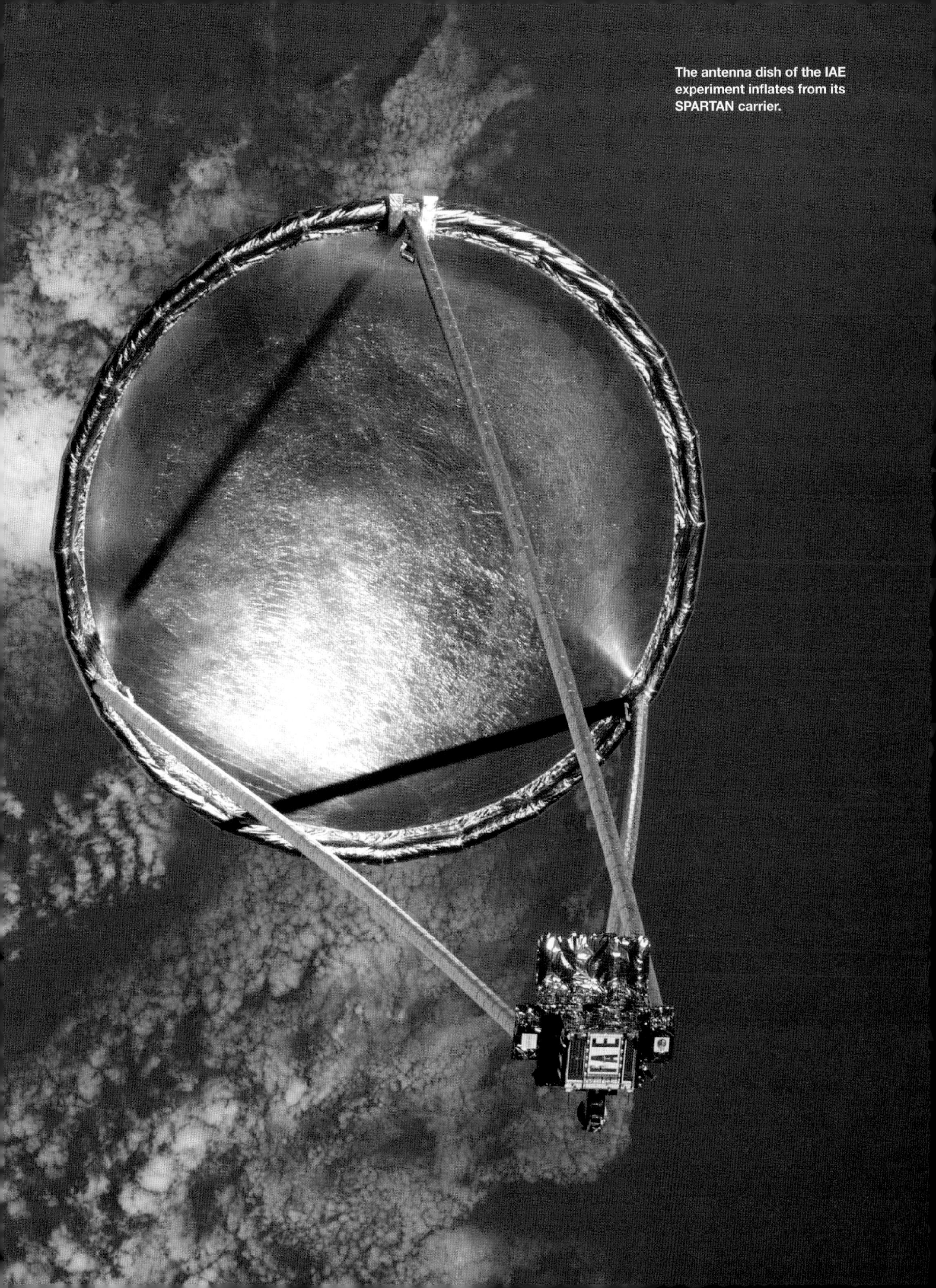

The antenna dish of the IAE experiment inflates from its SPARTAN carrier.

Mission no. **78**

STS

78

Orbiter	Columbia
Launch	June 20, 1996
Landing	July 7, 1996
Duration	16 days, 21 hrs., 47 mins., 35 secs.

The STS-78 crew in Spacelab: (clockwise from left) Rick Linnehan, Tom Henricks, Susan Helms, Jean-Jacques Favier, Bob Thirsk, Kevin Kregel, and Chuck Brady

Crew	Terence T. Henricks, Kevin R. Kregel, Susan J. Helms, Richard M. Linnehan, Charles E. Brady Jr., Jean-Jacques Favier, Robert Brent "Bob" Thirsk

Mission: STS-78's astronauts operated more than forty experiments aboard the international Life and Microgravity Spacelab (LMS), conducting research in life sciences, human physiology, fluid physics, and materials science. LMS helped refine science operations plans for the coming ISS.

Tom Henricks, Commander:

The primary mission was life and microgravity science with astronauts as the primary test subjects. We proved that humans could do good research on themselves in microgravity. We were an international mission with a Frenchman and a Canadian on the flight crew, and their backups were a Spaniard and an Italian. The cultural diversity forced us to find ways to work together. For example, if Frenchman Jean-Jacques Favier, who spoke English well, was challenged by technical jargon in English, bilingual Canadian Bob Thirsk could help.

We had to keep the two backup crewmembers engaged in case they had to fill in to fly at the last moment. Fortunately, Susan Helms was an outstanding payload commander, and I trusted her to handle the challenging experiment task assignments and scheduling. She coordinated with the Huntsville payload ground team to develop a training flow that was very productive and worked effectively on orbit.

We had two MDs, a Doctor of Veterinary Medicine, and a PhD as the primary test subjects. Those four doctors did amazing stuff. They attached electrodes to their calves to induce painful, maximum effort muscle contractions. Every now and then, I heard screams from the lab! They did blood draws twice a day. MDs are not the best people to do "sticks"—any doc will admit that. Rick Linnehan, the vet, was our best "sticker." One morning I went back to the lab, and the four "guinea pigs" were doing their blood draws, and there were blobs of blood floating in the lab—it was like a horror movie. Sometimes an intravenous catheter wouldn't stay in—maybe that's what happened.

For fun, we all went to the back of the lab and pushed off in attempts to fly through the Spacelab tunnel to the shuttle middeck without contacting anything. We had to be careful not to look up at the wrong time and crash into a hatch. At the end of one long day all seven of us settled around the opening of the Spacelab tunnel in the middeck. We'd turned all the lights off in the lab and tunnel so it was just a dark hole. Sitting around the opening of the tunnel, we imagined that our feet were dangling "down" into a well. We all sat there and chatted about what we missed from home. It was like a campfire experience around this well. Everyone imagined it was a deep hole—and then Kevin fell in and disappeared. Good one!

On Flight Day 14, I asked everyone how many more days they wanted to continue working at the pace we were on, with sixteen-to-eighteen-hour workdays. I asked because everyone seemed comfortable, rested, and productive. No one wanted to return as planned on Flight Day 16. The consensus was another week: a twenty-one-day mission. Regrettably, *Columbia* didn't carry enough consumables for such a stay.

Of course, as the commander, I found it very satisfying that on STS-70 and -78, Kevin and I had two of the better landings in the shuttle program. I attribute that to Kevin, who once was a shuttle training aircraft instructor pilot. Before he became an astronaut, he helped train me to land the orbiter. I told him, "Okay, Kevin, this is another training mission, and I want you to speak up if you see anything that's not right down the middle." We touched down as planned at 205 knots (105 meters per second), at less than a foot per second (0.3 meters per second) descent rate, 2,500 feet (762 meters) down the runway.

ABOVE: **Tom Henricks inspects computer hardware behind the Spacelab module control panels.**

LEFT: ***Columbia* launches from Pad 39B with the LMS containing 40 experiments.**

BELOW: **Rick Linnehan grimaces as he delivers a leg-strength measurement in Spacelab.**

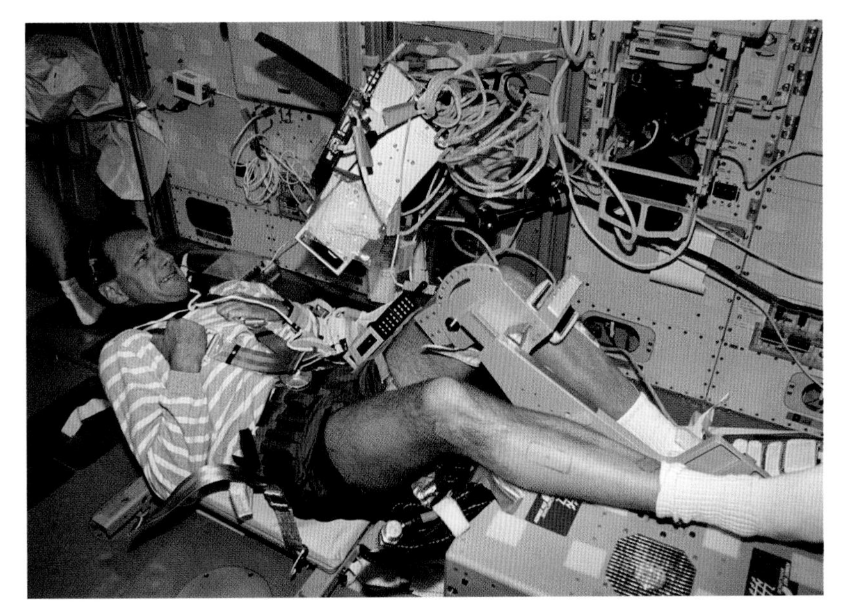

Mission no. **79**

STS

79

Orbiter	Atlantis
Launch	September 16, 1996
Landing	September 26, 1996
Duration	10 days, 3 hrs., 18 mins., 24 secs.

The STS-79 and Mir crews (in blue) in the station Base Block: Alexander Kaleri, Tom Akers, Jay Apt, Carl Walz, John Blaha, Valery Korzun, Bill Readdy, Terry Wilcutt, and Shannon Lucid

Crew	William F. Readdy, Terrence W. "Fencer" Wilcutt, Thomas D. Akers, Jay Apt, Carl E. Walz Up: John E. Blaha Down: Shannon W. Lucid

Mission: The *Atlantis* crew docked with Mir and delivered two tons (1.8 metric tons) of supplies from the SPACEHAB double module. Shuttle supply runs were vital to sustaining and equipping Mir. John Blaha replaced Shannon Lucid, who completed 188 days in space.

Carl Walz, Mission Specialist:

Bill Readdy and our crew worked hard to put together an ambitious plan. We worked together well and had a lot of fun. It helped that we had a great pair of hosts on Mir with Valery Korzun and Sasha Kaleri.

The whole back half of the SPACEHAB double module was logistics—monster batteries and all sorts of gear for Mir—and the front half carried experiment hardware. One experiment was an ISS rack called the Active Rack Isolation System (ARIS). The ARIS active control system used magnetic fields to "float" the rack and isolate sensitive experiments it carried from any spacecraft vibration. I was the ARIS "mother," and the darn thing kept breaking. I used up all the spare hardware and then some, trying to keep the thing running.

It was an amazing experience going to someone else's spaceship. Mir was very different from the ISS; when I lived on the latter it was brand-spankin' new. Mir had been up there for a while and was well lived-in. Different modules had different sensations and different smells. None of them were bad, but the Kristall module was loaded with old equipment, like an attic, and smelled very musty. Every old experiment that ever flew on Mir was jammed into this module. If you've read *The Lion, the Witch and the Wardrobe*—where characters push through all the coats to step into Narnia—that's what it was like going from the shuttle and swimming up through all this stuff.

We brought some gifts up to the Mir crew. I had lived in Las Vegas prior to Houston, and I knew Ethel M chocolates were really good—and some were liqueurs. I got a big box, and we brought them along as one of our gifts to the Russians. Bill and I and the rest of our crew floated over to the Mir to join Sasha and Valery in the Base Block, just floating, eating chocolates, and watching the world go by. Then I got this call from the shuttle that the ARIS rack had "blown up," and bits and pieces of the rack were flying everywhere. I instantly went from relaxing with our Mir friends to safing the experiment, collecting all the pieces and figuring out what to do next.

I was impressed by the Russian hardware and the capabilities of their specialists and engineers. They came at things from a different viewpoint, but they put a lot of thought into everything and came up with perfectly good operational solutions. If I learned one thing on Mir, it's that there's more than one way to fly in space.

I had already gotten word that I would be training for ISS, so talking to Shannon and John Blaha about their Mir training experience was helpful. "Shannon, are all the classes at Star City going to be in English?" She shook her head dismissively: "Nyet." "Do you think we will only be speaking English aboard the International Space Station?" "Nyet." Their answers made it pretty clear that what we were hearing from management was overly optimistic and probably wasn't going to happen

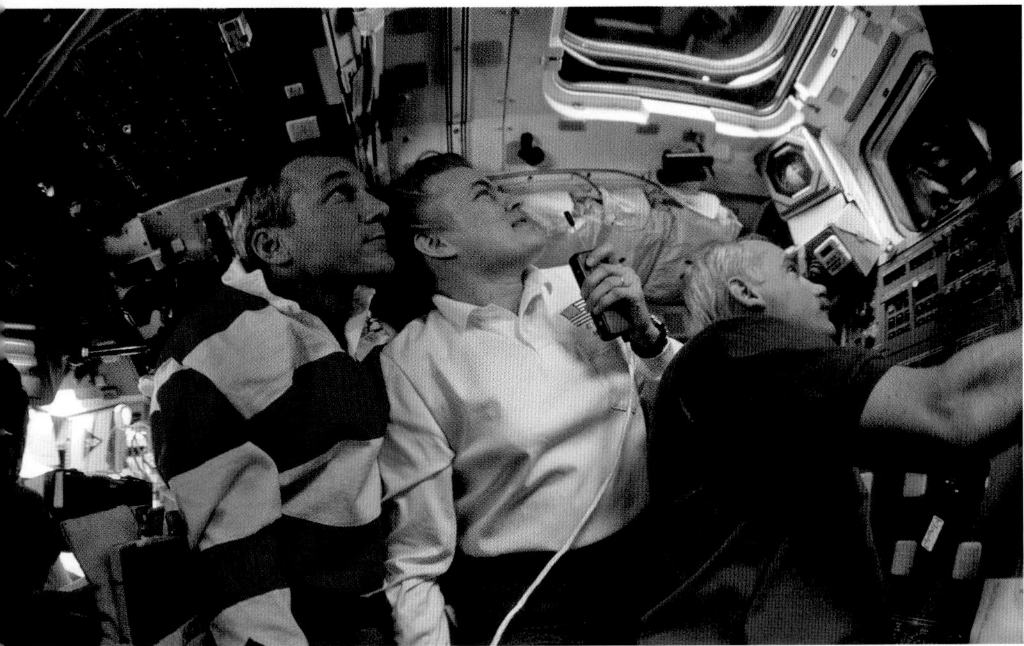

ABOVE: **An *Atlantis* IMAX image of Mir after docking, high above a storm system in the southern Indian Ocean.**

LEFT: **On the flight deck, Terry Wilcutt, Shannon Lucid, and Bill Readdy watch Mir recede after undocking.**

Mission no. **80**

STS

80

Orbiter	Columbia
Launch	November 19, 1996
Landing	December 7, 1996
Duration	17 days, 15 hrs., 53 mins., 17 secs.

The STS-80 crew on the middeck: (front row) Tom Jones and Story Musgrave; (back row) Ken Cockrell, Tammy Jernigan, and Kent Rominger

Crew	Kenneth D. "Taco" Cockrell, Kent V. "Rommel" Rominger, Tamara E. "Tammy" Jernigan, Thomas D. Jones, F. Story Musgrave

Mission: On launch day, *Columbia*'s crew used the RMS to deploy the ORFEUS-SPAS astronomy satellite, and later deployed the Wake Shield Facility 3 satellite. Both were retrieved after completing their science missions. Two spacewalks were cancelled when an errant screw jammed the airlock hatch mechanism.

Ken Cockrell, Commander:

I call STS-80 a transition from independent shuttle missions to when shuttles began to dock with, construct, and support other human spacecraft. We were also trying to gain experience with long-duration science, prior to attempting experiments on the Station that would last much longer. Our extended-duration orbiter flight was meant to last sixteen days, but we ended up setting a record, just short of eighteen days. Our mission had a lot going on: two satellites, two spacewalks, and traveling a lot during training to understand those tasks.

I think STS-80 was the first time that two free-fliers from the same vehicle were out there in orbit at the same time. When we were doing the rendezvous to pick up the Wake Shield, our trajectory took us up between the two satellites. As we approached the T_i burn, the rocket firing that would initiate the final rendezvous phase, we had a free-flier about ten miles (sixteen kilometers) on each side of us. We could see ORFEUS out one window and Wake Shield out the opposite one.

Rendezvous becomes a test of skills—to try to do it with the least amount of propellant. It's also a battle for patience within yourself to not make a correction before its time. You always have orbital mechanics in the back of your mind: what forces will slow you down and speed you up. On STS-80, coming up from nadir made the approach the easiest in the world. You have to force the orbiter to go up to the satellite, so it's easy to control closure rate. You should never want to brake—you want the laws of physics to do the braking for you.

The digital autopilot is keeping your attitude stable, and as the pilot, all you're doing is translation—left and right, up and down, while you look up through an optical alignment sight every now and then to gauge how well the satellite is centered in the payload bay. If you handle left, right, up, and down, the arm operator can go in and get it even if it's not perfectly centered in the bay. Rommel and I had a lot of fun trying to get the satellite's relative motion to totally zero. We joked that we could fly the orbiter so as to put the satellite's grapple fixture right into the arm's end effector. Then the arm operator would only have to pull the trigger to grapple it.

Tom Jones, Mission Specialist:

On reentry, Story Musgrave came up from downstairs to watch and film the flickering, incandescent plasma streaming out behind the orbiter. Braced against the left-side panels and the commander's seat, he never left! It was his last flight, and he was going to experience the landing *his* way. Every time I glanced left from my flight engineer's seat, he was still there—sweating, but standing all the way through landing.

Ken Cockrell:

The most fun came right after touchdown when someone started patting me on the shoulder. I had been completely absorbed with landing up until that point and hadn't realized that Story had not gone downstairs.

When he patted me on the shoulder and said, "Good landing, Taco," and then "I gotta go sit down," I thought, "This is one for the history books."

The joy, for me, of flying STS-80 (and because of my pilot's approach to the flight) was the satisfaction of training for and executing the myriad flying tasks to enable the mission. There were a lot of them, and each was a team effort. STS-80 gave us all a rich flying experience on the space shuttle *Columbia*.

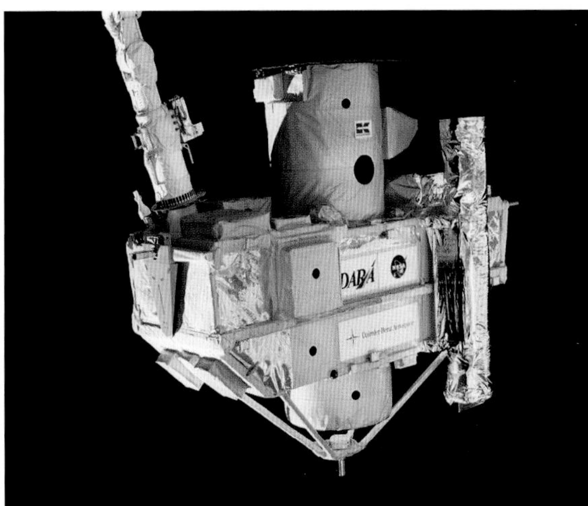

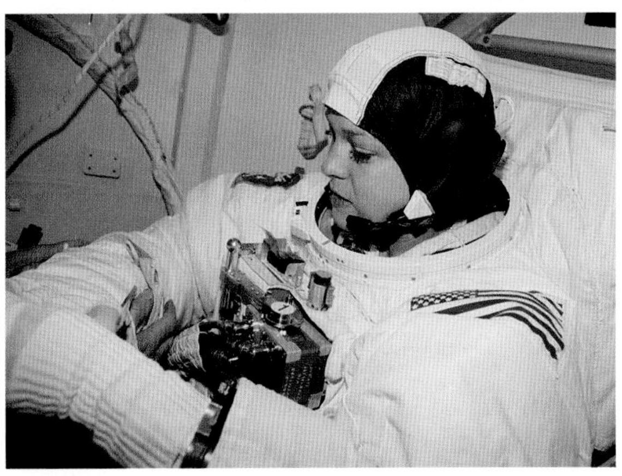

ABOVE: **Ken Cockrell and Tammy Jernigan award Story Musgrave (center) his "Master of Space" badge, celebrating his one thousand hours in orbit.**

ABOVE RIGHT: **The ORFEUS-SPAS observatory in the grasp of the RMS shortly before release.**

RIGHT: **Jernigan suits up in the airlock for an EVA, which was eventually foiled by a jammed hatch.**

BELOW: **In 1996, Cockrell and Kent Rominger fly *Columbia* to a smooth finish to the longest shuttle mission in history.**

Mission no. **81**

STS

81

Orbiter	Atlantis
Launch	January 12, 1997
Landing	January 22, 1997
Duration	10 days, 4 hrs., 55 mins., 21 secs.

The STS-81 and Mir crews in *Atlantis*'s middeck: (clockwise from top) Brent Jett, Jerry Linenger, Mike Baker, Valery Korzun, Alexander Kaleri, Marsha Ivins, John Blaha, Jeff Wisoff, and John Grunsfeld

Crew	Michael A. Baker, Brent W. Jett Jr., John M. Grunsfeld, Marsha S. Ivins, Peter J. K. "Jeff" Wisoff Up: Jerry M. Linenger Down: John M. Blaha

Mission: *Atlantis*'s astronauts docked with Mir, swapped Jerry Linenger for John Blaha, and transferred nearly 6,000 pounds (2,722 kilograms) of water, US science equipment, and Russian equipment. About 2,400 pounds (1,089 kilograms) of science samples, obsolete gear, and trash returned to Earth in SPACEHAB.

Marsha Ivins, Mission Specialist:

STS-81 was the first time I watched us approach another spacecraft in space. We had completed the final rendezvous burn in the dark, and when we hit sunrise suddenly there was this big thing out there in front of us, getting closer. It didn't feel normal, not at all like formation flying in an airplane. When we pulled up into Mir, we literally had solar arrays and pieces of Mir on either side of us, down where I had the feeling that they shouldn't be.

I realized that our slow approach to Mir was just relative motion, but somewhere in the back of my head, I was hearing, "That thing is going 17,500 miles (28,164 kilometers) per fricking hour!" Think about that! I was stunned, as I was again on STS-98, by how smoothly and elegantly this boat of a space shuttle performed the rendezvous maneuver.

John came back with us, and we left Jerry up there. They were both transfer items on my stowage list. I don't know if we were the first Mir crew to list the astronauts that way, but every crew after us would have the transfer astronaut as a line item on the cargo list.

The cosmonauts said to us, "Come over anytime." Well, I went wandering over one day, and one of the guys was bathing—completely naked! I turned around and slunk off in another direction, thinking, "Okay, you actually meant 'Come over any time—unless you're a woman.'" I always knocked after that. The first time Valery Korzun, the Mir commander, came over to the shuttle, I had just washed my hair and it was all out in a ball. He came over, saw me, and said, "Ahhh!" He turned around, flew back to Mir, and came back with a camera: "Take picture!"

I was taking video in the Kvant module, where they staged all their old gear. It's also the berthing point for the Progress supply ship. The stuff in Progress was bolted down, and to get to it, the had to unscrew metal brackets. Valery was looking for something in there, and it sounded like a cat fight inside some trash cans. Out he came, and he had a syringe stuck to him on one side and stains on the other sleeve, but he'd found whatever it was he wa looking for. Aye-yi-yi! We had all our clean, precise standards for shuttle: If we flew a bag of Wheat Thins, we'd have them stacked flat with all the corners precisely aligned. Meanwhile, a cat fight in trash cans was what was happen ing on Mir.

The Russians intentionally ran with higher CO_2 levels than we dc When we did our air-to-ground TV interviews in the Base Block, the air in there for three people was now being sucked up by ten Our instructions were, "Grab you oxygen molecule early, because you're going to need it!" We had the TV lights on and were doing the event, and I started falling asleep. By coincidence a master alarm sounded, coming from the shuttle, and Brent and I went zorching down into the orbiter to take care of it. Turned out to be a nuisance alarm, but entering the shuttle airlock from Mir made a noticeable difference—literally a breath of fresh air.

I was pleased at how well we got along with the Russians. Whe training with a Russian crew, we were limited to the five days we spent in Star City. Not much time We paid a lot of attention to what previous Mir crews had said abou what worked and didn't work.

ABOVE: **Marsha Ivins and her free-falling hair on *Atlantis*'s flight deck.**

RIGHT: **Ivins photographed *Atlantis*'s nose grazing Earth's twilight horizon.**

BELOW: **Jeff Wisoff, John Grunsfeld, and Mike Baker (far right) celebrate the arrival of shuttle-delivered fresh fruit to Sasha Kaleri and Valery Korzun.**

Mission no. **82**

STS

82

Orbiter	Discovery
Launch	February 11, 1997
Landing	February 21, 1997
Duration	9 days, 23 hrs., 37 mins., 7 secs.

The STS-82 crew on *Discovery*'s middeck: Joe Tanner, Steve Hawley, Mark Lee, Ken Bowersox, Steve Smith, Scott Horowitz, and Greg Harbaugh. The crew wear Hubble images on their T-shirts while holding an insulation blanket emblazoned with the catchphrase from the TV show *Home Improvement*.

Crew	Kenneth D. "Sox" Bowersox, Scott J. Horowitz, Mark C. Lee, Steven A. Hawley, Gregory J. Harbaugh, Steven L. Smith, Joseph R. Tanner

Mission: *Discovery*'s crew rendezvoused with Hubble for the second servicing mission, the first since STS-61 three years earlier. During four EVAs, astronauts installed advanced instruments and upgraded sensors, reaction wheels, and data recorders, adding a fifth EVA to replace deteriorating thermal insulation.

Steve Smith, Mission Specialist: STS-82 will be remembered as the first normal, non-crisis Hubble repair flight. As an agency, we still didn't have much spacewalking experience, and we didn't have four experienced spacewalkers available to send to Hubble. So what NASA decided to do was an experiment: find two experienced spacewalkers—Mark Lee and Greg Harbaugh—and send up two well-trained people who had flown previously but not done a spacewalk.

The Hubble team had said, "If they're going to be inside Hubble to change out the gyros, we need some tall people, because the gyros are really a long reach." I was super-thrilled when Mark Lee told me I was on the flight. I thought "Oh, my God—I'm so capable, they picked me!" He said, "Wait a second; one of the reasons we picked you is because you're tall." Mark deserves a lot of credit for developing the spacewalking techniques that we've used since the late 1990s: how we talk to each other, how we prepare our tool lists, how we prepare for contingencies. I was lucky to go out with such a capable person.

Scott Horowitz and Mark, both Air Force fighter pilots, were our STS-82 "pretend" doctors. We talked ahead of time about space motion sickness, and I said, "Hey guys, I'm probably gonna need a Phenergan shot to counter nausea when we get to orbit." Mark, a farmer, said, "If you let me do it, I just want you to know that my experience is with giving injections to cows, and my mantra is to 'bury the needle.' I'm gonna just jam it in there and make sure it's all the way in." I looked at Scott—I knew he had some experience giving actual shots to people—and said "I'll take *you*."

During rendezvous, we could see that the multi-layer insulation (MLI) on the outside of the Hubble was disintegrating. Having free-floating particles around the telescope—that's bad. So, they added a fifth EVA with Mark and me. Scott built all the blankets we would place over the damaged MLI out of anything he could find inside the shuttle.

Approaching Hubble, I thought "Wow, that thing's huge!" Just absolutely gorgeous and mesmerizing, one of the most amazing machines that humans have ever built. Touching it, going inside it, was kind of a spiritual thing, especially recalling that in some ways it's a time machine. The images it produces are of events that happened twelve billion years ago of objects that maybe don't exist anymore, so it's magical.

Grabbing Hubble for the first time, I put my face right up to the skin and it looked like the Moon, absolutely pockmarked with small craters. It was just pitted with holes after only seven years in orbit. At that moment I really came to understand the seriousness of space debris and was amazed that Hubble still worked.

I've got drawings I did when I was five years old of me doing spacewalks. In the summer of 1965, I watched Ed White go out the door. That was the first time an American astronaut walked in space. For me, actually doing an EVA decades later was the ultimate adventure. If we had screwed up Hubble, we would've ruined the jobs of thousands of engineers and researchers. Not long before flight, Frank Cepollina, the Hubble Space Telescope "main guy," said "Hey, just a reminder: It's worth six billion dollars." No pressure.

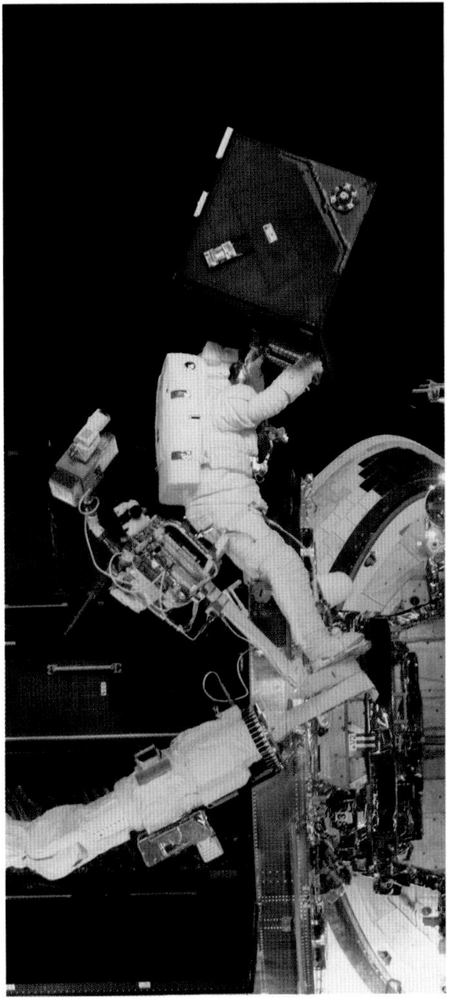

LEFT: **Steve Smith works near Hubble above the Australian coast.**

ABOVE: **Smith removes the Faint Object Spectrograph from Hubble.**

BELOW: **Scott Horowitz fabricates Hubble thermal insulation blankets from scrounged shuttle materials in the middeck.**

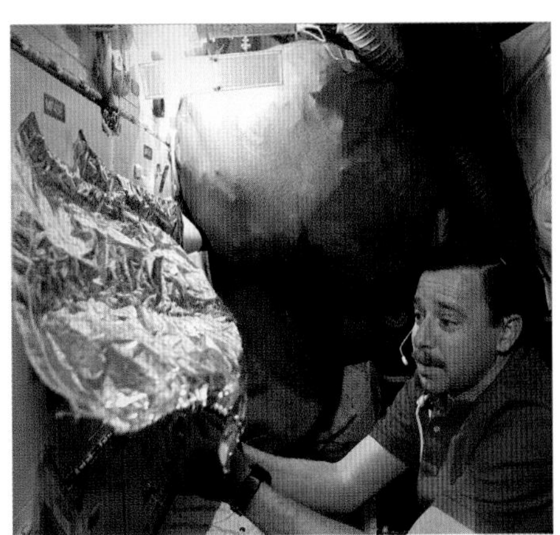

Mission no. **83**

STS

83

Orbiter	Columbia
Launch	April 4, 1997
Landing	April 8, 1997
Duration	3 days, 23 hrs., 12 mins., 39 secs.

The STS-83 crew in Spacelab: (front) Janice Voss, Jim Halsell, and Don Thomas; (back) Roger Crouch, Mike Gernhardt, Susan Still Kilrain, and Greg Linteris

Crew	James D. Halsell, Susan L. Still Kilrain, Janice E. Voss, Donald A. Thomas, Michael L. Gernhardt, Roger Crouch, Gregory T. Linteris

Mission: *Columbia*'s crew operated the first Microgravity Science Laboratory in the Spacelab, carrying an international suite of experiments in combustion physics, materials science, and microgravity disciplines. A malfunctioning fuel cell dictated an early landing just short of four days in orbit.

Susan Still Kilrain, Pilot:

As we took the elevator down from crew quarters, heading for the launch pad, the spouses of the married crewmembers were at the launch control center. For those of us who weren't married yet, Jim Halsell had arranged for our significant others to meet us at the bottom of the elevator, going out to the Astrovan. So, the elevator doors opened, I walked out, and there was my boyfriend, Colin! I gave him a hug and a kiss goodbye, and off I went. It might have been Dave Leestma, the chief of flight crew operations, who asked Colin, "Who are you and how do you know her?" Colin answered, "I have no idea who she is!" I was in the doghouse the whole time I was on orbit for breaking quarantine protocol until we returned and I could explain to Dave, "Yes, he is my boyfriend, and he was just trying to be funny."

Houston flight controllers noticed the fuel cell problem. We didn't see a signature of any kind onboard. Mission Control told us about it and said we would have to shut it down. That was a big deal: If you shut the wrong one down, you couldn't start it back up. But it all went fine. We later learned that the problem ended up being a harmless sensor failure. We weren't in any danger, but nobody knew that at the time.

We were in a "two-thirds-normal electricity" situation. We were trying to get whatever bits of science we could. Most of our lights were off, and we were working in the dark. The scramble threw off our whole sleep shifting plan; Jim and I were far off the proper circadia rhythm to land when we were most alert . By the time we land and got to crew quarters, I can't even tell you how many hours w had been awake.

Still, looking out the window was unforgettable. One highlight was Comet Hale-Bopp—we watched the Sun set, and then t Moon shortly thereafter, and we were left with just the comet out there. Out one window I could se the North Star, and out the other at the same time was the Southern Cross—spectacular.

We got to space, orbited a few times, saw Earth, took a few pictures, and came back. From the pilot's point of view, it was a big success. But from the payloa crew's reaction, I could tell that it was heartbreaking—until the ground let us know that they wer probably going to fly us again. W thought perhaps management was just saying that to keep our spirits up until we landed. Chief Astronaut Bob Cabana told us right after landing that they had approved us to fly a second time.

STS-83 was a dream come true, one I'd had twenty years before. I was working as an aerospace engineer for Lockhee doing wind tunnel testing, and I was bored. I knew I didn't want t do that for the rest of my life. My boss was a friend of Dick Scobee's and put me in touch with him. Dick was very helpful, willin to mentor, and said the best way into NASA was to become a test pilot. He recommended the Navy even though he was Air Force, because there were more Navy astronauts than Air Force at that time. I was impressed that he took the time.

LEFT: **Powered by seven million pounds (31.1 million Newtons) of thrust from its five engines, *Columbia* rises from Pad 39A.**

ABOVE: **Greg Linteris and Don Thomas at work in the roomy Microgravity Science Laboratory Spacelab.**

BELOW: **Comet Hale-Bopp and a sunset glow as observed from *Columbia.***

Mission no. **84**

STS

84

Orbiter	Atlantis
Launch	May 15, 1997
Landing	May 24, 1997
Duration	9 days, 5 hrs., 19 mins., 55 secs.

STS-84 and Mir crews in SPACEHAB: (front row) Jerry Linenger, Vasily Tsibliev, Charlie Precourt, Aleksandr Lazutkin, and Mike Foale; (back row) Ed Lu, Eileen Collins, Jean-Francois Çlervoy, Yelena Kondakova, and Carlos Noriega

Crew	Charles J. Precourt, Eileen M. Collins, Carlos I. Noriega, Edward T. Lu, Jean-Francois Çlervoy, Yelena V. Kondakova Up: C. Michael Foale Down: Jerry M. Linenger

Mission: STS-84 docked with Mir, exchanging Mike Foale for Jerry Linenger, who survived a dangerous fire onboard. *Atlantis*'s crew transferred a Russian oxygen generator and 7,500 pounds (3,402 kilograms) of water, experiments, and supplies from SPACEHAB.

Charlie Precourt, Commander:

We were struggling through the trials and tribulations of the emerging Russian-American relationship in space. STS-84 flew largely to recover from the aftereffects of the fire that broke out on Mir. We wanted to determine whether Mike Foale should stay on the Russian station and to ensure that the shuttle-Mir program would help us build a better space station and a stronger international partnership to support it.

Our docking was close to orbital noon, and at about 100 feet (thirty meters) out from Mir, the Sun overhead saturated the cameras and obliterated my vision. The whole station was lost in the glare, and it was spooky, because I had no navigation whatsoever. The best thing is to do nothing, right? I had to just let go and watch for about five minutes until the Sun passed through our field of view. I did not expect that.

We knew going there that the Mir crew was in distress because Sasha Lazutkin and Vasily Tsibliev with Jerry had been tasked with a couple of post-fire EVAs, trying to figure out what to do to recover. They had been putting all of their heart and soul into keeping things going at Mir—working crazy hours constantly. It was inspiring.

I asked our control center to give us a couple of unscheduled hours after docking where we could have a crew meal unbothered by calls from either center. At one point, Vasily called me over into his little bunk area and gave me a glove from a previous EVA, saying, "This is for you, Charlie," representing fond memories of our time on orbit working together to recover the space station. Building that relationship was pretty darn cool. What stuck in our minds was that we were both in Germany before the wall came down; I was flying F-15s and he was on the east side flying MiG-23s.

Ours was just a nicely put-together crew, and we had a great time knowing that we were breathing new life into the Mir folks with energy from seven new crew members. We had a truly international crew. Mike Foale had a British background, and Carlos Noriega's parents are from Peru. Ed Lu is a first-generation American with Chinese heritage, and we had Jean-Francois Çlervoy from France and Yelena Kondakova from Russia. Eileen Collins was fantastic in helping integrate Yelena into our crew. After we docked and could relax over dinner with the Russians and Jerry, every one of those folks brought something from their heritage to the meal.

All of these things we learned—fortunately, without hurting anybody on that Mir station—dramatically improved our ability to manage the ISS. I remember undocking from Mir near Australia, going out over the Pacific. We were headed northeast into the sunset, and the Moon was rising in the far east, with Mir in the foreground and Earth below. I had this real sense of the three-dimensional reality of being in space, as opposed to 2-D pictures taken from orbit. The Sun itself in the blackness of space doesn't occupy a large amount of the sky, so unless I turned around and looked for that light source, I could almost convince myself that Earth was glowing from inside. It made me wonder about our place in the universe and what all that means.

ABOVE: **Eileen Collins and Charlie Precourt organize thermal printer paper with flight plan updates.**

LEFT: ***Atlantis* approaches for docking carrying a double SPACEHAB and the tunnel connecting it to the crew cabin.**

Mission no. **85**

STS

94

Orbiter	Columbia
Launch	July 1, 1997
Landing	July 17, 1997
Duration	15 days, 16 hrs., 44 mins., 33 secs.

The STS-94 crew in Spacelab: (front) Susan Still Kilrain and Janice Voss; (middle) Mike Gernhardt, Jim Halsell, and Greg Linteris; (back) Don Thomas and Roger Crouch

Crew	James D. Halsell Jr., Susan L. Still Kilrain, Janice E. Voss, Donald A. Thomas, Michael L. Gernhardt, Roger Crouch, Gregory T. Linteris

Mission: NASA quickly turned around STS-83's Microgravity Science Lab 1 experiments, orbiter, and Spacelab, reflying the mission as STS-94. Astronauts operated twenty-five primary experiments, four glovebox investigations and four accelerometer studies covering the fields of combustion physics, biotechnology, and materials processing.

Don Thomas, Mission Specialist:
Between 1983 and 1998, we flew twenty-two of these Spacelab science missions. They're important because they opened the door for doing science in space. Everything we've done on the ISS is based on that experience.

We turned around in eighty-eight days, a crew record for shortest time between launches. After getting three days of science on STS-83, we just repeated the whole mission for STS-94. I think the key to success was, in a word, teamwork. We had the Mission Control Center in Houston, the Payload Operations Center at Marshall Space Flight Center in Huntsville, and scientists from ten nations around the world. Huntsville got better and better at creating experiment procedures and working with the scientists. The mission went really smoothly, and I would give them a lot of credit for that.

In orbit it's a familiar routine: You get up in the morning, get ready for your shift, look at the morning mail, get breakfast, brush your teeth, then check on the schedule to see if they altered the crew activity plan at all. Then you get to working your shift. It's very busy, with not many breaks in the day at all. The work could be more of a grind than a blast.

But the work is still satisfying. About the second day of the mission, Jim Halsell and I were just looking out the window—just the two of us up on the flight deck, looking out at Earth, and Jim said, "I forgot how great it is." I just smiled, thinking, "You work so hard, you train for maybe two years, with a lot of time away from home. Then all of a sudden, you're in space. It's worth all the effort, time, and sacrifices that you and your family make."

I sat up on the flight deck for the landing. On my first two missions, I was down on the middeck, so I had never seen the reentry plasma. I'd always wanted to see that fireball. On STS-94, we had a nighttime reentry, and I was spellbound looking out the overhead window with the mirror on my knee, watching the plasma develop. Its structure seemed to change constantly: It would flow out and seem to open up like an umbrella, changing over time, pulsing like a camera flash going off. The plasma at its most intense was bright white, and out the front windows it glowed pink-orange, as if you were looking into a blast furnace. The plasma vapor trail behind the shuttle seemed to go on for hundreds and hundreds of miles.

After we landed, I saw our suit technician, Bill Todd, and said, "I have a question for you. When you open the hatch after landing, does it stink?" The shuttle cabin had the toilet right there, and all the garbage inside; after being up sixteen days, it had to stink. He looked down and just shook his head, and said, "No, no, no." Astronauts can't stink, right? But I knew he was lying, so I pushed him—we had worked together on several flights, and I knew him pretty well. "C'mon, Bill. When you open the hatch, what does it smell like?" He looked at me and said, "It stinks like hell." I said, "I knew it!"

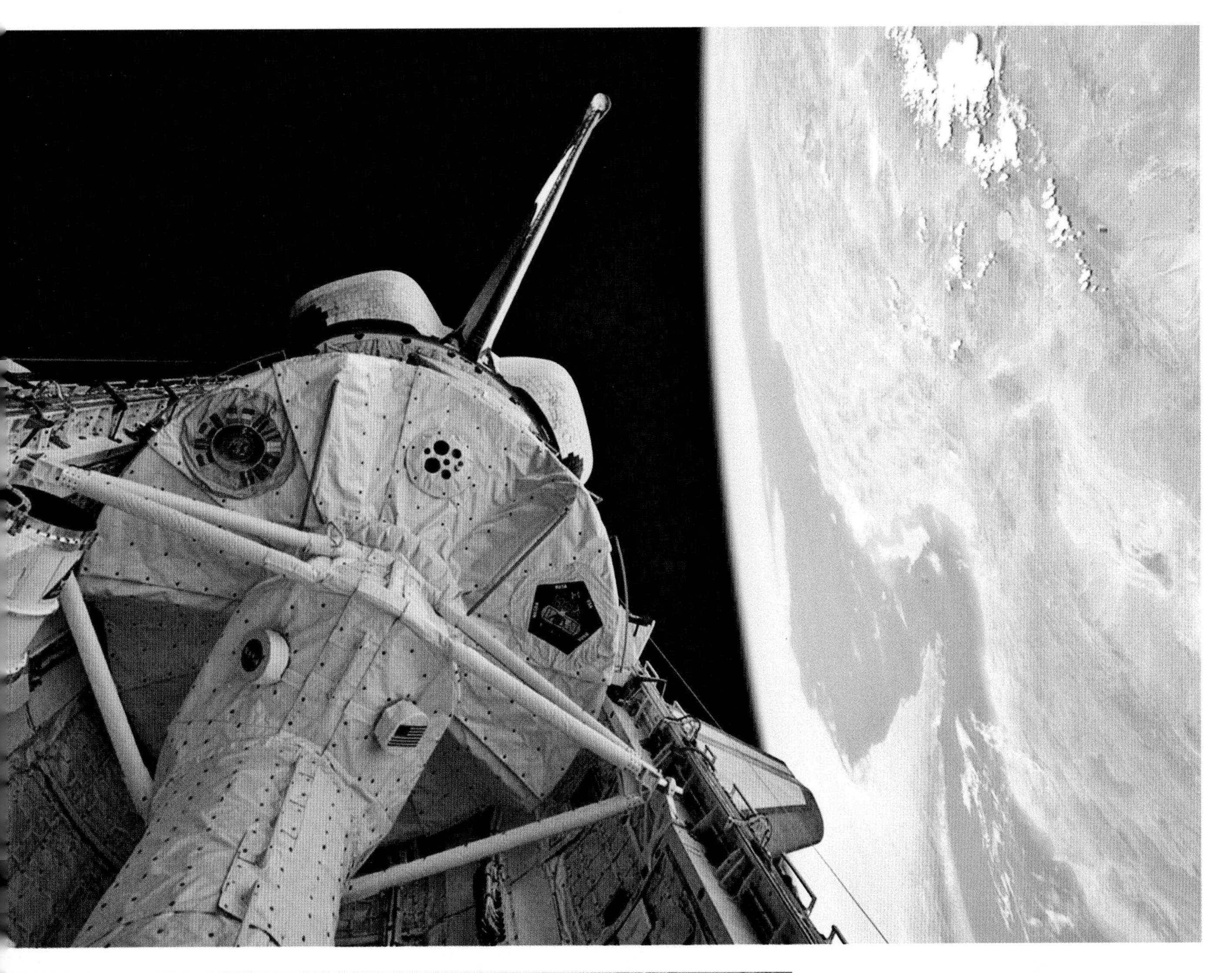

ABOVE: **The Microgravity Science Laboratory Spacelab module over the Persian Gulf.**

LEFT: **Susan Still Kilrain, Janice Voss, and Greg Linteris at work in Spacelab.**

BELOW: **Don Thomas records video results for a Spacelab experiment's investigators.**

Mission no. **86**

STS

85

Orbiter	Discovery
Launch	August 7, 1997
Landing	August 19, 1997
Duration	11 days, 20 hrs., 26 mins., 58 secs.

The STS-85 crew in the middeck: (front row) Bob Curbeam, Jan Davis, and Kent Rominger; (back row) Bjarni Tryggvason, Steve Robinson, and Curt Brown

Crew	Curtis L. Brown Jr., Kent V. "Rommel" Rominger, N. Jan Davis, Robert L. Curbeam Jr., Stephen K. Robinson, Bjarni V. Tryggvason

Mission: *Discovery*'s crew deployed CRISTA-SPAS 2 for nine days of atmospheric trace-gas research. Scientists on the ground operated the TAS-1 and IEH-02 payload bay experiments, while the crew joined ground controllers in operating the Japanese MFD arm destined for the ISS.

Jan Davis, Payload Commander: STS-85 was an international mission involving the United States, Germany, and Japan. How the mission integrated the ground, the training, the crew, and the international partners was a great example of what we have been doing since on the space station.

When we began training, our crew had four rookies out of six. As payload commander it was really fun shepherding those guys and teaching them the difference between training in the simulator and what they were actually going to see in space. A highlight for me was testing the Japanese Manipulator Flight Demonstration (MFD), a six-foot-long (two-meter-long) arm that became the Small Fine Arm on the ISS. The Japanese manipulator system at the Station has the ten-meter arm, and then the smaller arm is on the end of it and does the intricate movements needed.

Steve Robinson and I tested the MFD, that fine arm, in orbit, while controllers in Japan tested the tele-operated portion of the arm. It worked great and was very precise. One test was to grip a little doorknob and turn it to open a door. The controls and displays are very similar to flying the shuttle's RMS, the Canadarm, but with the MFD, we got our camera views on a laptop screen back on the flight deck. That's the same situation as on the space station—we don't have the luxury of looking out the window at the arm, so we rely solely on camera views.

The CRISTA-SPAS satellite had an infrared telescope package that looked at the composition of the ozone layer. NASA knew that I had never been space-sick, so we were able to deploy the satellite right away, on Flight Day 1. That and the retrieval on the tenth day went well. I had flown the shuttle arm on my second flight, working with the Wake Shield Facility, so I was familiar with the arm's oscillations. On retrieval, as I was approaching the grapple fixture, the arm motion coming into the pin on the CRISTA-SPAS was a little more than I saw when I was training. That was nerve-wracking: I didn't get it as dead-on as I would have liked with the arm exactly perpendicular. However, the capture and retrieval were fine.

This was my first flight where we didn't have a module back in the payload bay. But we did have the external airlock, opening up that area in the back of the middeck under the spacesuits. That's where I would change clothes. A couple of times I told the guys, "Okay, I'm going back to change," and a few minutes later while I was in the middle of changing, they'd call, "Okay, I'm coming back there!" They joked around with me like that—it was like camping out with a bunch of big brothers.

We had so many different things going on: payloads operating on the middeck as well as in the payload bay, a lot of robotics, and the Hale-Bopp telescope set up on the side hatch. All of these experiments involved a lot of training, with thirteen laptops to handle all of the payloads. We worked twelve-hour days, sixteen hours or longer with housekeeping and meals. By the last day, we were what I would call "slap-happy." We were doing a school interview from orbit, and one of the kids asked, "Does your hair grow faster in space?" We had a couple of—shall we say—balding crewmembers, and we all just lost it, laughing uncontrollably. Curt Brown was laughing so hard that he couldn't even answer.

Jan Davis and Bob Curbeam strapped in on *Discovery*'s flight deck for the STS-85 countdown rehearsal.

Mission no. **87**

STS

86

Orbiter	Atlantis
Launch	September 25, 1997
Landing	October 6, 1997
Duration	10 days, 19 hrs., 20 mins., 51 secs.

The STS-86 crew at Pad 39B: Mike Bloomfield, Dave Wolf, Jim Wetherbee, Wendy Lawrence, Scott Parazynski, Vladimir Titov, and Jean-Loup Chretien

Crew	James D. Wetherbee, Michael J. Bloomfield, Vladimir G. Titov, Scott E. Parazynski, Jean-Loup J. M. Chrétien, Wendy B. Lawrence Up: David A. Wolf Down: C. Michael Foale

Mission: A cargo ship collision on June 25 nearly depressurized Mir, so STS-86 carried four tons (3.6 metric tons) of equipment to replenish the outpost. Dave Wolf replaced Mike Foale, and Scott Parazynski and Vladimir Titov spacewalked to retrieve external experiments and install a leak repair cap on Mir.

Scott Parazynski, Mission Specialist:

STS-86 flew at a critical juncture in the US-Russian space relationship. With the recent fire and Progress collision, we became something of a rescue and resupply mission, with a lot riding on its success.

First, we changed out Mike Foale for Dave Wolf. I should have been Mike's replacement, but I turned out to be too tall. Wendy was my replacement, but she was too short to do spacewalks to help fix the Spektr module, which had been damaged by the collision with the Progress resupply ship. So, I was "Too Tall," Wendy was "Too Short," and we called Dave Wolf "Too Average"—which he really hated! Wendy and I also styled ourselves the "Russian Rejects."

The second thing that we—the "cavalry"—were there to do was deliver a critical replacement computer. Mir had experienced frequent "rolling brownouts," a bunch of hardware issues, and at least one and maybe two failures of Progress resupply ships, which had been unable to dock at Mir. The replacement primary computer that we brought up was vital in solving these problems.

The first time I floated into Mir, I thought, "This place is dark, it's dank, and it has a mildewy smell to it." Within thirty seconds I realized how lucky I was to be spending five days up there instead of five months.

Kristall, the first Mir module entered through the docking module, was really dark. Because the Russians had staged there everything planned for shuttle return, lashed against the walls, I remarked to Dave that "you really have to suck in your gut to squeeze through." But passing through Kristall, I saw this well-lit, little plant-growth greenhouse, glowing a beautiful, verdant green—a color you don't typically associate with spaceflight. Making a sharp, 90-degree right turn from there, I was in Mir core module, called the Base Block. It looked very much like the Service Module on the ISS, with the walls covered in drab, off-green Velcro, holding cameras and all sorts of gear. At the far end of the module was the dining room table, and on the bulkhead the picture of Gagarin and religious and spaceflight icons. Mir had a warm atmosphere, with Russian pop music playing in the background. We were visiting the crew's home, and it was neat to be invited into it.

On our Flight Day 7 spacewalk we went out through a hatch in the SPACEHAB tunnel adapter, facing directly up toward Mir. It was awesome—I asked the crew to keep the payload bay lights off as we egressed in orbital night. It was extraordinary having my eyes dark-adapted and seeing the stars that way without lights. I saw Anatoly Solovyev looking down from the Base Block module down into the payload bay. It felt unreal!

The cool thing was that Volodya—Vladimir—and I used a mix of Russian and English as we were doing the spacewalk, so some of the banter was bilingual. I heard Volodya saying "Titicaca" with wonder and a heavy Russian accent while we were flying over that alpine-blue Andean lake. He repeated it two or three times: "Lake Ti-ti-ca-ca," savoring the beautiful views of the lake and mountains.

The high point of STS-86 amid the Shuttle-Mir program was that we were good partners for the Russians when they were down and out; NASA stepped up and helped them in their time of crisis. When the tables turned in the aftermath of *Columbia*, they did the same for us.

ABOVE: **Mir with damaged Spektr solar array (short array on left), as seen from a departing *Atlantis.***

LEFT: **An out-of-control Progress cargo ship collided with Mir and punctured the hull and this Spektr solar array.**

RIGHT: **Scott Parazynski in the airlock following the joint US-Russian EVA.**

Mission no. **88**

STS

87

Orbiter	Columbia
Launch	November 19, 1997
Landing	December 5, 1997
Duration	15 days, 16 hrs., 34 mins., 4 secs.

The STS-87 crew on *Columbia*'s middeck: (front row) Steve Lindsey, Takao Doi, and Winston Scott; (back row) Kevin Kregel, Kalpana Chawla, and Leonid Kadenyuk

Crew	Kevin R. Kregel, Steven W. Lindsey, Winston E. Scott, Kalpana Chawla, Takao Doi, Leonid K. Kadenyuk

Mission: STS-87's astronauts worked with ground experimenters to operate the USMP-4 payload. The crew deployed the SPARTAN-201 astronomy satellite, but its control system didn't activate—a regrapple attempt with the arm set SPARTAN spinning, preventing retrieval. Spacewalkers later retrieved SPARTAN by hand, but its scientific mission was lost.

Winston Scott, Mission Specialist:

We carried several science and microgravity experiments, but the overarching mission of STS-87 was to evaluate tools, equipment, and techniques to prepare for assembly of the space station.

We delayed the planned portion of our first EVA to perform a satellite capture. Our SPARTAN satellite had malfunctioned, and in attempting to retrieve it we inadvertently tipped it. As a result, we had this 3,000-pound (1,360-kilogram) SPARTAN drifting in a very slow, complex spin, unable to be retrieved with the robot arm. After several days of consultation among the crew, the planners, the astronaut office, and the divers in the Neutral Buoyancy Lab in Houston, we decided that Takao Doi and I would manually retrieve SPARTAN on EVA 1.

I was told by Mission Control that this whole task was really up to me—I could accept or refuse. I was the lead EVA guy, having flown and spacewalked before. Takao was on his first flight and EVA, with English as his second language, and we were asking him to go outside and help catch a satellite. I thought, for about two tenths of a second, that perhaps we shouldn't do this, and then I said, "No, absolutely, we're going to go outside and get this dog-gone thing if it's the last thing we ever do. We are the A-team."

SPARTAN was spinning and wobbling very slowly in complex motion. We couldn't perceive the subtle details of its motion with the naked eye. We exited the shuttle airlock during the final phase of our SPARTAN rendezvous and perched ourselves on opposite sides of the payload bay while Kevin and Steve flew us toward the satellite. We monitored its position and motion to be sure that its electromagnetic torquer bars had dampened the spin enough so that we could catch it.

I said, "OK, when we get ready, Takao, this is what I'm going to say: 'Standby, standby, capture.' Do you understand?" Takao answered "Yes, I understand." With a GO from Mission Control, I said "Standby, standby, capture," and we each grabbed our part of the satellite. The amazing thing was that in space I was able to observe the difference between weight and mass. The huge mass of SPARTAN was floating up there, weightless, but when I grabbed it, I could feel all of that mass, resisting our movements.

We now had the spacecraft. We rotated it to the proper orientation for berthing in the payload bay. Several things went wrong, but we managed to orient it appropriately, lock it into the payload bay, and bring it home.

Two days later, we performed a second EVA to check off the planned spacewalking tasks. Takao was all over the place—his confidence was rock solid. Working with the space station construction crane, Takao was cranking it around at what appeared to be Warp Factor 8, moving it all over the place.

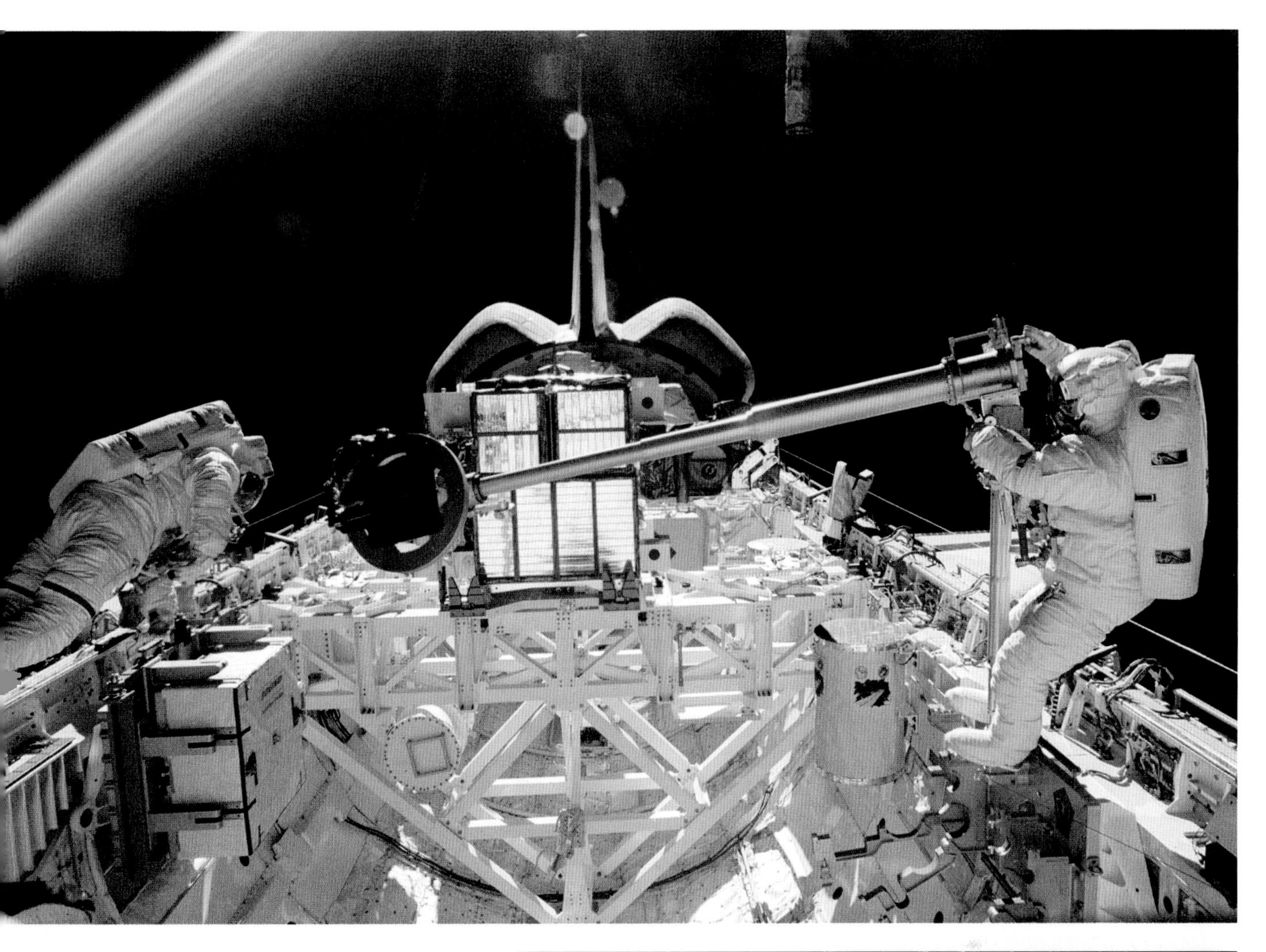

ABOVE: **Winston Scott and Takao Doi test the handling of a US Space Station construction crane.**

RIGHT: **SPARTAN ready for release from *Columbia*'s robotic arm.**

BELOW: **Scott ready to retrieve the AerCAM free-flying robot in the payload bay.**

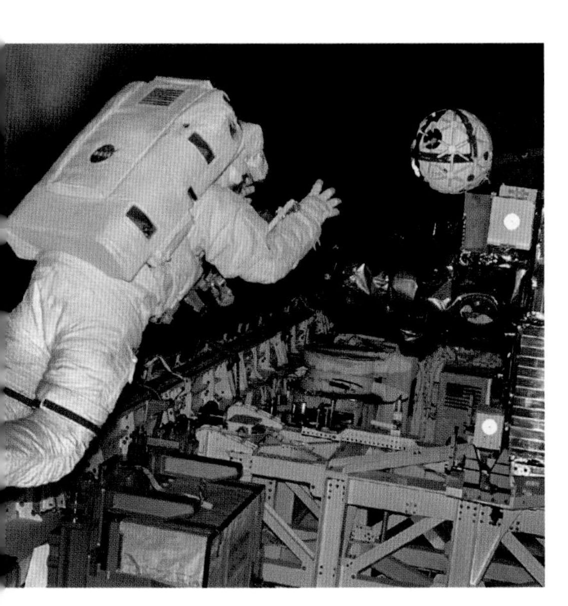

Mission no. **89**

STS

89

Orbiter	Endeavour
Launch	January 22, 1998
Landing	January 31, 1998
Duration	8 days, 19 hrs., 46 mins., 54 secs.

The STS-89 and Mir crews in the station's Base Block: (clockwise from left) Dave Wolf, Pavel Vinogradov, Salizhan Sharipov, Jim Reilly, Andy Thomas, Joe Edwards, Mike Anderson, Bonnie Dunbar, Anatoly Solovyev, and Terry Wilcutt

Crew	Terrence W. "Fencer" Wilcutt, Joe F. Edwards Jr., Bonnie J. Dunbar, Michael P. Anderson, James F. "J. R." Reilly, Salizhan Shakirovich Sharipov Up: Andrew S. W. Thomas Down: David A. Wolf

Mission: The STS-89 crew transferred more than four tons (3.6 metric tons) of experiments, water, and spare parts from SPACEHAB to Mir. Andy Thomas, the last American to serve aboard Mir, replaced Dave Wolf after his 119 days at the outpost.

Jim Reilly, Mission Specialist:

The eighth shuttle docking mission to Mir was a logistics flight, moving about 9,000 pounds (4,082 kilograms) of gear between the two vehicles, about 1,500 separate manifest items. On STS-89 we had four rookies, including Salizhan Sharipov, who was actually making his very first flight as a Russian cosmonaut but going up as a shuttle crew member.

Two of my missions had night launches. We were accelerating through first stage and had just gone supersonic, through Max Q, or maximum dynamic pressure, when Fencer told us on the flight deck, "Look out the window." We looked ahead at a thin layer of stratus; when we went through that cloud layer, a ripple spread out across its surface as we went through it. Our friends saw that shock wave on the ground. What we saw was this orange flickering light from the trailing plume that caused that cloud to glow just as we went shooting through it. For an instant we got a sense of how fast our orbiter was moving.

Terry Wilcutt was terrific as a commander: very quiet and supportive. He gave me the job of dealing with the logistics while we were docked. I thought that was brave because I was an inexperienced commodity as far as he was concerned. I don't know how he figured out that I could do it, but I managed to get it done. We came back with everything we were supposed to come back with, and left everything we were supposed to leave, and even ended up with the right crew members on the right sides of the hatches.

The Mir was an interesting place that had no real way to get obsolete equipment down, so it was just packed everywhere insi the station. When we first went ir we floated past an Orlan spacesuit and a whole series of older European experiments that were still strapped to the bulkheads ar the module interior. It was neat tc go into the node and look down the various arms of the station, all oriented differently. You could see somebody who appeared vertical in one module and then i another module, somebody who was horizontal. I got the sense, "Wow, there is no up or down." I compared Mir to a friend's garag a friend who kept everything. If you needed it, he probably had i in there somewhere.

The one thing everybody remembers about Mike Anderson is his smile. He had a million-wat smile, and he used it a lot. He just enjoyed everything. He lovec flying. He loved his kids. He love his wife. He was a very religious man. He was just really fun to work with, and we became really good friends.

As we were on landing approach in *Endeavour*, slowing to transonic, we heard this buzz throughout the orbiter. It must have been the shock wave movir forward as we were transitioning to subsonic, and it caused a rising-frequency buzz that was very loud. All of us were just kinc of looking at one another, asking, "What the hell is that?" Terry chuckled and said, "Oh, yeah, I forgot to tell you guys about that That's the buzz that happens when we go subsonic." We were silent for about a ten-count, and then either Reb Edwards or Mike Anderson said, "Damn, Fencer, I thought something was coming off." Those orbiters were so different. Each one was unique and custom-built.

LEFT: **Mir after undocking, with the docking module at top and damaged Spektr at left.**

ABOVE: **Jim Reilly transits the SPACEHAB tunnel into *Endeavour*'s crew cabin.**

BELOW: **Terry Wilcutt and Joe Edwards celebrate a successful mission by *Endeavour*'s nosewheel after landing.**

Mission no. **90**

STS 90

Orbiter	Columbia
Launch	April 17, 1998
Landing	May 3, 1998
Duration	15 days, 21 hrs., 49 mins., 59 secs.

The STS-90 crew in Neurolab: (bottom row) Kay Hire , Rick Searfoss, Scott Altman, and Rick Linnehan; (top row) Jay Buckey, Jim Pawelczyk, and Dave Williams

Crew	Richard A. Searfoss, Scott D. Altman, Richard M. Linnehan, Dafydd Rhys Williams, Kathryn P. Hire, Jay C. Buckey Jr., James A. Pawelczyk

Mission: STS-90 was an internationally funded Spacelab mission called Neurolab. During nearly sixteen days of intense experiments targeting the human nervous system, astronauts served as laboratory test subjects along with rats, mice, crickets, snails, and two kinds of fish.

Rick Linnehan, Payload Commander:

Of all the Spacelab Life Sciences missions, nothing even comes close to what happened on Neurolab. It was pure life sciences, looking at the central and peripheral nervous systems as part of the National Institutes of Health-designated Decade of the Brain. Nothing even remotely like it has been done on the ISS.

We flew more than 2,000 lifeforms, seven of which were large primates—us! We carried fish, insects, and rodents, and we interacted intimately with some of those animals. Others—the oyster toadfish and crickets—were sealed in standalone experiments housed in the Spacelab.

We all worked together: single, sixteen-hour shifts every day. Because of the sleep shift required to be at our best for landing, we were getting up earlier and—the way it turned out—working longer, always losing sleep in the process—tiring and sometimes stressful.

The Spacelab rat cages weren't designed well for microgravity, so the dams—the mothers—weren't able to hold on while attempting to cuddle and nurse their babies. The babies were getting separated from their mothers, floating away inside the cages, marooned in space. We took out every single one of those neonatal rat pups and gave them warm, subcutaneous fluids and antibiotics to keep them alive, treating the dams when needed, too. This payload was supposedly self-contained, but as we transferred and treated the rats, the faulty cages allowed urine and feces into the cabin. The repetitive clean-up was very labor-intensive. We had no choice but to stay up for five nights with little sleep, taking shifts when we could.

We became a neonatal intensive-care unit for about a week, in addition to executing all our other scheduled experiments. Despite the habitat problems, we saved about 60 percent of the neonatal pups, and the experimenters were able to get almost all their critical, planned science. Our Neurolab crew was exhausted by the end of the mission, but we were also happy that things had worked out the best they could.

We also flew adult rats with "hyperdrive units" implanted directly into their hippocampi and attached to their skulls. The goal was to study place cell navigation in space, or how we figure out how to move around in unknown environments or in the dark.

One of these "rats with hats" dislodged his "hat" in orbit and was eliminated from the experiment. When I took him out of his cage to treat him, he figured things out really fast. He wouldn't just float around in the cabin; instead, I put him on my shoulder and he held on. He spent enough time with me that I saw the animal actually enjoyed being out and interacting with us. He survived: His skull healed, and his scalp started to regrow through his veterinary treatment. We ended up calling him Curly, and he became our Neurolab mascot. It was a good thing that we saved him, because he became very useful as a flown survivor control, an ace in the hole that no one had initially counted on. Because he couldn't be used for the "rats with hats" experiments, he was available to several other investigators who found him invaluable. It worked out as more good science—and a good space story.

The payload crewmembers' extra work and skill helped us wade through Neurolab's many challenges, making it arguably the most successful life sciences mission in NASA history.

ABOVE: ***Columbia* rolling over to the Vehicle Assembly Building for stacking.**

RIGHT: **Dave Williams exhales into a pulmonary gaseous exchange sensor in Neurolab.**

BELOW: **Rick Linnehan reviews data on an experiment computer while Jay Buckey works at rear of Neurolab.**

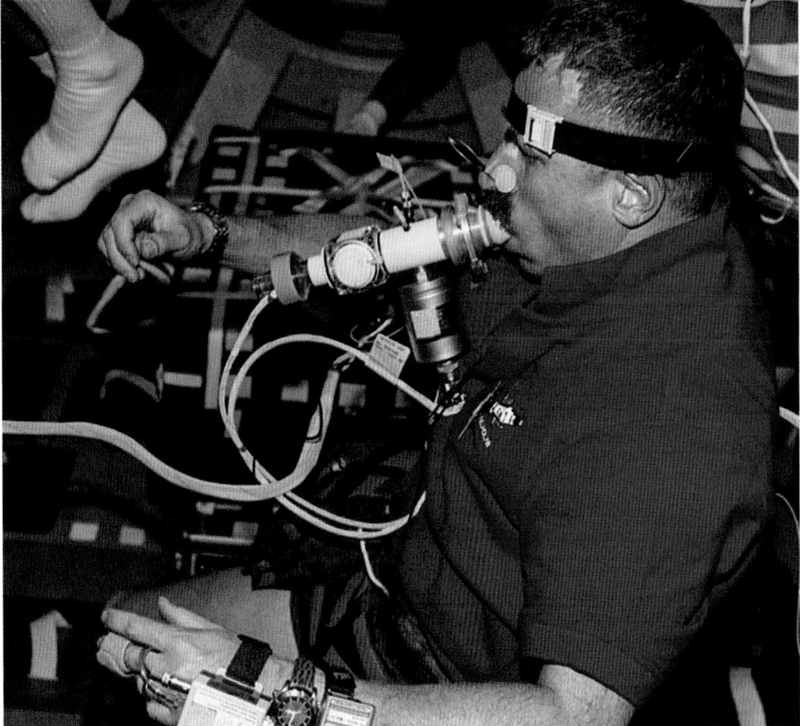

Mission no. **91**

STS

91

Orbiter	Discovery
Launch	June 2, 1998
Landing	June 12, 1998
Duration	9 days, 19 hrs., 53 mins., 53 secs.

The STS-91 crew in Pad 39A White Room: (front row) Franklin Chang-Diaz, Valery Ryumin, and Janet Kavandi; (back row) Dom Gorie, Wendy Lawrence, and Charlie Precourt

Crew	Charles J. Precourt, Dominic L. Pudwill Gorie, Wendy B. Lawrence, Franklin R. Chang-Díaz, Janet L. Kavandi, Valery Victorovitch Ryumin Down: Andrew S. W. Thomas

Mission: *Discovery*'s astronauts docked at Mir to deliver supplies and returned with Andy Thomas, marking 907 days spent at the station by seven US resident crewmembers. AMS detectors measured the cosmic ray energy spectrum above the atmosphere.

Janet Kavandi, Mission Specialist: I had wanted to be an astronaut since I was about five, when people were first flying in space. Back then, there were no women astronauts, and it wasn't a realistic, attainable goal until the shuttle, when NASA first selected scientists, females, and non-military astronauts in significant numbers.

I was in high school then and I studied hard, put a picture of Bruce McCandless's untethered EVA over my dorm table and wrote the words "Never Give Up!" across the top. I just kept going to school, kept trying, and worked in the space business at Boeing until I ultimately made it in.

The final countdown started, my heart rate went up, and I felt the engines ignite and the orbiter "twang." I was so exhilarated and so happy that tears flowed—I was lying on my back and they ran right into my ears! I wasn't afraid, but I just couldn't believe I was truly fulfilling my lifelong dream of going to space.

We knew we were going to rendezvous and that Mir would be big in front of us, but we were in the dark. When sunrise came and Mir revealed itself, it was gigantic. Mir showed a lot of age, and it had stains and discoloration from previous dockings and thruster pluming. But it was just exciting to realize that we had our little pressurized shuttle volume here, and over there was a bigger pressurized volume in space, and we were both going 17,500 miles (28,164 kilometers) per hour a couple of hundred miles above Earth. We were bringing together these two tiny places where life can exist, from two different countries that were mortal enemies just a few years before, to do this cooperative event as friends and technical equals. It's amazing what humans can do when they want to do things together for a positive reason.

It was disorienting to go into a module upside down, because for an instant, I felt I was going to fall—the references were all wrong. Logically I knew I wouldn't fall, but it was startling for a second to realize I was upside down in the module when there is no "up" and "down." I laughed at the fact that I could come in on the ceiling and still be okay.

We were in the Mir Base Block on a night off, and we were all having a social dinner together. The Russian commander Musabayev pulled out his guitar, and his favorite music was from the Beatles. He started to play and sing "Yellow Submarine." I don't know that he understood what all the words were. I looked out one of their windows, down at Earth, and thought, "Nobody down there realizes that there are people passing overhead in a Russian space station with a Russian cosmonaut singing 'Yellow Submarine.'" It was surreal!

The *Discovery* shuttle stack silhouetted by sunrise and enroute to the launch pad

Mission no. **92**

STS 95

Orbiter	Discovery
Launch	October 29, 1998
Landing	November 7, 1998
Duration	8 days, 21 hrs., 43 mins., 56 secs.

The STS-95 crew on *Discovery*'s middeck: (clockwise from bottom left) Pedro Duque, Chiaki Mukai, Scott Parazynski, John Glenn, Curt Brown, Steve Lindsey, and Steve Robinson

Crew	Curtis L. Brown Jr., Steven W. Lindsey, Scott E. Parazynski, Stephen K. Robinson, Pedro Duque, Chiaki Mukai, John H. Glenn Jr.

Mission: While operating SPACEHAB experiments, STS-95's astronauts deployed and retrieved SPARTAN for solar observations. Mercury astronaut John Glenn returned to space for biomedical experiments on aging. *Discovery*'s drag chute compartment door fell off during launch, preventing use of the parachute after landing.

Curt Brown, Commander:

John Glenn was assigned to the mission early in our training. I was proud to be the commander of the flight taking John up. Everybody was excited, and John and Annie, his wife, were the nicest folks I think I've ever met. The crew became a kind of family, with John and Annie being the grandparents, the crew being the kids, and the crew's kids being like their grandkids.

The drag chute door failed because of a buildup of tolerances: Everything was within limits, but the tolerances on the pins, the door, and the opening were on the, let's say, skinny side. Vibration caused the door to come off, and it hit the center main engine nozzle and caromed off it, leaving a visible impact scar on the nozzle. When we got back on the ground, we could see where it hit. If that door had punched a hole in the nozzle, I wouldn't be talking about it today. It was serious.

But ascent was nominal, so we were oblivious to it. Once we were in orbit, Houston said, "We've got a video we want to send to you." Steve Lindsey and I looked at it and said, "Holy shit!" We looked at each other for a few moments because we realized what could have happened. Fortunately, after ascent, we don't use those main engines again.

We spent a lot of time with the Canadian arm trying to look behind the orbiter to see if there was a drag chute trailing us. Would it come out during all those rendezvous burns and maneuvers, or during reentry? The folks on the ground spent a lot of time in the simulator, analyzing what would happen if the drag chute deployed early during reentry at different Mach numbers or energy levels.

We didn't use the drag chute after landing because the engineers wanted to see if it was still there. When we touched down, I think we had the biggest crosswind any orbiter ever landed in—about thirteen knots (seventeen meters per second). Big deal! We touched down crabbed slightly into the wind to kill the drift, and de-rotated and stopped. When we did a walk-around, that old drag chute was stuck right in there like it never knew anything had happened.

Having John onboard made the flight special. It was a big PR event, obviously. But we also had Pedro on board, the first Spaniard to go into space. We had the first American into orbit and we had the first Spaniard on the same flight. Chiaki Mukai was onboard too, part of our international crew. Chiaki was so full of energy and life. She's so much fun and quite an impressive woman. Those are the little things I think about as I look back at my missions.

We were worried about John getting too excited before launch and during ascent because of the heat inside the suit and the stress on your body. I'm just happy we got approval for a defibrillator on the pad and on the shuttle during the flight. And we had two doctors: Scott Parazynski was there along with Chiaki. I just didn't want to be a guy whose name ended up in the game show *Jeopardy.* The *Jeopardy* answer would be, "Astronaut Curt Brown." I didn't want the matching question to be, "Who killed John Glenn?"

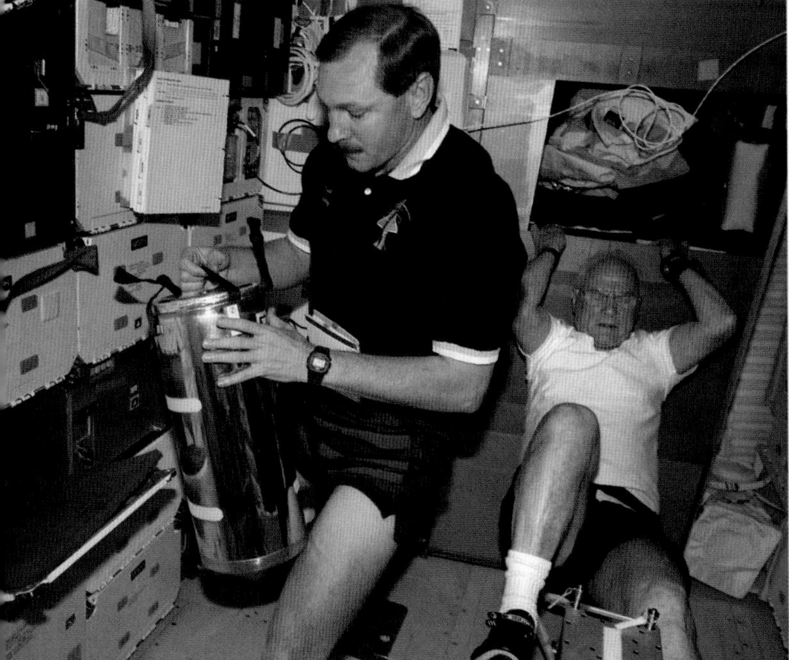

ABOVE: **John Glenn enjoying research work in the SPACEHAB module.**

LEFT: **Glenn pedals the middeck cycle ergometer while Curt Brown secures a sample container.**

BELOW: **Brown and Steve Lindsey guide *Discovery* to a landing on Kennedy Space Center's Runway 33.**

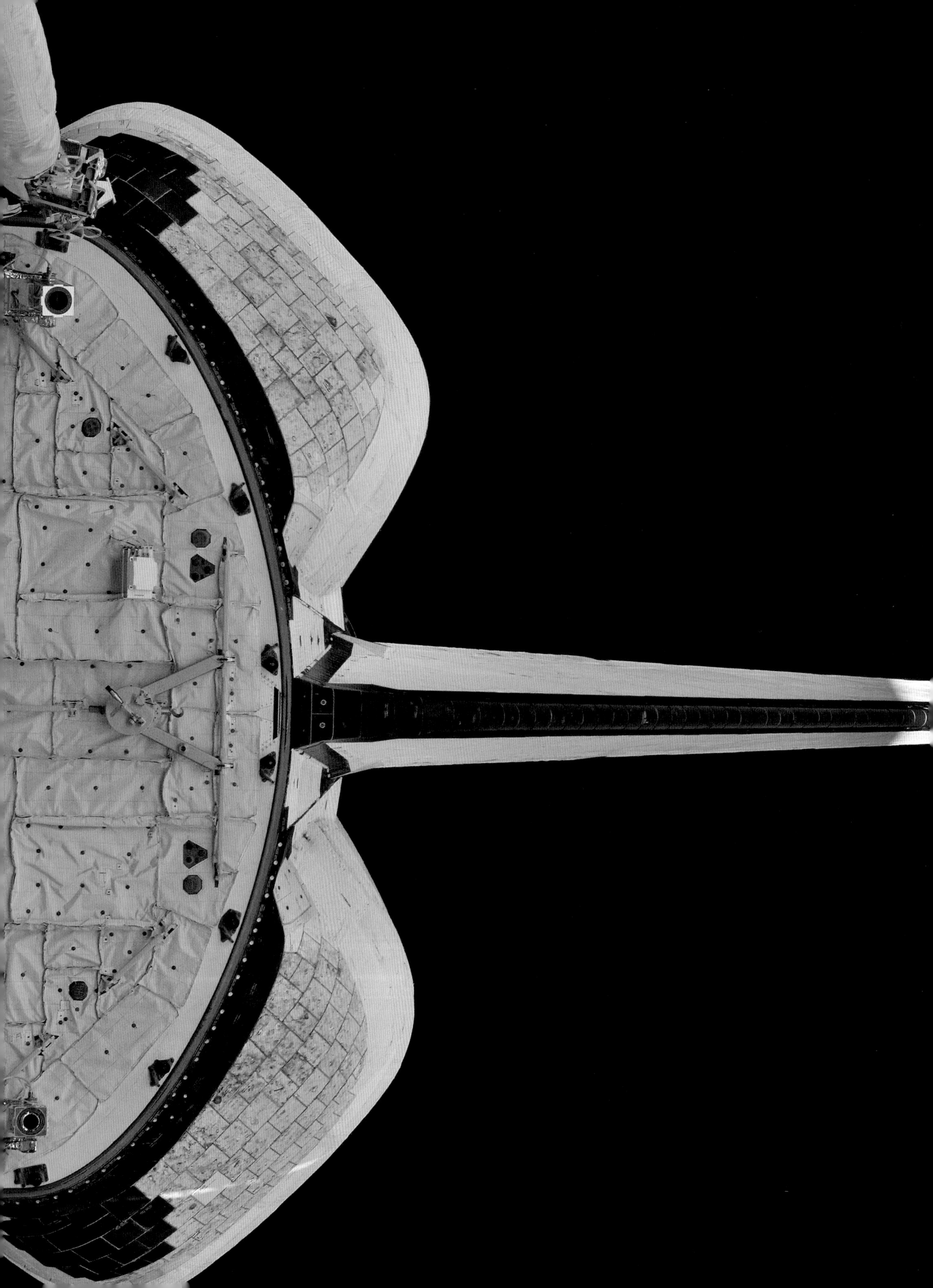

Section 3

BUILDING THE SPACE STATION

Losing *Columbia*

1998–2011

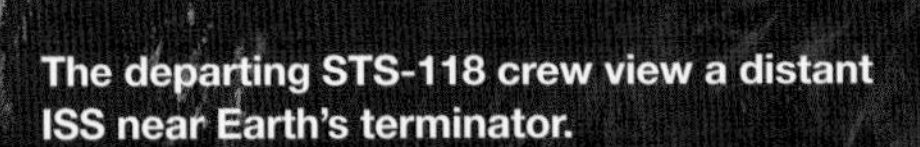
The departing STS-118 crew view a distant ISS near Earth's terminator.

To gain long-duration space experience and develop the newly formed American-Russian space partnership, shuttles supported the mid-1990s Shuttle-Mir program, often termed "Phase I" of the ISS. From February 1994 to June 1998, space shuttles made ten flights to the Russian station, and seven American astronauts logged long-duration stays there.

Shuttle visits required development of a docking system compatible with both Mir's docking port and the shuttle's payload bay. Engineers moved the orbiter's cylindrical airlock out of the middeck to just behind the crew cabin. The Mir-compatible docking hardware sat atop the airlock; NASA purchased the docking mechanism and its control panels from Russia. First tested at Mir on STS-71, the Orbiter Docking System (ODS) performed flawlessly on all nine Mir dockings and on dozens of trips to the International Space Station.

The orbiters' payload bay dimensions and mass limitations drove the design of all western Space Station components, from habitats to labs to truss segments. The first module, Unity, with six berthing ports, was launched by STS-88 in December 1998, and astronauts using the robot arm linked Unity to the first Russian module, Zarya (FGB). The shuttles would fly thirty-seven missions to deliver components, supplies, and crews to the new outpost.

Four years into Station assembly, on January 16, 2003, NASA launched a final stand-alone research mission aboard *Columbia*—STS-107. At eighty-two seconds into ascent, a briefcase-sized piece of insulating foam broke free from the external tank at the base of the bipod structure linking the tank to the orbiter's nose. The five-pound (two-kilogram) chunk of foam struck the left wing's leading edge at a relative speed of some 500 miles (805 kilometers) per hour, cracking or puncturing the eighth leading edge panel, made of reinforced carbon-carbon (RCC).

In orbit, the crew could not see the damaged wing leading edge, and flight controllers, after analysis of launch imagery, decided any damage to the wing or its heat shield tiles was superficial. Houston briefed the crew that the impact was not a safety concern and cleared STS-107 for a normal reentry on February 1, 2003.

During the reentry, beginning at Mach 23, superheated plasma penetrated the damaged left wing leading edge, melting and weakening the wing's internal structure. When it buckled under reentry loads, *Columbia* tumbled out of control and was torn apart by the hypersonic slipstream over northeast Texas at 181,000 feet (55,169 meters), traveling at Mach 15, nearly 12,000 miles (19,312 kilometers) per hour. The seven crewmembers, even in their pressure suits, could not survive exposure to near-vacuum, cabin breakup, frictional heating, and punishing aerodynamic forces.

NASA convened the *Columbia* Accident Investigation Board to determine the cause of the tragedy and recommend corrective actions. Telemetry data radioed to Houston documented the progressive destruction of the left wing and the orbiter's struggles to retain control. On the ground, searchers in east Texas and Louisiana recovered the crew's remains and 84,000 orbiter fragments, constituting 38 percent of *Columbia*'s reentry mass. Both the data and retrieved wreckage pointed to RCC panel 8 as the foam impact site, and a data recorder recovered nearly intact from the debris field confirmed the sequence of onboard events leading to *Columbia*'s destruction.

The Board pointed once again to NASA leadership, communication, and operations failures similar to those that led to *Challenger*'s loss seventeen years earlier. The most serious was management's lack of appreciation—over decades—of the catastrophic damage that external tank foam loss could wreak on the orbiter's fragile heat shield. Foam damage was a regular event, but was usually, though not always, minor. Over time, management gained confidence that foam loss posed an acceptable risk and continued to fly without addressing this dangerous external tank flaw. During the flight, mission managers failed to hear and investigate flight controllers' concerns that the impact damage might be serious.

During recovery, the agency focused much effort on reducing tank foam loss but could not eliminate it completely. Eventually, the shuttle program did limit foam loss to small pieces, below the mass threshold for causing impact damage.

In 2004, President George W. Bush announced that after fulfilling its ISS assembly mission, the shuttle would retire. The implications for the US human spaceflight program were serious. First, without the shuttle the US was forced to rely on Russia for astronaut transport, at least until the planned Orion crew vehicle or commercial spacecraft became available. That US orbital transport gap and the Russian crew launch monopoly embarrassed NASA and the nation for nearly ten years. Second, without the shuttle's large cargo capacity, NASA had to revamp its ISS logistics strategy, using more frequent launches of ISS partner cargo ships, and later, commer-

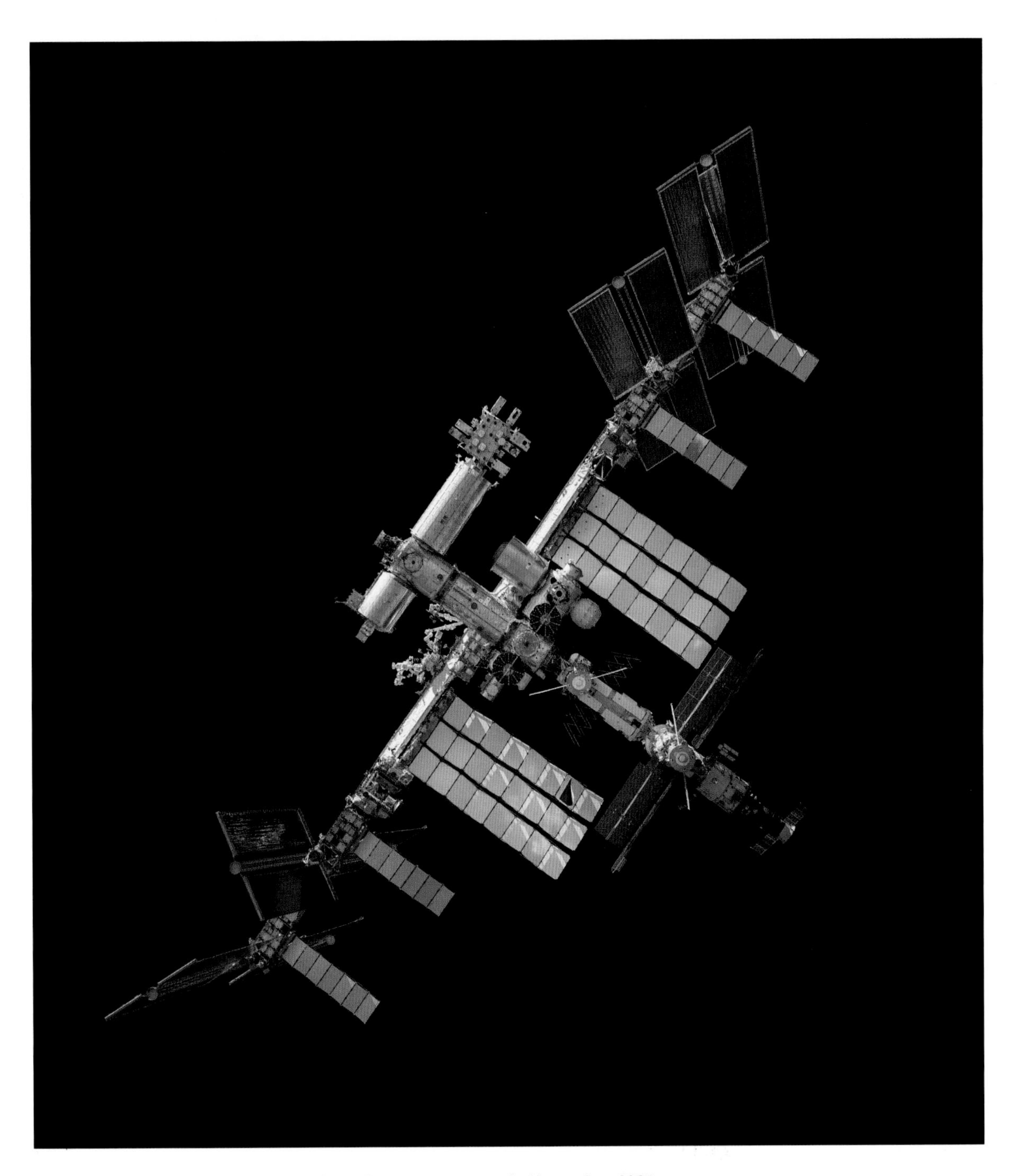

The completed ISS as seen from a Crew Dragon transport in November 2021.

ial vehicles to haul spare parts and supplies. NASA's rst commercial contract cargo launch came in 2012, but paceX's Crew Dragon astronaut transport didn't enter ervice until 2020.

The shuttle's retirement in 2011 might have been xpected to constrain the outpost's future research apabilities. But the 2012 addition of Dragon commercial cargo delivery and return enabled NASA and its partners to expand the crew size to seven and gradually increase the Station's scientific output. That outcome capitalized on the shuttle's successful fulfillment of its most important technical assignment, building an engineering wonder that has continually and safely hosted crews for nearly twenty-five years.

Mission no. **93**

STS 88

Orbiter	Endeavour
Launch	December 4, 1998
Landing	December 15, 1998
Duration	11 days, 19 hrs., 17 mins., 56 secs.

The STS-88 crew displaying flags of ISS partners in the middeck: C. J. Sturckow, Jerry Ross, Jim Newman, Nancy Currie, Bob Cabana, and Sergei Krikalev

Crew	Robert D. Cabana, Frederick W. "C. J." Sturckow, Nancy J. Currie, Jerry L. Ross, James H. "Pluto" Newman, Sergei K. Krikalev

Mission: *Endeavour*'s crew began ISS assembly by using the RMS to berth the Unity module atop the orbiter docking system. After rendezvous with the Zarya module, the crew linked it to Unity, creating the nascent ISS. Astronauts executed three EVAs to connect the new Station systems.

Bob Cabana, Commander:

This was the first ISS assembly mission, setting the stage for the whole assembly sequence. It was critical that it went well. We had to get those first two modules together to establish the Space Station and show progress.

During the rendezvous and proximity operations, I had Jim Newman in my ear, saying, "What do you think about a couple of 'in' pulses? How about an 'up'?" We were getting ready to grapple the FGB to bring it down into the payload bay. With the Node blocking it, I couldn't see the FGB (Unity) out the windows, but I watched our camera views through two little eight-inch (twenty-centimeter) monitors, trying to keep the module perfectly centered. While we were waiting to pass over a Russian ground station for permission to grapple, the autopilot was holding our attitude, easing the pilot workload. But the orbiter drifted and hit an attitude limit, and the Digital Autopilot (DAP) fired the jets to center it back up. The firing induced an "up" translation, and all of a sudden this 45,000-pound (20,411-kilogram) mass was moving down into the payload bay and toward the arm. I started firing thrusters to move away, but nothing happened—the FGB was still going to hit us.

I was using the "B" DAP for fine, precise control and had enough sense to select the "A" DAP, where I had more thruster power. I backed away from the FGB and got stable, then moved back in. Everything was fine again, but there was just dead silence in the cockpit; nobody said anything. It was pretty tense. When stable again, I said, "Hey, Pluto! You wanted to help during the rendezvous. How come you didn't have any words of advice for me when *that* happened?" Jim said, "Oh, I know when to keep my mouth shut."

Before launch, the media kept asking who was going to be the first person to enter the station, and I wouldn't tell anybody. When it came time to open the hatch after docking, I said, "Sergei, get up here." We pulled the hatch open, and the pictures show that as we went through every hatch from the beginning of the Pressurized Mating Adapter (PMA) all the way through into the FGB, Sergei and I entered all the hatches side by side. So there was no "first person" to enter the Space Station. I felt that if we had an International Space Station, we needed to enter as an international crew.

Later in Unity, I called Nancy Currie to join me. I had my feet anchored under this handrail when she came over, and I held her and placed her as still as I could in the middle of the Unity Node, and then gently released her. She could not reach anything—she was stuck there! She did the first thing everyone tries to do: swim in space. And you can't—you just rotate. So she could not propel herself in any one direction. But because it's impossible to release someone absolutely still with no momentum, eventually that slow drift and the air currents moved her to where she could grab something. But for a while, she was stuck in the middle of the Node. It was fun to watch.

Over the last twenty-five years, we have learned how to truly live in space on the ISS. That has prepared us for going back to the Moon in a sustainable way, and prepared us for the long trips to Mars.

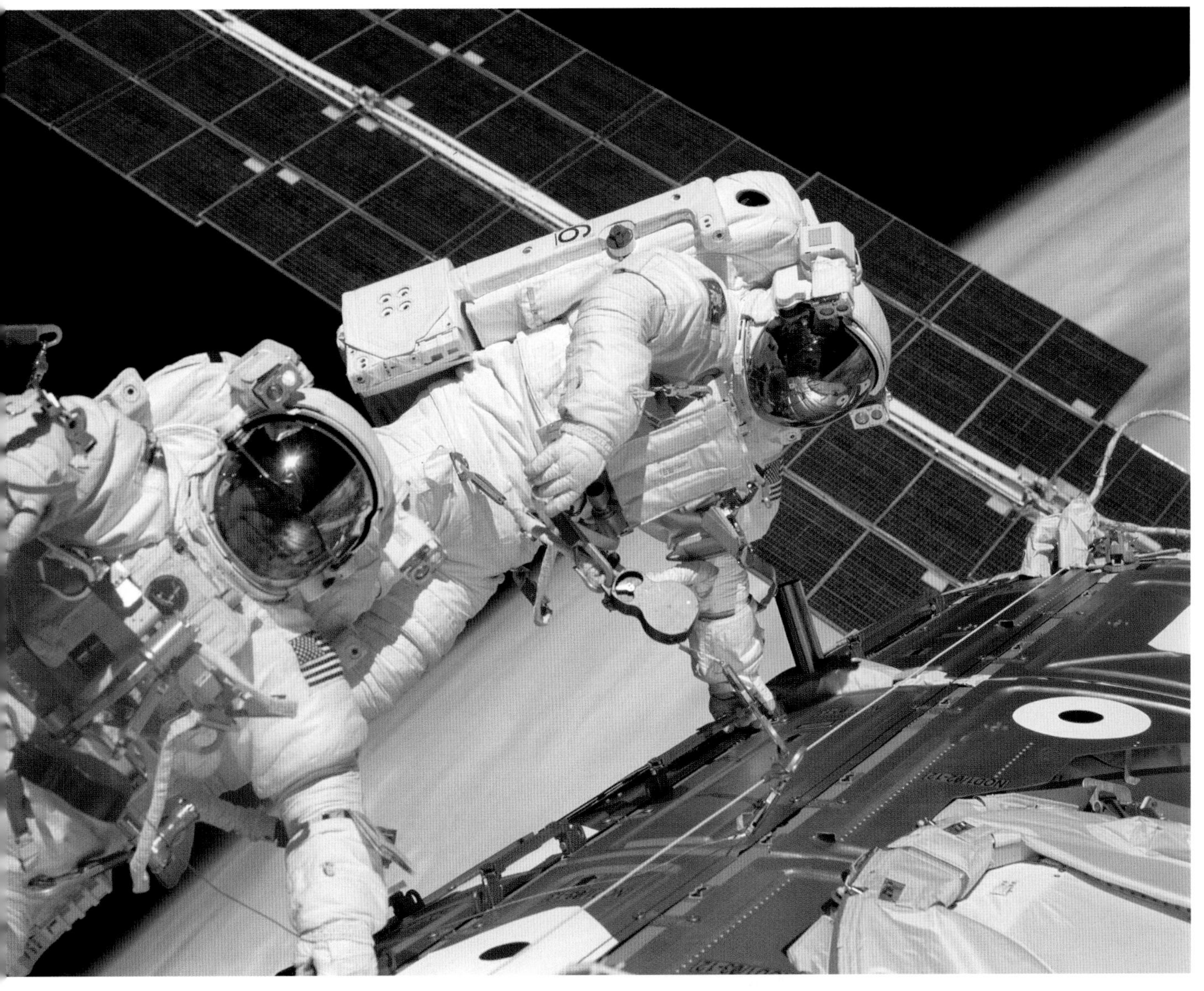

ABOVE: **In 1998, Jerry Ross and Jim Newman, outside Unity, perform the first EVA on the ISS.**

RIGHT: **Zarya (at left) and Unity formed the newly inaugurated ISS.**

BELOW: **Sergei Krikalev and Bob Cabana working just inside the Russian FGB, Zarya.**

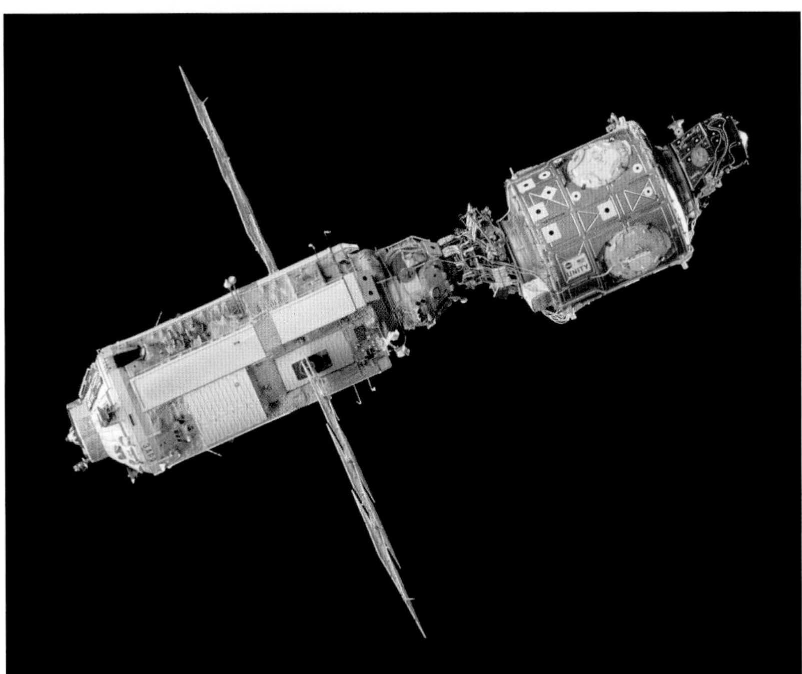

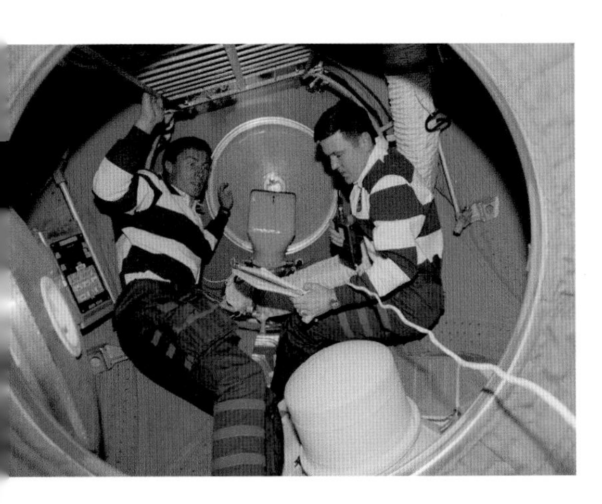

Mission no. **94**

STS

96

Orbiter	Discovery
Launch	May 27, 1999
Landing	June 6, 1999
Duration	9 days, 19 hrs., 13 mins., 1 sec.

The STS-96 crew in Unity: (clockwise from upper left) Valery Tokarev, Tammy Jernigan, Kent Rominger, Rick Husband, Ellen Ochoa, Julie Payette, and Dan Barry

Crew	Kent V. "Rommel" Rominger, Rick D. Husband, Ellen Ochoa, Tamara E. "Tammy" Jernigan, Daniel T. Barry, Julie Payette, Valery Ivanovich Tokarev

Mission: STS-96 docked with the ISS to transfer 3,567 pounds (1,618 kilograms) of supplies from SPACEHAB to the Station. Barry and Jernigan's spacewalk installed a crane and work platforms outside. The crew launched the highly reflective STARSHINE satellite as a tracking target in student science lessons.

Tammy Jernigan, Mission Specialist:

STS-96 was part of the suite of flights required for Space Station construction. Our main goal was to attach construction equipment outside the ISS and to outfit the inside in preparation for the first crew coming aboard. We delivered everything from scientific apparatus to clothing, filters, and backup electronics for future missions.

We were the first docking mission to the Station, and we docked to the US-built Unity node. With our docking mechanism lined up, we hit the Station at just the right velocity, which engaged latches to join the two spacecraft in what we call a "soft capture." Next, I sent commands from the flight deck to create a "hard capture"—a rigid, airtight connection—between the docking mechanisms. Then we could equalize the pressure between the shuttle and the ISS to open the hatches into Unity, which was attached to Zarya, the Russian FGB. At the time, there was no one on the Station to greet us, so we docked and opened up the hatches ourselves.

On a previous mission, STS-80, astronaut Tom Jones and I were in the airlock about to begin our spacewalk, but a stray screw that had lodged in the airlock hatch mechanism kept us from opening the hatch. On STS-96, the hatch opened with such extraordinary ease—it was shocking to me how effortless it was to open a properly operating hatch. Many people from the Cape had called and sent me emails, reassuring me that the hatch would open this time.

Floating outside to see this spectacular view of Earth—that full-field view outside—is my most memorable experience from STS-96. I saw this beautiful blue ball suspended in the darkness of space, wrapped in the thin atmosphere that protect life on Earth. As I looked at this beautiful blue orb, there were nc boundaries visible, and I though "We are all in this together. We need to be good stewards of thi planet with which we've been entrusted."

The spacewalking experience was so much easier than the trai ing exercise in the pool had been because we were not operating against the water's viscosity and the difficulties of achieving neutr buoyancy. Only in the airlock did the suit feel like an encumbrance within the airlock's tight confines the suit felt very bulky. But outsic the ship, riding on the arm or operating on the Station structur the suit was comfortable and didn't hinder my work at all.

The US crane and Russian Strela crane each weighed a couple of hundred pounds, but manipulating them was easy. My training in the water tank had gotten me accustomed to movin slowly, to avoid fighting the water's drag. In space, all I had to do was remember that once I gc something going, I had to be in position to put in enough force t counteract its inertia.

Throughout the mission, our international crew worked well to gether, and I credit our commanc er, Kent Rominger, with instilling the sense that "we'll always have each other's back." I also had tre mendous admiration for our pilot Rick Husband. He was just one c the finest human beings I'd ever met. Sadly, we lost my dear frien four years later when *Columbia* broke apart on reentry.

ABOVE: **The rising sun silhouettes *Discovery*'s exhaust plume as the shuttle heads toward the ISS.**

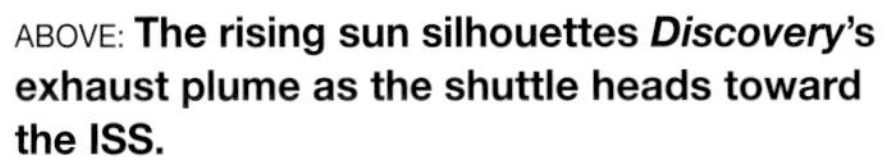

RIGHT: **Anchored on the RMS foot restraint, Tammy Jernigan installs the Russian Strela crane on the ISS.**

BELOW: ***Discovery*'s crew photographed Unity and Zarya as they departed the growing station.**

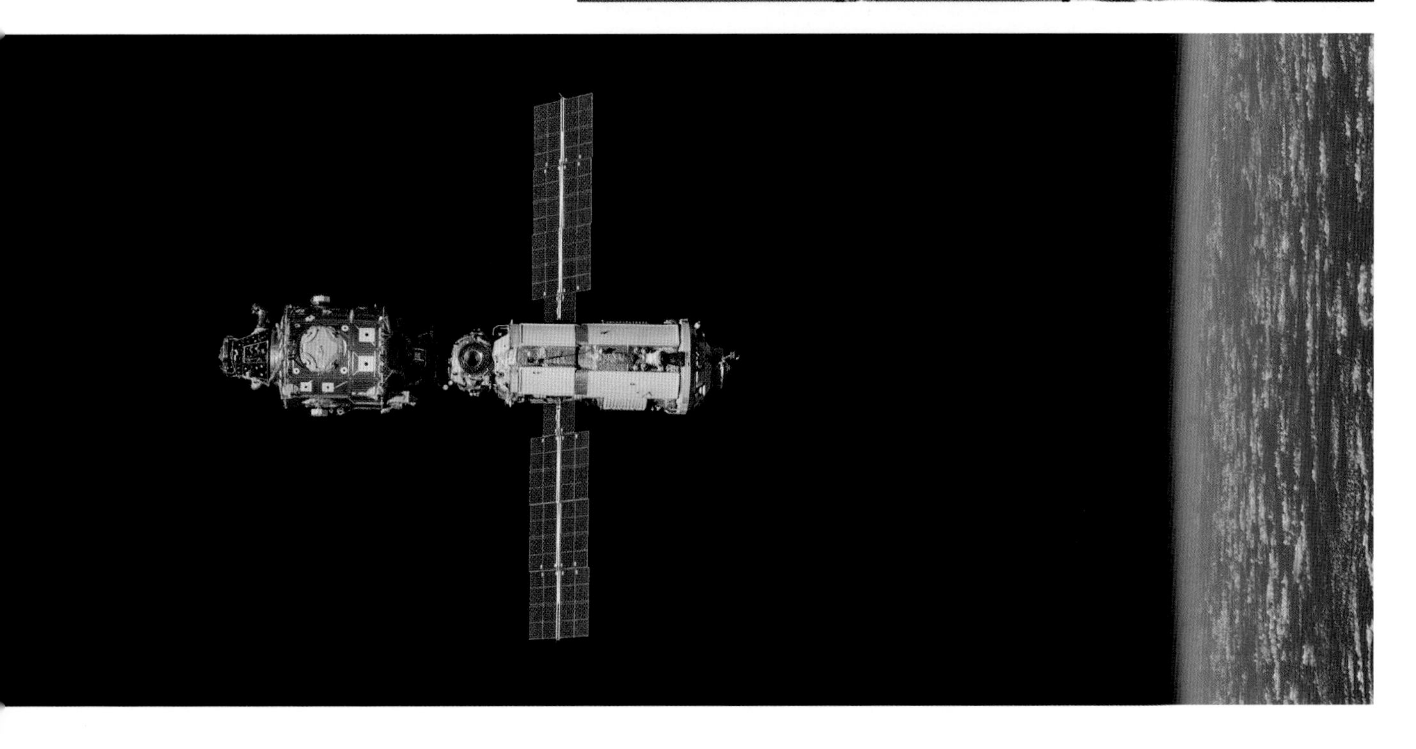

Mission no. **95**

STS

93

Orbiter	Columbia
Launch	July 23, 1999
Landing	July 27, 1999
Duration	4 days, 22 hrs., 49 mins., 35 secs.

The STS-93 crew with *Columbia* at Pad 39A: Steve Hawley, Michel Tognini, Cady Coleman, Eileen Collins, and Jeff Ashby

Crew	Eileen M. Collins, Jeffrey S. Ashby, Steven A. Hawley, Catherine G. "Cady" Coleman, Michel Tognini

Mission: Five seconds after liftoff, *Columbia* suffered an electrical short, nearly aborting the ascent. A hydrogen leak caused a premature engine cutoff into a lower than planned orbit, but the astronauts deployed the twenty-five-ton (22.6-metric-ton) Chandra X-Ray Observatory toward its distant observing orbit. Eileen Collins became the first woman to command a shuttle mission.

Eileen Collins, Commander:

The Chandra observatory took more than five years to build, and it is still operating perfectly after twenty-three years in orbit, returning images and spectra in X-ray wavelengths. It is a sister observatory to Hubble, one of the four Great Observatories, and I'm really proud of the people who built it and our team who launched it. We had a few challenges getting Chandra to orbit, however.

About five seconds after liftoff, an electrical short circuit interrupted power for a microsecond. We should have gotten a master alarm, but we didn't because the brief blip wasn't long enough to trigger the alarm. We learned after the flight that a burred screw head had been rubbing on a wire bundle, abrading the wire and causing the short circuit. The short circuit didn't cause an immediate abort, but we completely lost two engine controllers.

We had no insight onboard into the controller problem, but we found out when Scott Altman, our capcom, called us after the roll maneuver the shuttle executes just after liftoff: "*Columbia*, AC bus sensors off. We see your controllers failed—primary on the center, backup on the right." Jeff Ashby turned off the AC bus sensors to prevent an inadvertent engine shutdown, and that's the only action we had to take through the rest of the ascent.

I remember thinking during ascent, "You don't want to waste too many brain cells trying to figure this out," because as the commander, I was looking at our vehicle performance. You cannot let that failure distract you. We were so heavy with Chandra that if we lost an engine, we were on the very edge of what was certified for safe flight with our weight and center of gravity. This was a serious short circuit, so I was extremely happy when we finally made it to MECO.

Because the space shuttle ma engines were reusable, some pa would have to be inspected and repaired before the next flight. Feed posts that sprayed oxygen into the combustion chamber through the injector plates were among the parts that engineers scrutinized, and if they thought a post was worn enough that it could leak excess oxygen, they would block that post with a gol pin. I think three of these posts were pinned on the right engine oxidizer injector plate.

When that engine started up at launch, one of those pins poppe out of the injector plate and hit three adjacent liquid hydrogen cooling tubes lining the engine nozzle, causing them to leak hydr gen all the way to orbit. Replayed launch video showed a glowing plume of hydrogen streaming fro the right engine nozzle.

We were leaking hydrogen, bu we actually ran out of oxygen first, because the main engine controller compensated for the power loss caused by the leak b increasing the oxygen-to-hydrogen burn ratio. We had no insight into that problem all the way to orbit until at MECO, we saw a fifteen-foot-per-second (five-meters-per-second) underspeed. I'd seen that underspeed in the simulator before, caused b guidance system errors, maybe up to nine feet (three meters) per second. This was fifteen feet (fiv meters) per second. I said, "Oh, that was early. I wonder what tha was?"

But we'd made it safely. I kne we had an underspeed, and an electrical short problem, but we were not in the ocean! We had Chandra onboard, and we could have lost that payload, not to mention the crew.

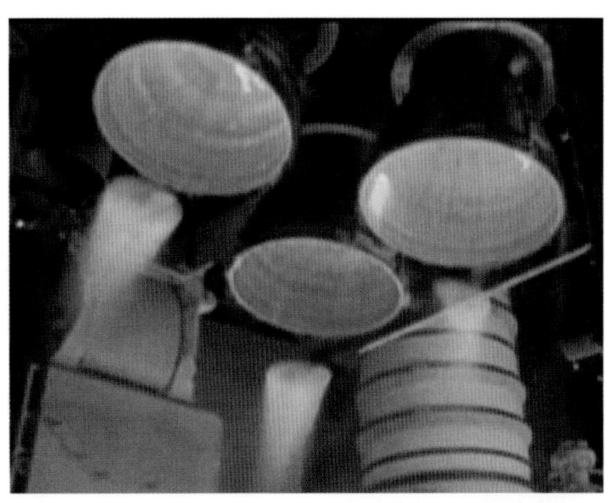

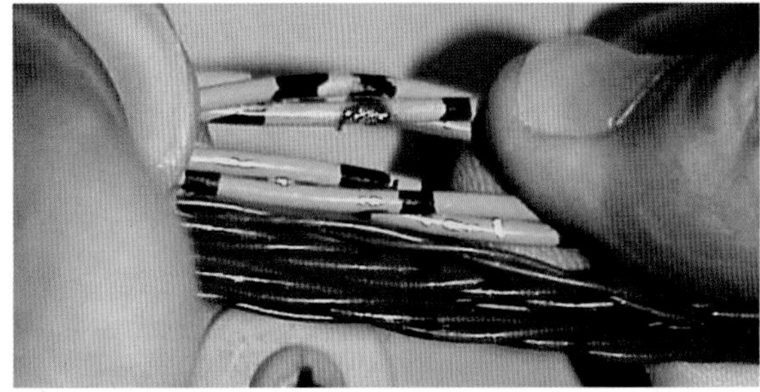

TOP: ***Columbia*'s right engine nozzle interior shows the light blue trace of a hydrogen leak after liftoff.**

ABOVE: **Just after liftoff, exposed abraded wiring in *Columbia*'s payload bay allowed electric current to short-circuit to the adjacent screwhead.**

LEFT: **The Chandra X-Ray Observatory nearly fills *Columbia*'s payload bay at the launch pad.**

BELOW: **Eileen Collins guides *Columbia* to a smooth touchdown at Kennedy Space Center's Shuttle Landing Facility.**

Mission no. **96**

STS **103**

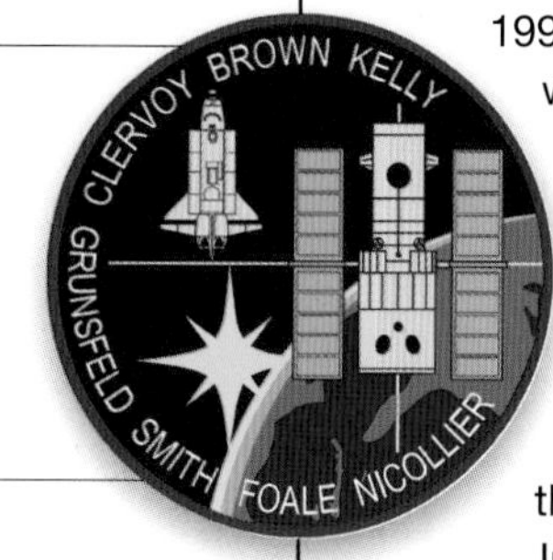

Orbiter	Discovery
Launch	December 19. 1999
Landing	December 27, 1999
Duration	7 days, 23 hrs., 10 mins., 47 secs.

The STS-103 crew on the flight deck: (front row) Claude Nicollier, Scott Kelly, and John Grunsfeld; (back row) Steve Smith, Mike Foale, Curt Brown, and Jean-Francois Çlervoy

Crew	Curtis L. Brown, Scott J. Kelly, Steven L. Smith, C. Michael Foale, John M. Grunsfeld, Claude Nicollier, Jean-Francois Çlervoy

Mission: NASA moved up this third servicing mission to the HST after Hubble lost a fourth attitude control gyro. STS-103's astronauts performed three spacewalks to replace six gyros and the 550-pound (249-kilogram) fine guidance sensor. *Discovery*'s peak altitude of 378 miles (608 kilometers) was the highest reached during the program.

Claude Nicollier, Mission Specialist:

We lifted off on December 19, 1999, and two days later we met with a telescope that was not working. Our goal was to make Hubble work again. We succeeded in recovering Hubble's operational capability just before the turn of the century.

In the spring of 1999, Hubble suffered three gyro failures. NASA then decided that we were going to fix Hubble earlier than the planned date in mid-2000. Our flight then became a Hubble "Launch-On-Need" mission, because we were one failure away from losing the capability to use Hubble. Then in November 1999, during the final preparations for the mission, the telescope lost a fourth gyro. The two remaining gyros were not sufficient to point the telescope; we could only orient the telescope's solar arrays to the Sun.

On this third Hubble-servicing mission, the main goal was to replace three Rate Sensor Units, each containing two gyroscopes, for six gyros total. We planned also to replace the computer along with one of the fine guidance sensors.

During my only EVA on this mission, my first contact with Hubble was a very special moment. Grasping the yellow handrail with my gloved hands on the lower part of the telescope for the first time, I felt something very special flowing through my body and mind. I stayed there for a while, "hanging" onto the Hubble that had given us so much and would continue doing so, but only if we had mission success.

On this EVA, we had a problem with the insertion of the Fine Guidance Sensor 2 (FGS) into the telescope. It's a pretty large box, like a grand piano, about fi feet (one-and-a-half meters) in length, about sixteen inches (for centimeters) thick, and more tha three feet (one meter) in width. I was inserting the new FGS2, an it had gone in maybe two feet (h a meter) out of the total four or f feet (one or one and a half mete needed when I felt a lot of frictio I could not insert it further.

On each side of the FGS was bar, square in cross section. The alignment bars were intended to slide inside U-shaped rails on th telescope, but there was simply too much friction. I tried to wigg the FGS2 gently in yaw. I felt tha any slight error in yaw during insertion was causing the exces friction, stopping everything.

There were many failure mod we were trained to face so despite problems we could still reach the goal, but this was one we had not really thought about Finally, Mike Foale held one sid of the instrument, and I held the other. We had markings on the square lateral-alignment bars— and we inserted the FGS inch b inch, pushing in a coordinated manner from both sides, doing our best to null any yaw error. It worked, but it took a long time. Our movements were controlled and precise, but slow. The rest the spacewalk went fine, but the full duration was eight hours an ten minutes.

After each of the three spacewalks, the Hubble team at Goddard checked the result of th fix by telemetry, and each time they gave us a thumbs up. I was extremely happy when, on Chris mas day, Jean-François Çlervoy released a telescope that was again fully functional. The scient ic value of Hubble makes it dear my heart, and also to the hearts astronomers and the public.

ABOVE: **John Grunsfeld and Steve Smith work inside Hubble's open instrument compartment.**

LEFT: **Scott Kelly photographed Jean-Francois Çlervoy, Mike Foale, Claude Nicollier, Curt Brown, Grunsfeld, and Smith after undocking.**

Mission no. **97**

STS 99

Orbiter	Endeavour
Launch	February 11, 2000
Landing	February 22, 2000
Duration	11 days, 5 hrs., 38 min. 44 secs.

The STS-99 crew on *Endeavour*'s middeck: (bottom row) Mamoru Mohri, Janice Voss, and Dom Gorie; (top row) Gerhard Thiele, Kevin Kregel, and Janet Kavandi

Crew	Kevin R. Kregel, Dominic L. Pudwill Gorie, Janet L. Kavandi, Janice E. Voss, Mamoru Mohri, Gerhard P. J. Thiele

Mission: STS-99's Shuttle Radar Topography Mission extended a radar antenna boom 196 feet (sixty meters) beyond the orbiter's port side to create a digital elevation map of Earth between 60 degrees North and 56 degrees South—a terrain database widely used by the military and civilian aviation.

Gerhard Thiele, Mission Specialist:

As much as I admire all the work constructing the Space Station or repairing the Hubble and so on, and as much as I'd love to be on one of those missions—because I really consider these things important—I'm a scientist, and getting a challenging scientific mission was for me richly rewarding.

It was even more rewarding because we were extremely successful. Before the mission, we asked the lead scientist and the chief engineer, "How much do we need to get so that you'll still talk to us when we return?" They said they would be extremely happy if we came back with something like 70 percent of the planned coverage. We thought that was a stretch.

But we came back with 99.7 percent! We didn't even dream about this to begin with. The data's precision, as we discovered later, was not quite as high as we hoped, but it was still far beyond what was imagined in the beginning. As much as I enjoyed contributing to science, I also knew that no matter how many times I would fly, I would never get to be a part of a crew like this again.

For about half a year, we spent maybe three days a week out at JPL training with the folks there, some three hundred engineers and scientists who put this project together. When you see all this effort, all this heart that others are putting into this mission, you know you'd better not fail. And you'd better give everything you have if you don't want to disappoint them. That certainly helped me make doubly sure that I would do things exactly as they would, had they been in my place.

One night, we crossed the terminator approaching Europe. Ireland and France were covered with clouds. But as we came south, the cloud layer became thinner. We couldn't identify the big cities—Marseilles, Lyon—but we saw pearls of light shimmering through the cloud layer. It was so beautiful.

We crossed the Alps into Italy. And from there on, it was cloud-free. We saw the entire boot of Italy and its cities and villages, as if the Italians had lit candles to outline the boot. I couldn't believe this view by night. And in as long as it takes to tell all this, we were already over the Mediterranean and Crete. Ahead, there was a band of light coming across the horizon—the Nile River.

Later we were flying a night pass across the Pacific. At night over the ocean, you can't see Earth. You can only tell where the planet is because there is a certain area in your view where the stars are blocked by Earth. This was one of those long ocean passes, maybe thrity-five minutes where the radar was turned off. We dimmed all the lights and computer screens so we could barely see them. Our eyes adjusted, and Kevin Kregel, Janet Kavandi, and I were all hanging out by the windows and looking down at the pitch black. No one said a word for half an hour. No one. That was almost a holy moment.

I was looking out at the stars—the cosmos. If you ask the astronauts, eight out of ten would say that Earth was the special thing to look at—where they wanted to be. I wanted to be in the blackness. If I could have taken over command of the shuttle, I would have said, "That's where we are going now."

ABOVE LEFT: ***Endeavour*** **roars aloft carrying the Earth-mapping radars of the Shuttle Radar Topography Mission.**

ABOVE RIGHT: **The Galapagos Islands, Ecuador, with Fernandina (at left) and the seahorse-shaped Isabela.**

BELOW: **A 60-meter boom extended from *Endeavour*'s port side carrying radar receiving antennas.**

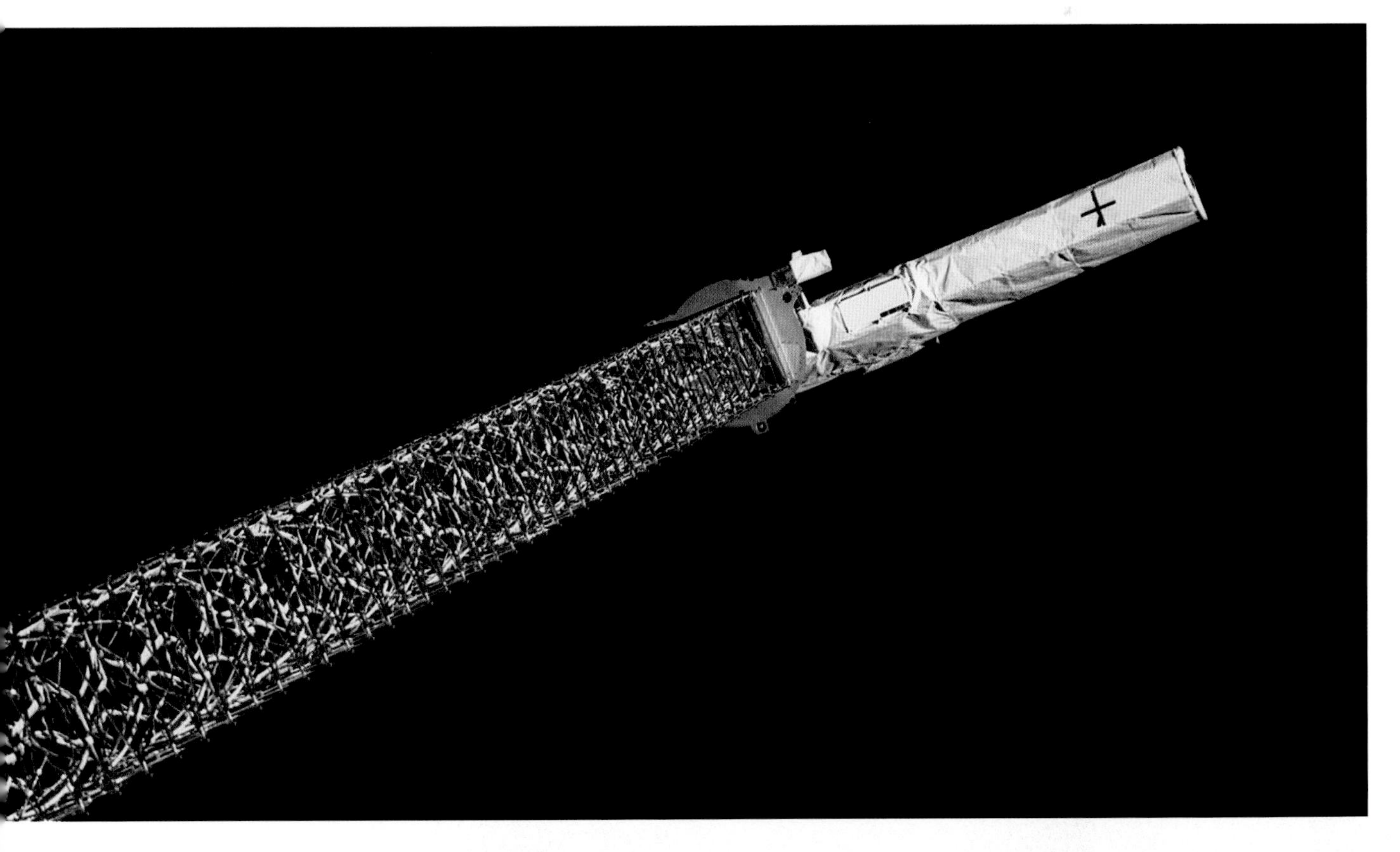

Mission no. **98**

STS

101

Orbiter	Atlantis
Launch	May 19, 2000
Landing	May 29, 2000
Duration	9 days, 20 hrs., 9 mins., 9 secs.

The STS-101 crew in the 39A White Room outside *Atlantis*'s hatch: (clockwise from left) Mary Ellen Weber, Jeff Williams, Jim Halsell, Jim Voss, Susan Helms, Yury Usachov, and Scott Horowitz

Crew	James D. Halsell Jr., Scott J. "Doc" Horowitz, Mary Ellen Weber, Jeffrey N. Williams, James S. Voss, Susan J. Helms, Yury V. Usachov

Mission: The shuttle's third docking with the ISS installed an antenna and camera cabling while upgrading solar array batteries. Russian delays forced STS-101's original mission to split into two parts: STS-101 delivered 3,300 pounds (1,497 kilograms) of crew supplies, and STS-106 outfitted the new Zvezda service module.

Susan Helms, Mission Specialist:

The ISS program was very worried that several batteries would fail on Zarya, the Russian FGB that was the first segment of the Station launched. If those batteries failed, the remaining power might not be enough to keep the ISS alive. Solving the power problem became a major goal of STS-101.

The mission was originally crewed by Jim Halsell, Scott Horowitz, Jeff Williams, Dan Burbank, Mary Ellen Weber, and Ed Lu. Management told Jim Halsell, "We're splitting your flight into two missions, and we're going to throw the second ISS expedition crew on with you because they have already been trained on how to change out those failing Russian batteries." Dan and Ed got moved to STS-106, and Jim, Yury and I joined STS-101. That swap happened eight weeks before launch.

Jim Halsell, Commander:

My initial reaction was about as parochial as you might expect. I told management that the current STS-101 crew had gelled into a strong unit that was 110 percent ready and that changing players this close to launch would add a small degree of additional risk best avoided if possible. But I remembered how Apollo 13 swapped out command module pilots just three days before heading to the Moon. Considering this and other factors, I told our chief of flight crew operations, "If the program feels strongly that the positives outweigh the negatives, we will make it work." Although we spent a couple of weeks mourning the absence of our good friends, we welcomed the new guys and got on with meeting the challenge.

Susan Helms:

It was the weirdest crew dynamic: Eight weeks before launch you had four astronauts who were grieving because their crewmates were reassigned, and you had three people joining who were ecstatic at getting a surprise shuttle flight. Jim Halsell scrambled, got his crew together, and said, "You're allowed to grieve about this for a week and then I don't want to hear any more about it. They are our crew now."

Jim Halsell:

I never heard anyone from the original crew complain about the replacements. It reminded me of how professional athletes are traded and expected to play seamlessly with their new team in the very next game. We felt proud at demonstrating that same versatility in America's space program.

Susan Helms:

Jim, Yury, and I felt that we had taken a three-month pause from our ISS 2 training just to squeeze a quick shuttle flight. It was rough going from one demanding training flow to another with no break. But we couldn't complain about the long hours or how hard we were working. Anyone would have said, "You assholes, you got an extra shuttle flight. Just shut up."

Soon after docking, Jim Voss, Yury, and I had entered the Space Station for the first time, and we found ourselves there alone. The others were over in *Atlantis*. Suddenly Yury said, "Okay, good! Now, close the hatch!" and the three of us laughed. We got what he meant: We'd been training for ISS 2 for almost three years, and Yury was ready to get started! Being up there together was really cool, knowing we would return in less than a year. Seeing the Unity node and the FGB in person was such a great preview—so much better than any training we could have gotten on the ground!

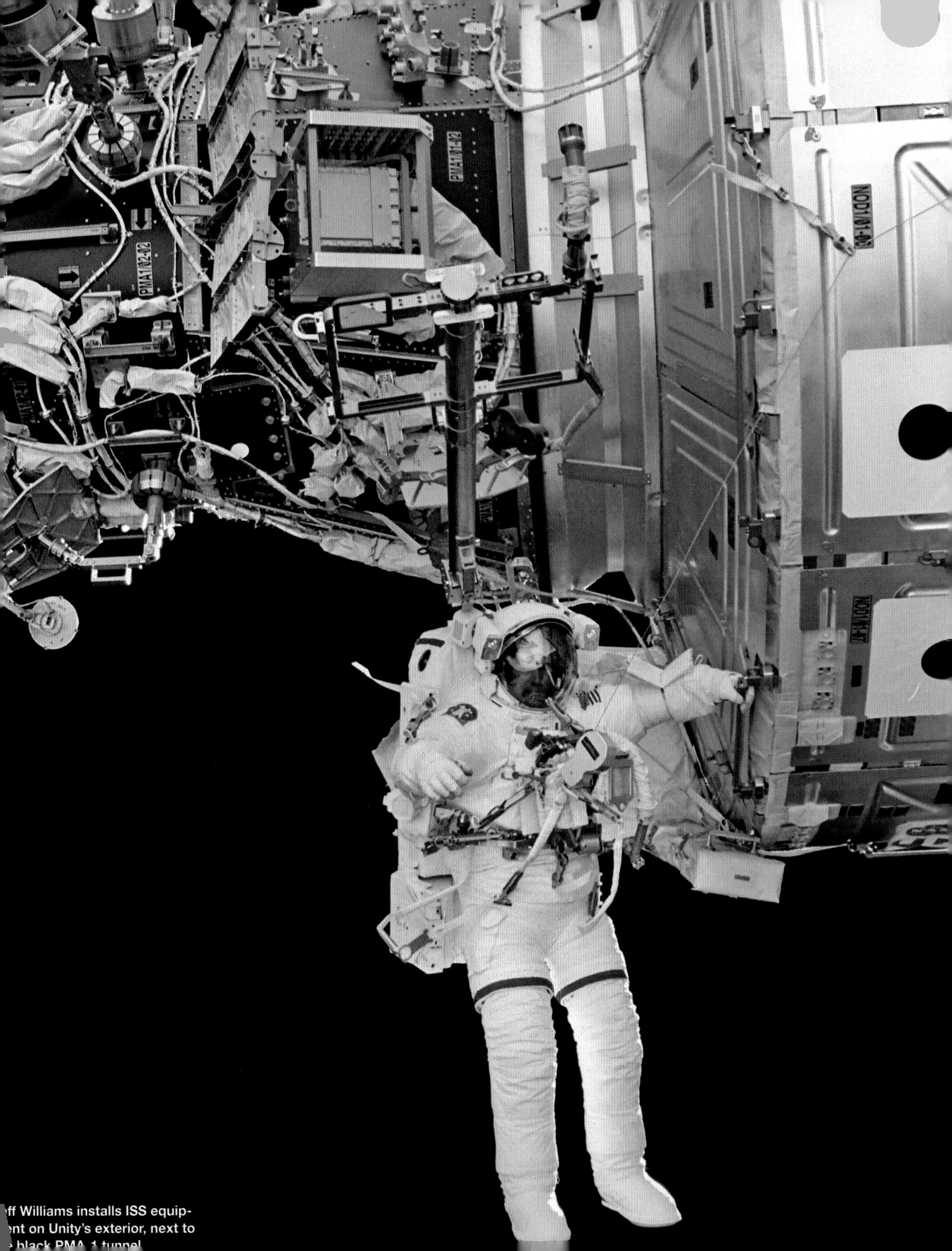

ff Williams installs ISS equip-
ent on Unity's exterior, next to
e black PMA-1 tunnel.

Mission no. **99**

STS

106

Orbiter	Atlantis
Launch	September 8, 2000
Landing	September 20, 2000
Duration	11 days, 19 hrs., 10 mins., 57 secs.

The STS-106 crew inside Zvezda: (front row) Yuri Malenchenko, Terry Wilcutt, and Scott Altman; (back row) Dan Burbank, Ed Lu, Rick Mastracchio, and Boris Morukov

Crew	Terrence W. Wilcutt, Scott D. Altman, Daniel C. Burbank, Edward T. Lu, Richard A. Mastracchio, Yuri I. Malenchenko, Boris V. Morukov

Mission: To prepare the ISS for the first expedition's arrival, *Atlantis*'s crew transferred more than 6,000 pounds (2,722 kilograms) of supplies. A six-hour joint US-Russian spacewalk connected power and communication cables between newly arrived Zvezda and the Zarya module.

Dan Burbank, Mission Specialist: The FGB, named Zarya, and the Service Module, called Zvezda, were both too massive to launch fully outfitted atop their Proton boosters. Because of that limitation, a lot of their systems testing had been done in Russia, and then a lot of the heavy stuff had been removed and shipped to the Kennedy Space Center to be brought up on shuttles. Our task was to take on some of the outfitting, logistics resupply, and life extension activities for the Zarya and Zvezda modules.

Early on the first morning in orbit, it was still dark in the middeck when I woke up in my sleeping bag in the airlock. I floated up in between the two wall-mounted spacesuits and looked up through the little circular docking hatch window—a window on this glorious, spectacular world scrolling by beneath me. Before the rest of the crew got up, I must have spent an hour with my nose pressed against that little window, marveling about being in this magical place.

One of the current conditioners for the electrical power system on the FGB had failed, so after docking, Boris Morukov and I had the job to remove and replace it. In the simulator building in Star City, we practiced this operation over and over again in the high fidelity FGB mockup. It was pretty straightforward: Remove a bunch of connectors on top of the thing, release a couple of captive bolts, slide it out, then put a new one in there and mate the connectors to the new conditioner box.

So, in orbit, we pulled up the FGB floorboards to gain access to the conditioner, and we found an additional brace, a secondary piece of structure, which went directly over the top of our worksite. This bar ran across an inch or so above the cable connectors, right on top of where the box was. And instead of threaded fasteners holding down the end of this interfering strut, there were Hi-LOK rivets—big, stout ones you would use on airplanes. There were four of these rivets on each end of this metal strut. I thought, "Oh, my goodness. What are we going to do with this?"

To get through this obstacle, we taped over the rivet heads, and got out a metal-cutting chisel and a Russian sledgehammer—not a huge sledgehammer, but also not the kind of thing you would expect to find on the Station. It turned out to be a fairly useful item! We just took the sledgehammer and chisel and cut off the heads of these four Hi-LOK rivets. The tape over the rivets helped contain any debris we generated. We pulled the bar out, and that was it.

STS-106 was an excellent, compressed example of all of the great things that made us proud about the ISS. The commitment to the excitement and importance of human exploration—independent of country, culture, and language—was extraordinary.

The thing that surprised me is that human beings can adapt to the space environment, so radically different than the one we evolved in. We leave behind gravity and terrestrial walking and jogging speeds—that kind of thing. The fact that we fly jet airplanes, that we operate on each other's brains, that we fly in space and are able to adapt to that environment—it's miraculous. To me, the wonderful plasticity of the human being, able to live in an environment that is so fundamentally foreign, is tremendously uplifting and just spectacular.

ABOVE: **Moisture condenses in shock waves as *Atlantis* goes supersonic.**

LEFT: **Yuri Malenchenko, in a US spacesuit, works outside the Russian segment of the ISS.**

BELOW: **Dan Burbank runs the EVA checklist from his station on the flight deck.**

Mission no. **100**

STS

92

Orbiter	Discovery
Launch	October 11, 2000
Landing	October 24, 2000
Duration	12 days, 21 hrs., 42 mins., 42 secs.

The STS-92 crew at Pad 39A: Pam Melroy, Brian Duffy, Mike Lopez-Alegria, Jeff Wisoff, Leroy Chiao, Bill McArthur, and Koichi Wakata

Crew	Brian Duffy, Pamela A. Melroy, Koichi Wakata, Leroy Chiao, Peter J. K. Wisoff, Michael Lopez-Alegria, William S. McArthur Jr.

Mission: After ISS docking, *Discovery*'s crew used the RMS to berth the Z1 truss atop Unity, and place the PMA-3 docking tunnel at Unity's lower berthing port for future dockings. Four spacewalks connected the new ISS elements and flight-tested the SAFER rescue jetpacks.

Brian Duffy, Commander:

On STS-92, we brought up the first piece of the structural backbone of the ISS, the Z1 truss. We were named as a crew two and a half years before we flew. Our long training timeline was not because we were slow learners, but because the Russians were late with their Zvezda module; we couldn't put our piece in place until theirs was up.

We had a blast training together. We liked one another and spent a lot of time together, not only at work, but also after hours. We'd do things with the families, have rotating movie nights, go to dinner together. During training, the Austin Powers movies with Dr. Evil and Mini-Me were all the rage. We called ourselves "The Evil Crew," and when we'd fly from one airport to another in NASA's T-38 jets, we'd check in on the radio as "Evil Flight 1, 2, 3, and 4."

On Flight Day 2, the radar failed. The way rendezvous is supposed to work is that when the shuttle emerges from darkness into daylight, there is supposed to be a space station out my overhead window, 2,000 feet (610 meters) away with me closing in at ten feet (three meters) per second.

But when we came into daylight, I looked up out the overhead window, and there was nothing there. One of the crew said, "Where is it?" Another said, "Look out forward. It's over there!" We were coming up short in the rendezvous, and we didn't have time to talk to Houston because the geometry was getting worse by the second. I had to make a real-time decision as to what to do. Either I blow off the rendezvous and try tomorrow, or I try to save this. I decided to try to save it, so I just started firing those four aft-firing thrusters—BOOM!

The crew was just watching, silent, no one saying a word. The ground couldn't see anything, because an electrical failure had taken out all of our high-data-rate telemetry. All they were seeing was jets firing and fuel quantity decreasing. They had no idea where the Station was. They didn't try to talk to me, and I didn't try to talk to them.

I had to fire all those thrusters to force our trajectory forward, speeding us that way like a bat out of hell. But as we passed underneath the Station, I didn't have to do anything more, because orbital mechanics carried us in front and over the top, and we stopped there without braking; I hardly had to fire the thrusters. By the time we looped up on top, we were right back on the fuel timeline, back to predicted fuel consumption. I got lucky, and it worked out just fine.

It was amazing how versatile the orbiter is, and how accurately you can control it, even though it weighs one hundred tons (90.7 metric tons). I had the camera looking up at the Station target, so for me it was like flying anything, right? You do it the way you've trained.

When we opened the hatch into the Node, it was like floating into a gymnasium. We went from the cramped and busy space shuttle middeck into a Station that was big, bright, clean, and warm—just a very comfortable place to go.

ABOVE: **Jeff Wisoff monitors Mike Lopez-Alegria as they take turns test-flying the SAFER jetpack at the ISS.**

LEFT: **An IMAX camera photographed Leroy Chiao working near Unity after installation of the Z1 truss.**

BELOW: ***Discovery*'s drogue chute begins to pull the main drag parachute from its compartment just after touchdown at Edwards Air Force Base.**

Mission no. **101**

STS 97

Orbiter	Endeavour
Launch	November 30, 2000
Landing	December 11, 2000
Duration	10 days, 19 hrs., 57 mins., 22 secs.

The STS-97 crew with the Expedition 1 crew (in blue) in Zvezda: (front row) Brent Jett, Bill Shepherd, and Joe Tanner; (back row) Sergei Krikalev, Carlos Noriega, Yuri Gidzenko, and Mike Bloomfield; (top) Marc Garneau

Crew	Brent W. Jett, Michael J. Bloomfield, Joseph R. Tanner, Marc Garneau, Carlos I. Noriega

Mission: *Endeavour*'s crew docked at the ISS, met by Expedition 1, which had lived aboard since early November. The STS-97 crew berthed the seventeen-and-a-half-ton (15.8-metric-ton) P6 truss atop the Z1 truss and deployed the P6 solar arrays. EVAs by Joe Tanner and Carlos Noriega prepared the Station exterior for expansion.

Carlos Noriega, Mission Specialist

STS-97 took up the first US solar arrays for the Space Station, a critical part of the assembly sequence. The ISS needed that power; the Russian segment didn't provide anywhere near what we were going to provide with the P6 arrays. In addition to this critical construction step, STS-97 enabled people to see the Station much more easily at night, because the solar arrays are so reflective.

I had little to do or worry about during ascent because I was strapped in by myself on the middeck. The suit techs wanted to get me out of the way, so they shoved me in there, sat me down, and I actually fell asleep waiting for the launch. When we got far enough down in the count, I heard them yelling from the flight deck to wake me up. That tells you my level of apprehension on the STS-97 mission.

During the deployment of the first solar array at the end of EVA 1, we started observing some unusual motion. The solar panels were supposed to accordion out evenly. Instead, the panels would stay bunched coming out, and then every so often they would "un-bunch." The panel faces were sticking together because they had been compressed in storage for too long. As the panels deployed farther and farther, the oscillations became more and more violent. By the time the array deployment was complete, we knew that we had a damaged array. Even though the array was fully deployed, the tensioning lines wouldn't tighten. Those slack cables meant that we had a potentially floppy array that would not be able to take future docking loads. For a while, they were even talking about jettisoning tha malfunctioning array.

Over the next three or four days, our Houston team flew out to Sunnyvale and with the engineers out there came up wit the repair procedure. Joe Tanne and I had some ideas ourselves because we too had intimate knowledge of the array.

The undersides of the array boxes were considered "no touch because they had sharp edges ar stored energy in the cable spools both things we had to avoid with our gloves. Yet the underside is where we had to go to fix this thing. The plan had me using a to with a little hook at the end to sna the tension wire and get it out of the way. Then Joe would manuall wind up the tension reel. Then as he let it unwind, I would feed the wire into the spool so we could again use the tensioning reels.

Our work position required us stand up in foot restraints with nc handholds for getting in and out of them. We had to use a T-bar, a short rod mounted on the foot plate, topped with a T-handle, to help get our boots into the restraint. Never a lot of fun in the water, that technique was not a trivial task on orbit. Joe got in, and then he was my ingress aid. would basically climb up on him, grab his hands, and stand up intc the foot restraint.

Once we got in place, people were amazed at how fast we got that whole reel-in procedure done. We were finished at each of four worksites in about twenty minutes. Because we got the reels fully wound up, we had regained full functionality and th tensioning worked as it was supposed to. For years after, when you spotted the ISS in the night sky, you were looking at our sola arrays going by.

ABOVE: **Carlos Noriega waves at Joe Tanner from atop the P6 truss and its solar array wing.**

RIGHT: **Tanner and Carlos Noriega suit up in *Endeavour*'s middeck for their third EVA.**

BELOW: **View from *Endeavour* of the ISS with extended P6 solar arrays.**

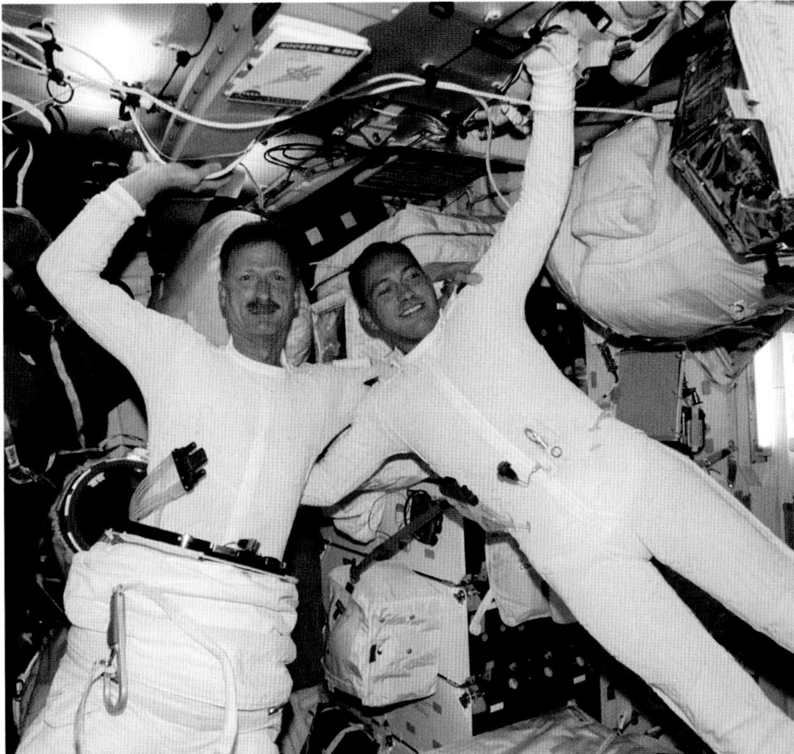

Mission no. **102**

STS

98

Orbiter	Atlantis
Launch	February 7, 2001
Landing	February 20, 2001
Duration	12 days, 21 hrs., 20 mins., 4 secs.

STS-98 astronauts (in dark blue) Mark Polansky, Ken Cockrell, Marsha Ivins, Bob Curbeam, and Tom Jones join the Expedition 1 crew—Sergei Krikalev, Bill Shepherd, and Yuri Gidzenko—in the Destiny lab.

Crew	Kenneth D. Cockrell, Mark L. Polansky, Robert L. Curbeam, Jr., Marsha S. Ivins, Thomas D. Jones

Mission: STS-98 delivered the US Destiny science laboratory to the ISS, using the robot arm to berth the sixteen-ton (14.5 metric-ton) Destiny at Unity's forward port. The crew's three spacewalks connected the lab to Station power, computer, and cooling systems while the astronauts transferred cargo into the outpost.

Tom Jones, Mission Specialist:

Delivery of the US laboratory would enable future scientific research at the ISS and switch prime ISS altitude control from Russian thrusters to the American lab's software and Z1's control moment gyros. The lab was the Station's brain. It was one of a kind, with no back-up. Our crew had to deliver the goods.

We docked on Flight Day 3 and readied for our first spacewalk the next day. Looking out the open airlock hatch for the first time, I could see it was a brilliant, sunny day outside. When that sunlight hit my suit, I felt its heat penetrate the suit's many insulation layers and warm my arms and legs, like on a pleasant spring day. My worries about the spacewalk evaporated. All that EVA training just clicked in.

The mission threw us several curveballs. A serious ammonia leak occurred as Bob Curbeam connected coolant lines to the lab, spewing ammonia overboard and frosting Bob's spacesuit with toxic ice crystals. He quickly reached an isolation valve to cut the flow, and with help from Mission Control, worked out a way to bypass the leak. Later, a stuck thermostat drove Destiny's internal temperature to nearly 100 degrees Fahrenheit (38 degrees Celsius), but Ken Cockrell and Marsha Ivins sped through activation of Destiny's cooling systems and prevented any damage.

Spacewalking is the peak astronaut experience. On EVA 1, I was high above the orbiter on the Station, releasing launch restraints holding down a P6 thermal radiator. Up there in the dark, I watched Marsha rotating the bus-sized, cylindrical lab gracefully through 180 degrees into berthing attitude. What an impressive sight—watching these giant modules come gracefully together. Not for the la time, I thought, "Do you realize, Tom, where you are, and what you're being allowed to see? Wh an amazing opportunity and gift!

That feeling returned near the end of EVA 3. We'd finished almost all our work outside, and was up at the Station's "bow" or the PMA-2 docking tunnel. Mark Polansky radioed, "Okay, come down to the payload bay and let do the incapacitated crew member demo." I thought, "Well, I'm leaving the lab for the last time." asked Mark, "Give me a moment out here." He answered, "You go it." I drifted to the lab's front-end handrail, facing the orbiter's tail twenty feet (six meters) away. The lab was behind and above me—nothing obscured the view Earth in any direction.

Lightly gripping the handrail w just a couple of fingertips, I spun slowly around, shifting my hands halfway to view both hemisphere Glorious Earth was visible below rolling silently beneath as I completed the circle. I gazed a thousand miles out to the blue horizo and 220 miles (354 kilometers) down to the Pacific beneath my boots. Looming above the lab wa the vertical P6 truss, holding tho golden solar arrays high against black emptiness. I was on the prow of a giant windjammer, falli swiftly, silently around Earth. The inky black of the cosmos arched above all.

It was *my* moment. Emotions swelled in my chest, flushed my cheeks, gave rise to irresistible tears. I knew only a few dozen humans had seen anything like this view, beautiful beyond words I felt personally grateful that God had given me this gift but humbl at the sheer scale of the scene and my own insignificance. Thos moments will never leave me.

LEFT: ***Atlantis*** **catches rays of the setting Sun to send a shadow lancing toward a full Moon.**

ABOVE: **Marsha Ivins uses the robot arm to rotate Destiny into perfect berthing alignment at the front of the ISS.**

BELOW: **The ISS crew photographed a departing *Atlantis*, payload bay empty, over the western desert of Morocco.**

Mission no. **103**

STS 102

Orbiter	Discovery
Launch	March 8, 2001
Landing	March 21, 2001
Duration	12 days, 19 hrs., 49 mins., 32 secs.

The STS-102 crew and Expedition 1 crew: (front) Yuri Gidzenko, Bill Shepherd, and Jim Voss; (middle) Sergei Krikalev, Susan Helms, and Yury Usachov; (back) Jim Kelly, Paul Richards, Jim Wetherbee, and Andy Thomas

Crew	James Wetherbee, James M. "Vegas" Kelly, Andrew S. W. Thomas, Paul Richards Up: James S. Voss, Yury V. Usachov, Susan J. Helms Down: Sergei K. Krikalev, William M. Shepherd, Yuri P. Gidzenko

Mission: *Discovery*'s crew exchanged Expedition 2 for Expedition 1 and delivered nearly five tons of supplies and equipment to the ISS via the Leonardo MPLM. Susan Helms and Jim Voss set an EVA duration record—eight hours, fifty-six minutes—to prep the Destiny lab for robot arm and truss installation.

Andy Thomas, Mission Specialist:

STS-102 was the first crew rotation mission to the ISS, with Yury Usachov, Jim Voss, and Susan Helms replacing Bill Shepherd, Sergei Krikalev, and Yuri Gidzenko. Bill, Sergei, and Yuri were the first resident crew, who had flown up on a Soyuz in November. This was also the first flight of the Leonardo Multi-Purpose Logistics Module (MPLM) to the Station. The MPLM demonstrated that we could ferry logistics up in the shuttle's payload bay, berth the MPLM, deliver the rack-loaded items, reload it with whatever we needed, and bring it back.

When I first went into the Russian segment at the ISS, in 2001, it had been three years since I'd flown on Mir. Inside the Russian modules, I was absolutely dumbfounded by the sense of déjà vu—how much like Mir they were. Zvezda looked like Mir, was lit up like Mir, and even smelled like Mir. The cosmonauts even displayed pictures of Gagarin and Korolev above the hatch, just as they had on Mir.

During the first EVA, Jim Voss and Susan Helms were outside, and Susan accidentally dropped a Ziploc bag filled with things being brought back to the airlock. She was on the arm at the time and of course reached for the bag now floating away. Jim "Vegas" Kelly was the arm operator: first-class pilot, first-class robotics operator. He leapt into action and flew the arm with Susan toward an intercept. The problem was he had this big honking space station between him and the airlock hatch near Susan. Without a direct line of sight, Jim could only use cameras, but he was able to do it because of his good situational awareness. His feat was one of the most impressive things I'd ever seen done in flight, enabling Susan to retrieve the tool bag.

STS-102 gave me my first chance for an EVA. It was a good experience on the whole, although we failed to get the pressure bladder inside my boot smoothed properly; not until we were through airlock depressurization and started outside did I begin to feel the heavy pressure on the top of my foot. I don't think the importance of smoothing the boot liner was given the necessary attention during suit-up training. I thought, "Well, I ca bear it." In hindsight, it was my mistake not to suspend the EVA and go back in, repressurize and fix it. I didn't because I thought we might postpone the EVA until the next day and lose a whole da of planned activities.

By the end of the EVA, I was really in pain. That much discomfort disrupts your concentration, and you can't focus on the task at hand. At least I was in the foot restraint for much of the EVA, which helped alleviate the pressure somewhat.

Late in our training flow a month or so before launch, the EVA branch asked me if I wanted to go up on top of the P6 truss and engage a latch there on the solar array, which had failed to engage during initial deployment. At the top of the truss, I had a view of the shuttle one hundred feet (thirty meters) below, and of Earth turning beyond. I was on the dark side of Earth looking down at the FGB, the truss, and the modules of the Station, suspended in nothing! Beyond was the deep black of infinity.

ABOVE: **Yuri Gidzenko hovers inside the expansive Leonardo multipurpose logistics module.**

RIGHT: **Vegas Kelly and Andy Thomas enter the Destiny lab from the PMA-2 docking tunnel.**

BELOW: **The Leonardo multipurpose logistics module arrives aboard *Discovery* on its first trip to the ISS.**

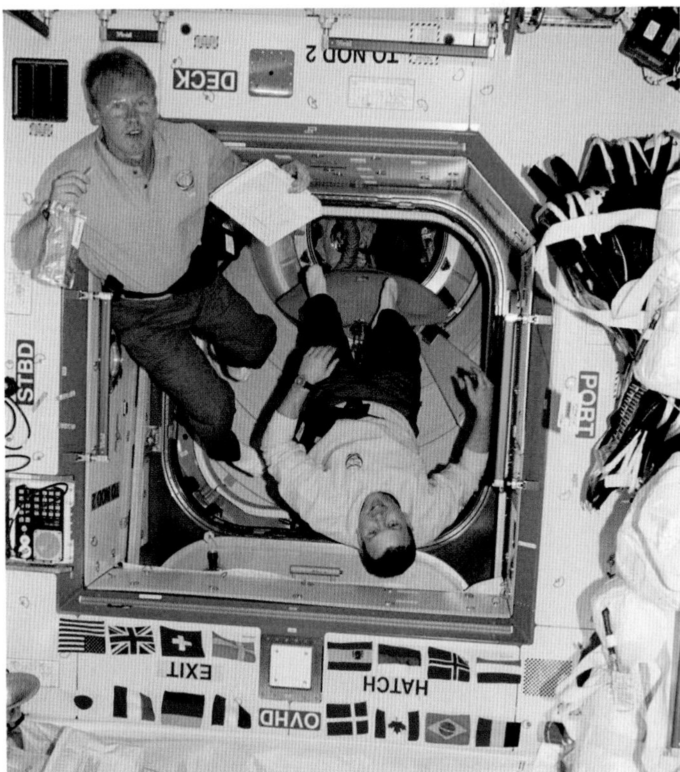

Mission no. **104**

STS

100

Orbiter	Endeavour
Launch	April 19, 2001
Landing	May 1, 2001
Duration	11 days, 21 hrs., 30 mins., 1 sec.

The STS-100 and Expedition 2 crews aboard the ISS: (clockwise from bottom middle) Jeff Ashby, Scott Parazynski, John Phillips, Chris Hadfield, Susan Helms, Kent Rominger, Yury Lonchakov, Yury Usachov, Umberto Guidoni, and Jim Voss

Crew	Kent V. "Rommel" Rominger, Jeffrey S. Ashby, Chris A. Hadfield, Scott E. Parazynski, John L. Phillips, Umberto Guidoni, Yury Lonchakov

Mission: *Endeavour*'s astronauts delivered Canadarm2, the robot arm for the ISS, and used the Raffaello MPLM to deliver 6,000 pounds (2,722 kilograms) of cargo and experiments. Two spacewalks helped activate Canadarm2, and the two crews overcame a Station computer crash to regain Canadarm2 and ISS control.

Kent Rominger, Commander:

In the sequence of Station construction, every mission added a puzzle piece. STS-100, or 6A, was very important because we were delivering the Canadarm2, the robotic arm needed for the rest of the assembly sequence. We also had a logistics module, Raffaello, full of three tons (2.7 metric tons) of spares, electronics, parts, and Station supplies.

As commander, I took time to sit back and think, "What's important? How do we really ensure this mission is successful?" Once we docked with the Station, the crew would be not just seven astronauts, but ten, and it would be important that we worked well together.

I went to both US and Russian mission control centers in advance and asked that they build an hour and a half into our schedule for a meal on the first evening we were together as a crew. I decided to make this a special meal: a luau. So we flew Hawaiian shirts for everybody, including the Station crew.

That first evening together, the meal provided a great bonding opportunity on the Station. The luau atmosphere in the service module gave us a chance to get to know each other on a more personal level and added to our excitement following the hatch opening.

That dinner paid dividends later in the mission when the Station had computer problems. The control center had uploaded new software for controlling the new Canadarm2 to give it its full capability. A problem in that upload crashed the Space Station's command and control computers; the Station was seriously crippled. With the computers down, not only could the ISS not command the arm, but it also couldn't communicate with the ground, and we couldn't release the latches to unberth the Raffaello logistics module. Even undocking the shuttle would require an emergency undock, which can be done only by firing pyrotechnics and leaving part of the space shuttle docking system attached to the ISS. For a time, I was thinking, "This is a big deal."

Using communications routed through the shuttle, Mission Control in Houston started working their magic. To get the Station back up and running, we cannibalized non-critical computers to restore those command-and-control machines that were in the critical communications path. We worked very well as an integrated crew getting those Station computers replaced, and I looked back and thought, "That luau was one of the smartest things I did."

On the first spacewalk, Chris Hadfield suffered eye irritation from the defog chemical applied to the helmet's inner surface. The mix of the defog and water leaking from his drink bag got into his eye and blinded him temporarily. The crew and Mission Control worked through this challenge together with great results. The flight controllers had Chris vent oxygen overboard from his helmet purge valve to reduce the eye irritation.

While working in the MPLM, we set up a bungee net—we called it a trampoline—so that when we were unloading cargo bags, we could stash the empty bags behind the bungees to temporarily restrain them. Those bungees created a great trampoline surface: You could come shooting into the MPLM, crash up against the bungees, and rebound back out.

I look back and take great pride in how well the teams worked together on the ISS and how well the crew worked with the control centers to pull off this complicated mission.

ott Parazynkski passes the ffaello cargo module as he nsfers a spare Direct Current ritching Unit to the ISS.

Mission no. **105**

STS

104

Orbiter	Atlantis
Launch	July 12, 2001
Landing	July 24, 2001
Duration	12 days, 18 hrs., 34 mins., 56 secs.

The STS-104 and Expedition 2 crews in the Quest airlock: Jim Reilly, Janet Kavandi, Steve Lindsey, Charlie Hobaugh (front), Jim Voss, Yury Usachov, Mike Gernhardt, and Susan Helms

Crew	Steven Lindsey, Charles O. "Scorch" Hobaugh, Michael Gernhardt, James F. "J. R." Reilly, Janet Kavandi

Mission: *Atlantis*'s crew delivered and activated the Quest airlock at the ISS, and performed three spacewalks to connect the new airlock to Station systems and install high-pressure gas tanks to resupply spacesuits. Inside, the joint crew overcame water and air leaks to put Quest into service.

Steve Lindsey, Commander:

Our job on STS-104 was to bring up and install the airlock, outfit it, then perform the first spacewalk from the ISS. We did our first two EVAs out of the shuttle airlock. We then did the last EVA out of the Quest airlock, "the New World."

Our mission saw the first "operational" use of the Space Station arm. EVA 1 was a hatches-closed EVA, with the ISS crew operating the Station arm, and us flying the shuttle arm. We trained for that task as much as we could, but it was still challenging. The dual-arm operations were not a good idea because of our small crew: we had Scorch managing the spacewalk, Mike Gernhardt and Jim Reilly outside on the EVAs, Janet Kavandi operating the arm, and I was doing all the camera work, providing views not just to Janet, but also to the ISS crew. It was complicated.

As a first-time commander, I hadn't flown the shuttle much on orbit or landed it. So my first memorable experience was coming up and manually flying the orbiter to a docking with the ISS. That wasn't particularly difficult, but it was a big mission milestone. That experience of docking, equalizing pressure, opening hatches, and transferring people from one vehicle to another in space was unique.

The Station impressed me as a lot quieter compared to the shuttle—slower, less frenzied. On a shuttle mission you just go, go, go—an all-out sprint. On ISS, the crew was busy; they were working hard. But they were pacing themselves as if running a marathon.

I worked hard to develop a "one-crew" relationship between Expedition 2 and ourselves, and I think it paid off. We went out to dinner together, and we planned a lot of events with them outside training—as much as we could before they flew—because I knew we would have to work together as one crew to be able to pull off this mission. They were few and we were few, and we faced dual-arm operations, complex EVAs, and no room for error. Although they owned the Station and we the shuttle, we tackled the airlock assembly tasks together. There were no "I'm in charge of this—No, we're in charge of this" arguments.

What amazed me during the EVAs to install Quest was how well the airlock and the oxygen and nitrogen tank installations went. I was particularly worried about the tank installations. The arm had to maneuver each tank up, get it in position, and then our EVA crew had to take the tank from the arm by hand—they were 1,200 pounds (544 kilograms) each—and berth and connect them manually. These were almost one-way operations: If the crew couldn't get the tanks berthed on the airlock, then returning them to the payload bay would have proved extremely difficult. Mike and J. R. had helped develop all those tank-to-airlock interfaces to ensure the berthing guide design could handle the tight installation tolerances. That tank task ended up going very smoothly.

If we had failed in those early assembly missions, it would have been a disaster for the Space Station program. My satisfaction came from watching the Station team and my own shuttle team come together and accomplish something that was intense and incredibly difficult, yet succeeding at every task.

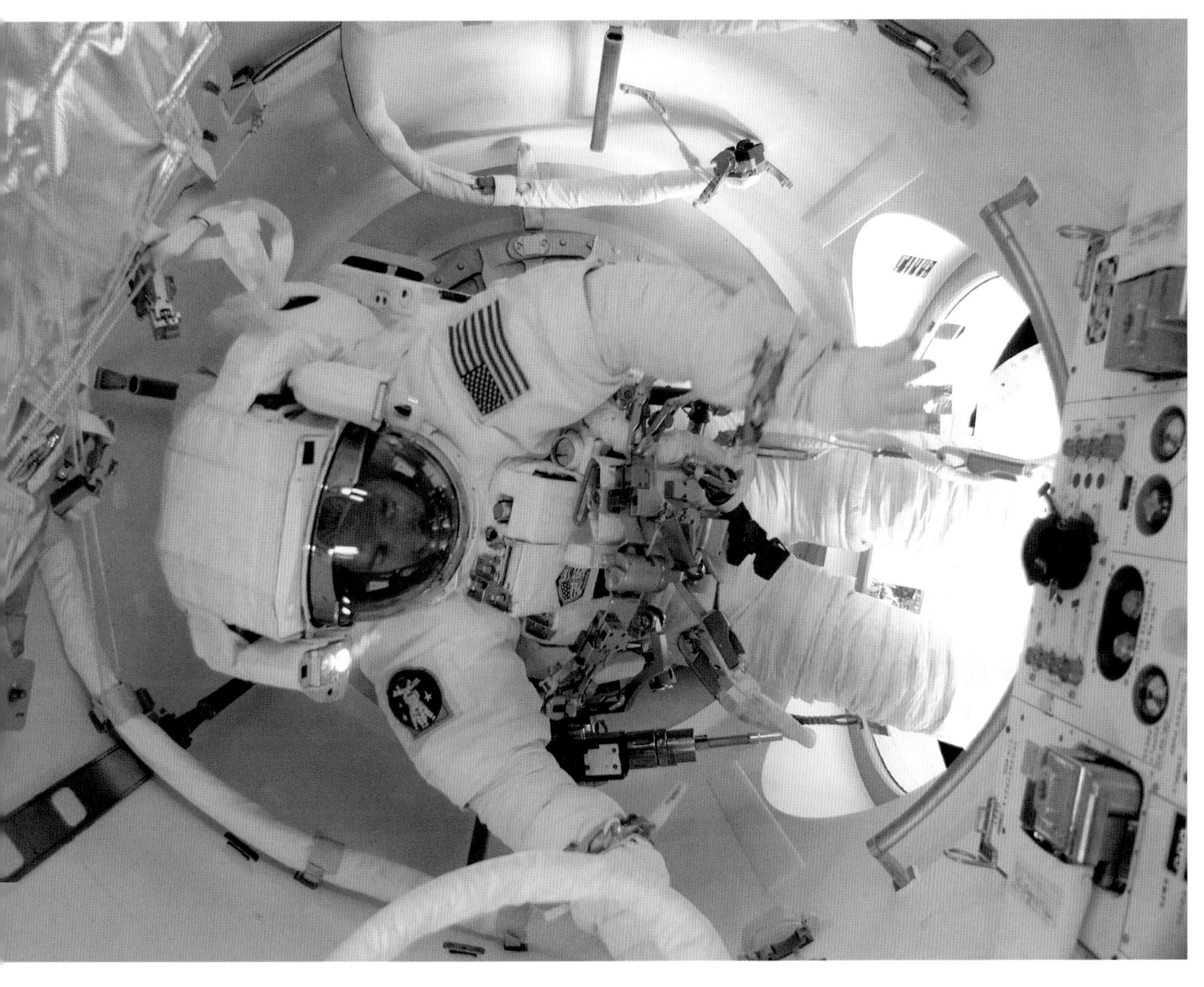

ABOVE: **Jim Reilly eases boots-first out of the Quest airlock hatch.**

BELOW LEFT: **The Quest airlock in the grip of Canadarm2, headed for berthing on Unity.**

BELOW RIGHT: ***Atlantis* departs the ISS with Quest delivered, the payload bay empty.**

Mission no. **106**

STS 105

Orbiter	Discovery
Launch	August 10, 2001
Landing	August 22, 2001
Duration	11 days, 21 hrs., 12 mins., 45 secs.

The STS-105 crew in Destiny: (clockwise from top) Pat Forrester, Dan Barry, Scott Horowitz, and C. J. Sturckow

Crew	Scott J. Horowitz, Frederick W. "C. J." Sturckow, Daniel T. Barry, Patrick G. Forrester Up: Frank L. Culbertson, Jr., Vladimir N. Dezhurov, Mikhail Tyurin Down: Susan J. Helms, Yury V. Usachov, James S. Voss

Mission: STS-105 exchanged Expedition 3 for Expedition 2. *Discovery*'s astronauts resupplied the ISS with 7,000 pounds (3,175 kilograms) of cargo using the Leonardo MPLM, prepared the Station for truss installation, and attached ammonia coolant tanks to the ISS exterior.

Pat Forrester, Mission Specialist:

STS-105 was a crew rotation flight: Expedition 2 went up on STS-102, and we brought them back on -105. They were the first ISS crew to go up and down on the shuttle. We were also a logistics flight, added by program managers to the original manifest between assembly flights 7A and 8A. Designated as 7A.1, we were called a "dot" flight.

From the time we were assigned until the time of our scheduled launch was not much more than six months. That meant an intense training load for the four of us on the flight deck. We did very little training with Expedition 3 (Frank Culbertson, Vladimir Dezhurov, Mikhail Tyurin) beyond emergency training at Johnson Space Center. They then joined us at Kennedy for countdown rehearsal and launch.

I remember being out at the Beach House, a NASA property where astronauts can gather privately with their families, for a pre-launch barbecue dinner when my phone rang. It was Jim Voss, calling from orbit. That was the first time I'd ever received a call from the Space Station. Jim was just calling to say "Hi" and that he was looking forward to seeing us. I thought, "Hey, this is real. We're part of a team, and some of that team is already in space."

In orbit, I slept on the flight deck along with Doc Horowitz. He had his sleeping bag across the pilot and commander's seats, and mine was strung across the aft flight deck. One morning I woke up early. My head was right by the overhead window, and I slid the sunshade cover aside to peek out. I saw we were headed right over the Mediterranean, coming across the Holy Land. It was suc an amazing sight first thing in th morning—the cradle of humanity where the entire redemptive stor played out.

During one of our spacewalks, we hung two Materials ISS Expe iments (MISSEs) outside. This w the first time we had placed an e ternal experiment on the Station and we installed one on the end the airlock. When I spacewalked alongside the structure of the IS I felt safe, as though I was being cradled by that structure. But there was something about bein out on the end of the airlock tha made me feel like I could fall off. Working out there at the very en I thought, "Hey, if I miss a handr I'm gone!"

For me, coming back from space was a whole lot harder th going. I was never sick in orbit, but coming back, I didn't feel we After landing, we finished with the medical staff, shared a meal in crew quarters, and were then released to stay in Cocoa Beach with our families. The next morning, I planned to see my parents for the first time since launch, bu my vestibular system was still messed up.

My wife, Diana, helped get me to breakfast with my parents, bu while we ate I felt that the room was spinning the entire time. When we returned to the beachfront condo, I felt proud of myse for getting to breakfast and back About that time, I opened the ba door, looked out from the balcon —and there was Doc in his gym clothes, running down the beach That was a humbling moment.

Worse moments were to come. We landed on August 22, 2001. Ours was the last flight before 9/1

ABOVE: **Above *Discovery* and a dusky Earth, Dan Barry leads an EVA.**

LEFT: **Using a Hasselblad camera, Pat Forrester photographed a docked *Discovery* above a sunlit Earth.**

BELOW: **Forrester installs a MISSE space exposure package.**

Mission no. **107**

STS 108

Orbiter	Endeavour
Launch	December 5, 2001
Landing	December 17, 2001
Duration	11 days, 19 hrs., 35 mins., 44 secs.

The STS-108 and Expedition 3 and 4 crews in Zvezda: (front row) Yuri Onufriyenko, Mikhail Tyurin, Dan Tani, Linda Godwin, Mark Kelly, and Frank Culbertson; (back row) Carl Walz, Dan Bursch, Dom Gorie, and Vladimir Dezhurov

Crew	Dominic L. P. Gorie, Mark E. Kelly, Linda M. Godwin, Daniel M. Tani Up: Yuri I. Onufriyenko, Carl E. Walz, Daniel W. Bursch Down: Frank L. Culbertson, Jr., Mikhail Tyurin, Vladimir N. Dezhurov

Mission: *Endeavour*'s STS-108 flight exchanged Expedition 4 for Expedition 3. Astronauts berthed the Raffaello logistics module at the ISS, transferring three tons (2.7 metric tons) of supplies to the station. Spacewalkers performed maintenance on the ISS solar array gimbal mechanisms The crew also deployed the Starshine 2 satellite.

Dan Tani, Mission Specialist:

We were the first flight after 9/11. The morning of September 11, we were in the Johnson Space Center's Building 9, using the Full Fuselage Trainer to rehearse the post-insertion phase, where we convert the rocket into an orbiting spaceship. The TV was on, and we watched all the awful real-time video from New York and Washington, DC. That was three months before we launched on December 5.

September 11 changed quite a few things, including the security status at the Kennedy Space Center. There were discussions about whether a shoulder-fired, surface-to-air missile could be launched from the causeway connecting Merritt Island to Cape Canaveral and hit the shuttle. It turns out that it cannot. We had armed guards during quarantine. If we wanted to go out for a run, there was an armed escort in a car following us down the road. This was comforting in a way, but also pretty alarming. And on launch day, our Astrovan's escort helicopter was doing low passes over the road, dipping below fifty feet (fifteen meters) about three hundred feet (ninety-one meters) in front of us. We joked that the biggest security risk we had was that damn Huey running into us.

Our EVA was added relatively late. Two of the Sun-tracking gimbals for the solar arrays had been acting up, and engineers designed some insulation blankets to wrap around them. Linda Godwin and I were tasked to go wrap these two beta gimbals up on the P6 truss. That was a fun task to perform.

Mark Kelly was driving the arm, and while riding it up to the work site, our job was just to hold on. The arm moves slowly—no great forces involved. I was just trying to make sure that I maintained decent body position so that I wasn't kicking equipment or my partner on the way. I had this fantastic view looking at all the structure we'd brought up to low Earth orbit, now orbiting Earth. Now I was a part of that scene as I moved from the shuttle up to P6. Those were the fun EVA moments when I was simply riding instead of doing, and I tried to take in as much as I could.

Being in space is so overwhelming that it's easy to lose track of where you are, or what you're doing, to lose your situational awareness. We call that disorientation "space brain." My first task in orbit just at the end of post-insertion was to set up the middeck computer network with three or four laptops. I had to hook up all the network power and cabling; back then it was a wired network with ethernet cards that plugged into the laptops. We went out to *Endeavour* the day before launch and inspected all that network wiring. It was all there, pre-positioned and taped down in place on the middeck wall.

In orbit, though, I completely forgot that all that cabling was already routed. From the flight deck locker, I dug out all the stuff I needed—cables, cards, everything—and went down to the middeck with my computers. I have an embarrassing photo of me with cables around my arm, holding all this equipment, getting ready to set up the network. I got down there, and I thought, "What? What was I thinking?" I didn't need any of that stuff. That's what I mean by "space brain."

It was a good, funny lesson that cost no one anything except a little embarrassment and time. So for my second flight, I wrote down absolutely everything I wanted to do after launch: "Take off helmet, put helmet in bag, put gloves in bag. . . ."

ABOVE: **NASA's UH-1 helicopter flies low and close to escort the Astrovan enroute to Pad 39B.**

RIGHT: **The ISS glides above Miami, Florida, and the Bahamas' Andros Island.**

BELOW: **Dan Tani sends birthday greetings to his wife, Jane, while on an EVA.**

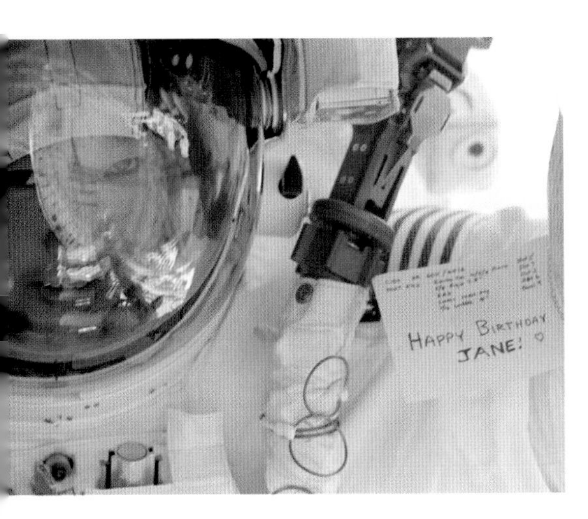

Mission no. **108**

STS

109

Orbiter	Columbia
Launch	March 1, 2002
Landing	March 12, 2002
Duration	10 days, 22 hrs., 9 mins., 51 secs.

The STS-109 crew on the middeck: (front row) Nancy Currie, Scott Altman, and Duane Carey; (back row) John Grunsfeld, Rick Linnehan, Jim Newman, and Mike Massimino

Crew	Scott D. Altman, Duane G. Carey, John M. Grunsfeld, Nancy J. Currie, James H. Newman, Richard M. Linnehan, Michael J. Massimino

Mission: On the fourth servicing mission to the Hubble Space Telescope, STS-109 astronauts installed advanced cameras, added cooling capability, and replaced both solar arrays. Two astronaut spacewalking teams completed five EVAs lasting a record thirty-five hours, fifty-five minutes.

Mike Massimino, Mission Specialist:

On STS-109, HST Servicing Mission 3B, we had to replace the solar arrays, install a cooling unit, and replace the Power Control Unit (PCU). Changing out the PCU was a hugely important task, because controllers actually powered down the telescope during its replacement. We also had to install extra thermal covers and sun shades to protect Hubble from temperature extremes and glare in orbit.

As on all the Hubble-repair missions, the plan was just to launch, do a rendezvous, grapple Hubble, spacewalk, spacewalk, then get rid of it, and come home. We didn't do much of anything else. On our first spacewalk, I was on the arm, moving up toward the telescope to catch the old solar panel that had already been folded up. I was getting high over the payload bay, and saw the wing of the shuttle, the American flag, the NASA meatball, and Earth in the background. It was an overwhelming view. You don't get that scenery in the Neutral Buoyancy Lab!

We were a bit higher on this mission than the Station—about 350 miles (567 kilometers) up. That gave us a different perspective on Earth, seeing a little more of its curvature. We were seeing the planet, through the suit visor, without the window in the way. It wasn't what I thought it would be. Earth isn't a protected place. I thought, "It's a planet, man! We're out here speeding through the cosmos, in the middle of all its extremes and dangers."

Replacing the Power Control Unit was the job everyone was most worried about. That PCU was a little scary—it had thirty-eight electrical connectors, some underneath, and all of them hard to reach. I think the telescope team hoped they would never have to change it out. But the reality was that it was failing, and we needed to replace it.

We were concerned at having to de-mate and mate all those connectors and weren't sure whether we could get them off and then get the new ones back on. We couldn't undo these big cannon connectors with our gloved hands—that wouldn't work. Instead, we had this special wrench that the team invented that gave us a lot more torque. The engineers redesigned the replacement Power Control Unit so that its connector panel slanted out at an angle, giving us a better look at what we were doing—a big improvement over the original, difficult geometry. Fortunately, that PCU task ended up being much easier to perform than we had thought. Not that it was easy, but John Grunsfeld and Rick Linnehan didn't have any real problems. It went as written.

On EVA 4, Jim Newman and I installed the Advanced Camera for Surveys. That instrument was used by Nobel Prize winner Adam Riess to help confirm the accelerating universe theory and investigate the existence of dark energy.

I was surprised at how prepared I was to do the work. Steve Smith, who'd been to Hubble, lived around the corner, and he was my mentor. He came around to visit the night before quarantine to see how I was doing. He told me, "You're trained and ready to go. Maybe you don't think you are, but you're ready to go. Remember—it's an open-book test. If you have any questions, just ask." That was pretty good advice.

LEFT: ***Columbia* erupts from Pad 39A at dawn to pierce the cloud deck overhead.**

ABOVE: **Nancy Currie runs the RMS from her perch on a work platform on *Columbia*'s flight deck.**

BELOW: **Mike Massimino (bottom) and Jim Newman replace a worn Hubble reaction wheel.**

Mission no. **109**

STS

110

Orbiter	Atlantis
Launch	April 8, 2002
Landing	April 19, 2002
Duration	10 days, 19 hrs., 42 mins., 39 secs.

The STS-110 and Expedition 4 crews dressed for a "rodeo" dinner on the ISS: (front row) Ellen Ochoa, Mike Bloomfield, and Yuri Onufriyenko; (middle row) Dan Bursch, Rex Walheim, and Carl Walz; (back row) Steve Frick, Jerry Ross, Lee Morin, and Steve Smith

Crew	Michael J. Bloomfield, Stephen N. Frick, Jerry L. Ross, Steven L. Smith, Ellen Ochoa, Lee M. E. Morin, Rex J. Walheim

Mission: *Atlantis*'s crew delivered the central S0 truss to its berth atop the Destiny module, where spacewalking teams anchored it permanently into place and connected electrical, data, and cooling lines. Inside, astronauts transferred cargo and experiments to ISS.

Mike Bloomfield, Commander:

Over the twenty years that we built the ISS, one of the essential things we needed to do was use the Station's own resources to help with its construction. STS-110 was the first mission to do that.

We were dependent on the ISS crew—Carl Walz, Dan Bursch, and Yuri Onufriyenko—for two reasons. First, we were going to do our spacewalks out of the ISS airlock. We were the first crew to use that airlock to build ISS.

The second resource we relied on was the Station's robotic arm. We handed off the S0 truss from the shuttle arm to the Station arm. That ISS arm actually put the truss in place, and we also used it to support the spacewalks, with some of the spacewalkers working out there on the end of the arm. After our mission, crews used the ISS arm and the ISS airlock to support every assembly EVA.

On ascent, when we accelerated through Mach 12, the vehicle rolled from "heads down" to "heads up." Looking out the left window from the commander's seat while heads down, I saw the Atlantic Ocean. But after we rolled heads up, I was looking left at the East Coast, and the first thing I saw was Cape Hatteras. When I was in the Air Force, I flew out of Langley, Virginia and was very familiar with that section of coastline. Looking down at Hatteras and the Outer Banks, I remembered that three years earlier I was down there at Kitty Hawk a hundred years after the Wright brothers had jumped off those sand dunes. I thought, "Now here we are—the human race—going through Mach 13 on the way to Mach 25. We've got a 40,000-pound (18,144-kilogram) payload in the back, a crew of seven, and we're going to visit three people who've been living in space for five months." I tried to comprehend the pace of that progress, that in just a hundred years, people had gone from jumping off the sand dunes to this technological achievement of expansive human spaceflight.

One night, we invited the ISS crew over to the middeck for dinner. Dinners are a fun time to get everyone together, get to know one another, chat about what happened that day, and talk about the next. Carl was a big rodeo fan and had missed that year's Houston Livestock Show and Rodeo, so we decided to take the rodeo to them. We brought up barbecue, some red bandanas to wear, country western music, and lemonade. For the space shuttle drink menu, lemonade was about as country western as we could get.

I was filling the lemonade-mix bags with cold water at the galley and Yuri was across the middeck from me. I took a lemonade, said, "Hey, Yuri!" and floated the drink bag through the air to him. As it got closer and closer, Yuri was all anticipation and focus. I'll never forget the look on his face when he finally reached out and grabbed it. His eyes widened—it was cold! On the Station they had no way at the time to chill things—no refrigerator for cooling drinks or food. Yuri, instead of drinking his cold lemonade, deliberately spent the next fifteen minutes just pressing the cold foil sack to his neck, his back, and his face, enjoying a sensation he apparently hadn't felt in about five months.

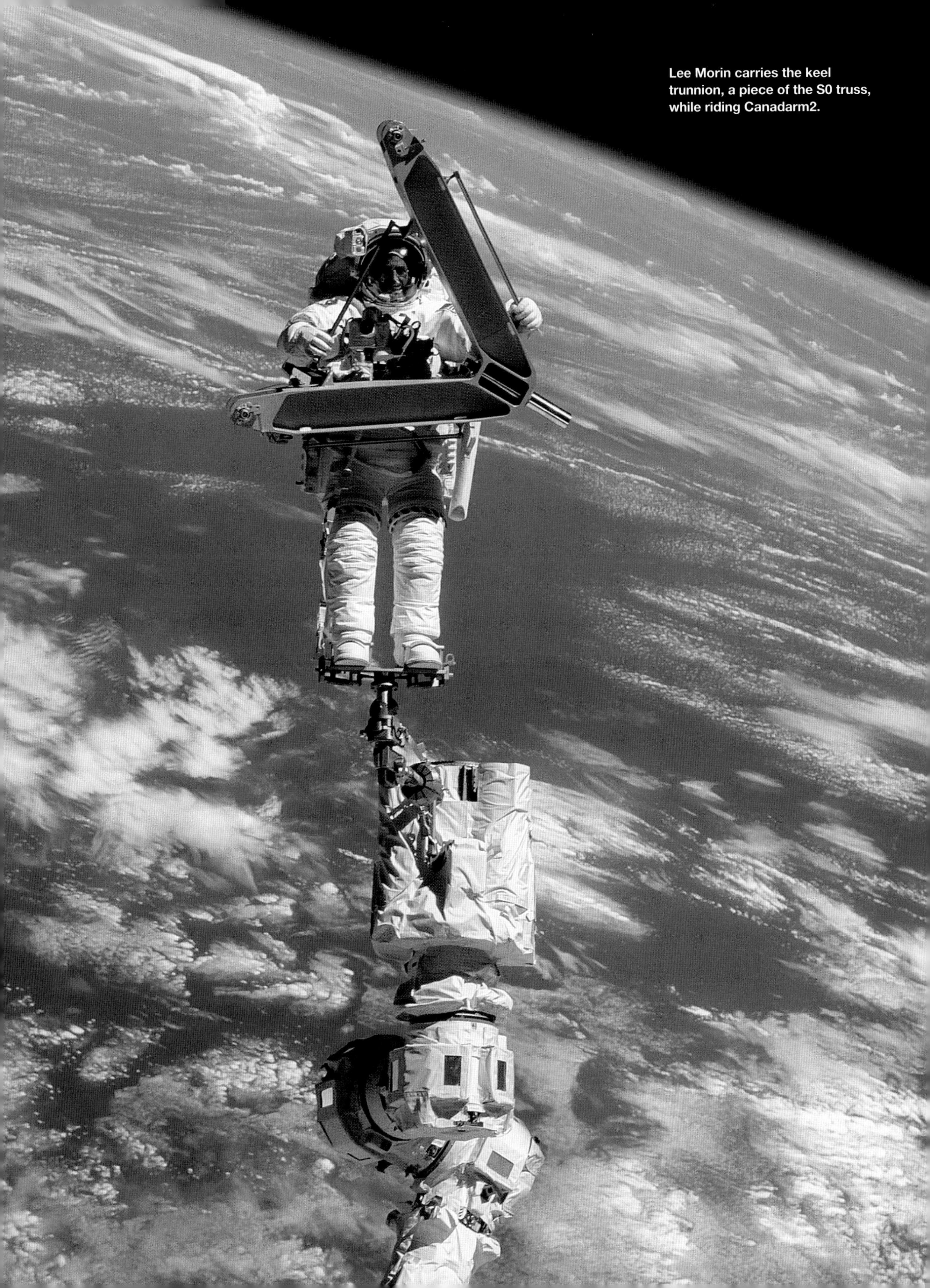

Lee Morin carries the keel trunnion, a piece of the S0 truss, while riding Canadarm2.

Mission no. **110**

STS 111

Orbiter	Endeavour
Launch	June 5, 2002
Landing	June 19, 2002
Duration	13 days, 20 hrs., 34 mins., 53 secs.

The STS-111 and Expedition 4 and 5 crews in Destiny: (front row) Yuri Onufriyenko, Franklin Chang-Diaz, Ken Cockrell, and Valery Korzun; (middle row) Dan Bursch and Paul Lockhart (back row) Carl Walz, Philippe Perrin, Sergey Treshchov, and Peggy Whitson

Crew	Kenneth D. "Taco" Cockrell, Paul S. Lockhart, Franklin Chang-Diaz, Philippe "Pepe" Perrin Up: Valery G. Korzun, Peggy A. Whitson, Sergey Y. Treschov Down: Yuri I. Onufriyenko, Daniel W. Bursch, Carl E. Walz

Mission: STS-111's astronauts installed the Mobile Base System, giving full mobility to the Canadarm2, repaired the arm's wrist joint, and transferred more than 8,000 pounds (3,629 kilograms) of supplies to the ISS.

Paul Lockhart, Pilot:

STS-111 was a blue-collar type of mission, a typical Space Station build-out flight. We brought up Expedition 5, carried home Expedition 4, ferried supplies, and performed three spacewalks to further construction of the ISS. Although we weren't bringing up any major pieces of the Station, STS-111 was representative of the grinding construction work that had to get done.

We had two rookie spacewalkers, Franklin Chang-Diaz and Phillipe Perrin. Philippe—"Pepe"—a French military test pilot, had adapted well to NASA and brought along extensive flight experience. Franklin, one of NASA's most experienced astronauts, had not yet performed a spacewalk, but had a wealth of spaceflight knowledge. Before each spacewalk, I conducted a walk-through briefing with them, much as I had done for Air Force military test flights. I believe that it added structure to the mission's complex spacewalks and that it helped meld Phillipe and Franklin into a strong EVA team. We completed three very complex spacewalks, one an unscheduled repair of the Station robotic arm, added to the flight plan just weeks before launch.

Reentry was thrilling. It is difficult to simulate the dynamic, intricate re-entry of the orbiter, especially in the final phases of approach and landing. Reentry is a ballet that ends in a mad rush of switch throws, computer checks, and hands-on-stick actions to bring the vehicle down safely to the runway.

I recall clearly the graceful "ballet rolls" the shuttle performed to stay on its desired trajectory. As the orbiter rolled left, I watched Earth fill the commander's windo while outside my window was th blackness of space. When the or biter rolled back right to maintain course, we had the opposite view Beautiful Earth filled my window, and I saw the ground and cloud formations far below.

Approaching Hawaii, the vehicl rolled again to the right. The clouds that had been far below u were now much closer, and I was suddenly aware of our extreme speed. Our Mach number becam visceral as I watched the clouds whiz by. Around then, I also bega feeling the onset of light g-forces which pushed me down into my cockpit seat and made it difficult to see over the glareshield.

I felt that I was behind on every action until I put my hands on the control stick. As soon as I took control of *Endeavour*, everything seemed natural and right. The vehicle came alive as we sped through Mach 5 into the lower atmosphere, and the orbiter transitioned into a glider. Descending around the heading alignment cone to line up on the runway, *Endeavour* flew beautifu ly; as we rolled out on final, Taco took command, and I became his cheerleader and apprentice, feeding him needed information and confirming the vehicle was configured correctly for landing. Taco landed beautifully, and soo we were "Wheels Stop" on the runway.

After the last switch was throw to safe the vehicle, I was free to savor the feeling that all our tasks were complete, done as expertly as we knew how. Our hands were dirty, but we got all the work don we put our tools away, and we came back. Then we all went out to Domingo's in Boron, California and had a beer.

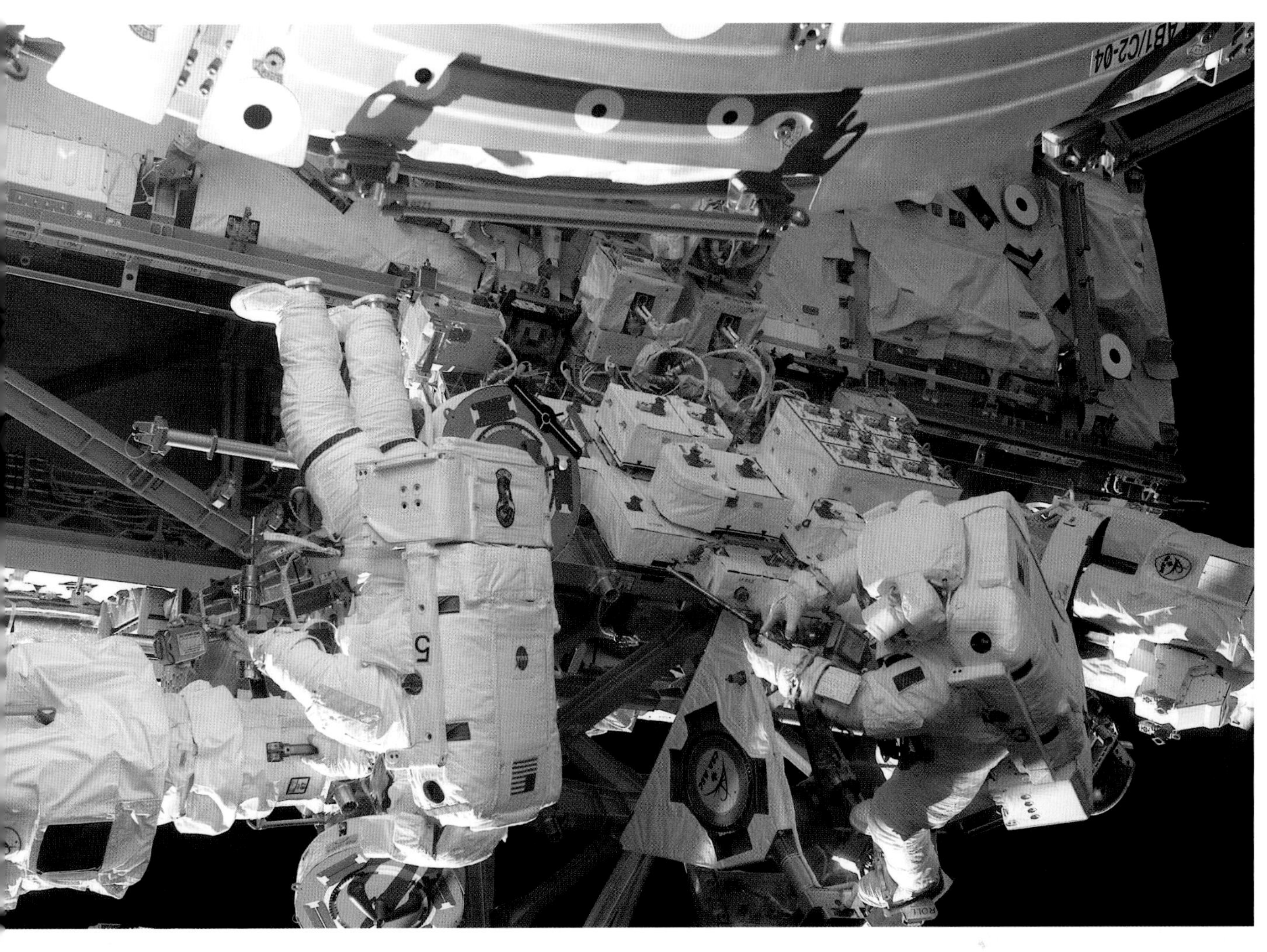

ABOVE: **Franklin Chang-Diaz and Philippe Perrin work to install the Mobile Base System on the S0 truss.**

BELOW LEFT: ***Endeavour* docked at Destiny's forward port, PMA-2, with the S0 truss at left.**

BELOW RIGHT: **Paul Lockhart working from *Endeavour*'s pilot seat on the flight deck.**

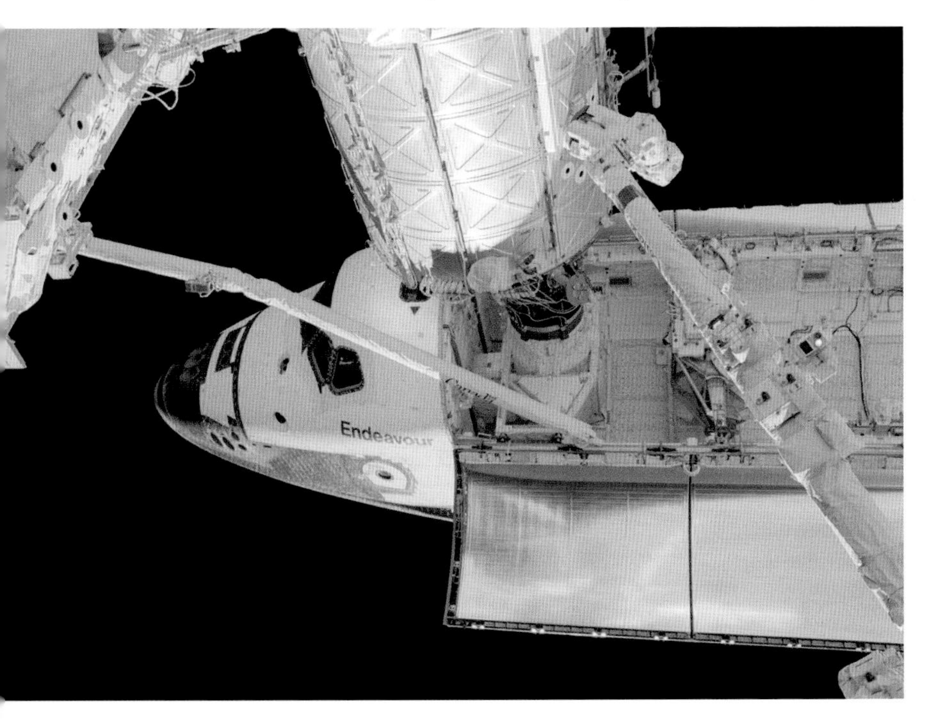

Mission no. **111**

STS 112

Orbiter	Atlantis
Launch	October 7, 2002
Landing	October 18, 2002
Duration	10 days, 19 hrs., 57 mins., 50 secs.

The STS-112 and Expedition 5 crews: (front row) Peggy Whitson, Valery Korzun, and Sergey Treshchov; (back row) Dave Wolf, Sandy Magnus, Pam Melroy, Jeff Ashby, Piers Sellers, and Fyodor Yurchikhin

Crew	Jeffrey S. Ashby, Pamela A. Melroy, David A. Wolf, Piers Sellers, Sandra H. Magnus, Fyodor N. Yurchikhin

Mission: At the ISS, the crew installed the sixteen-ton (14.5-metric ton) S1 truss and performed three spacewalks to install a mobile work platform. During launch, *Atlantis*'s external tank shed insulating foam that struck the attach ring on the left SRB, presaging the damage that would later doom *Columbia*.

Sandy Magnus, Mission Specialist:

We were an early part of the whole "Build-the-Truss" series of flights. We installed the S1 truss, put a bunch of big cameras on it, and deployed the starboard radiator.

I was Mission Specialist 2, assigned to logistics and robotics. My focus was on the Station arm because my role was to pull the truss out of the shuttle payload bay and get it over to the Space Station and lock it in for the EVA guys.

When we got to orbit and had main engine cutoff, my seatbelt was still on, but I was floating. My butt wasn't touching the chair. I let go of my ascent checklist, and it just stayed there and floated. I started giggling because it was so fun and unusual and magical. Pam Melroy and Jeff Ashby started laughing. They said, "Oh, we love flying with rookies, because you never know how they're going to react." I didn't giggle for long. I realized, "Oh, wait. I gotta go to work now."

I opened the payload bay doors, and there's Earth. I looked out and said, "Oh my God! Our atmosphere is so thin!" That's my longest-lasting impression of Earth: It's beautiful, but wow, it's fragile. You know how when you blow on dandelion fluff, it just scatters away into the breeze? That was exactly my impression of the atmosphere. Somebody could come along and blow on it, and it would just fritter away.

You could argue that we dodged a bullet because the foam piece that happened to hit our SRB came from the same tank location as the piece that broke free and struck *Columbia*'s wing. You'd think that a piece of foam couldn't do any damage to the wing leading edge, but it dented metal on our mission. Our foam incident should have raised som eyebrows, but NASA continued believe that foam loss just could pose a serious safety risk.

Once in orbit, we learned abou another close call. One of the redundant circuits on bolts holdi down the SRBs failed. In this system, there are two circuits th fire pyrotechnics to split the nuts on the hold-down bolts in half. Breaking the nuts apart frees the bolts, so the shuttle doesn't rip itself free of the launchpad when the boosters light.

One of those circuits failed on one of the SRBs. The back-up worked, and the nuts split, but with only a single string firing. If those nuts don't blow and the bolts rip the booster skirts off as the shuttle leaves the pad, we might have lost all SRB steering, and then the shuttle would have gone out of control. A range safe officer would have then had to hi the destruct button. When they got that information up to us, I looked at Ashby and said, "Huh. don't think that's something I'll te my mother." That incident worrie us more than the foam strike on the booster.

Despite those worries, it was really nice to fly for the first time. That first flight assignment seem to take forever to arrive. It's myst rious, and nobody quite knows how it happens. But I'd wanted t be an astronaut from around mid dle-school age, so I was just real happy I could finally fulfill my role

LEFT: ***Atlantis* approaches the ISS for docking, the S1 truss filling the payload bay.**

ABOVE: **Peggy Whitson, Sandy Magnus, and Pam Melroy aboard the ISS in 2002.**

BELOW: **Dave Wolf on Canadarm2 carries a TV camera for installation on the S1 truss.**

Mission no. **112**

STS

113

Orbiter	Endeavour
Launch	November 23, 2002
Landing	December 7, 2002
Duration	13 days, 18 hrs., 47 mins., 26 secs.

The STS-113 and Expedition 5 and 6 crews in Destiny: (front) Ken Bowersox, Jim Wetherbee, and Valery Korzun; (middle) Don Pettit, John Herrington, Mike Lopez-Alegria, and Peggy Whitson; (back) Nikolai Budarin, Paul Lockhart, and Sergey Treshchov

Crew	James D. "Wxb" Wetherbee, Paul S. Lockhart, Michael Lopez-Alegria, John B. Herrington Up: Kenneth D. "Sox" Bowersox, Nikolai M. Budarin, Donald R. Pettit Down: Valery G. Korzun, Peggy A. Whitson, Sergey Y. Treshchov

Mission: *Endeavour*'s crew exchanged Expedition 6 for Expedition 5. The astronauts delivered the P1 truss to the ISS, and their spacewalks helped activate P1 and outfit it for future expansion. STS-113 was the last mission before *Columbia*'s loss in 2003.

John Herrington, Mission Specialist:

STS-113's primary mission was to take three people to space and bring three people home. Our secondary mission was a major assembly task: installing the P1 truss. P1 has an ammonia cooling system for the Station, with three radiators, a huge ammonia coolant tank, and the Thermal Radiator Rotary Joint (TRRJ) on which the radiator section rotated.

We launched at night and we jettisoned the ET at night, so we couldn't take pictures of it because we couldn't see it. So we don't know if foam popped off the bipod ramp but missed us. I look back on it and think that we dodged a bullet, launching between STS-112, when foam struck an SRB, and *Columbia*, when foam loss caused a disaster.

We were the last crew to perform a Station crew rotation in combination with a major assembly task. And for a long time, we were the last mission to fly at night. After *Columbia*, shuttle program managers wanted to be able to image the vehicle going uphill.

Wxb pulled P1 out of the payload bay with the shuttle arm, and handed it off to Peggy Whitson, who used the Station arm to move it up into place. Once the ISS crew got the ready-to-latch lights, Don Pettit, running the computer, used motorized bolt assemblies to latch P1 in place. We spacewalkers couldn't go and hang out on the truss until that connection was made.

On the third EVA, while we were in the airlock, the mobile transporter was moving down the front of the truss and got stuck. The cameras couldn't see it, and no one could figure out what had happened. Mission Control finall told me to open the hatch and g inspect it.

The whole time while translating out there, I was thinking, "Please don't let me find a tethe there marked 'John Herrington, EV 2,' tangled in the transporter. I climbed up on top of the mobile base system and looked: An electronics box on the transport had bumped into the ultra high frequency (UHF) antenna on the truss. We had not deployed the UHF antenna yet; Mission Contr had told us to hold off until EVA I gave a sigh of relief that it wasn me! Then I said, "Houston, I foun the problem." I couldn't believe I actually said that.

Jerry Ross had told me that sometime during my spacewalk should pause, look out, and take it all in. I was on the back of the P1 truss, at the far left end wher the P3/P4 truss would go, hangi by a thumb and forefinger onto nothing but those little protrudin alignment cones.

I looked out over the limb of Ear to the vastness of the universe. I couldn't see the shuttle; I couldn see anybody. For the first time in my life, there was nothing betwe me and whatever else was out there. Hanging on out there, I fel like a rock climber on the ultimat cliff—220 miles (354 kilometers) straight down. I had goosebump

All of us enjoyed seeing the IS when we first arrived, this compl combination of modules floating the vastness of space. When we left, the ISS was different—we'd added to it. We participated in th amazing engineering achieveme that has been crewed since 200 If I were asked to point to a majo accomplishment in my life, it's that. I'm very proud I got to take very small part in the constructio of the ISS and do well what I wa trained to do.

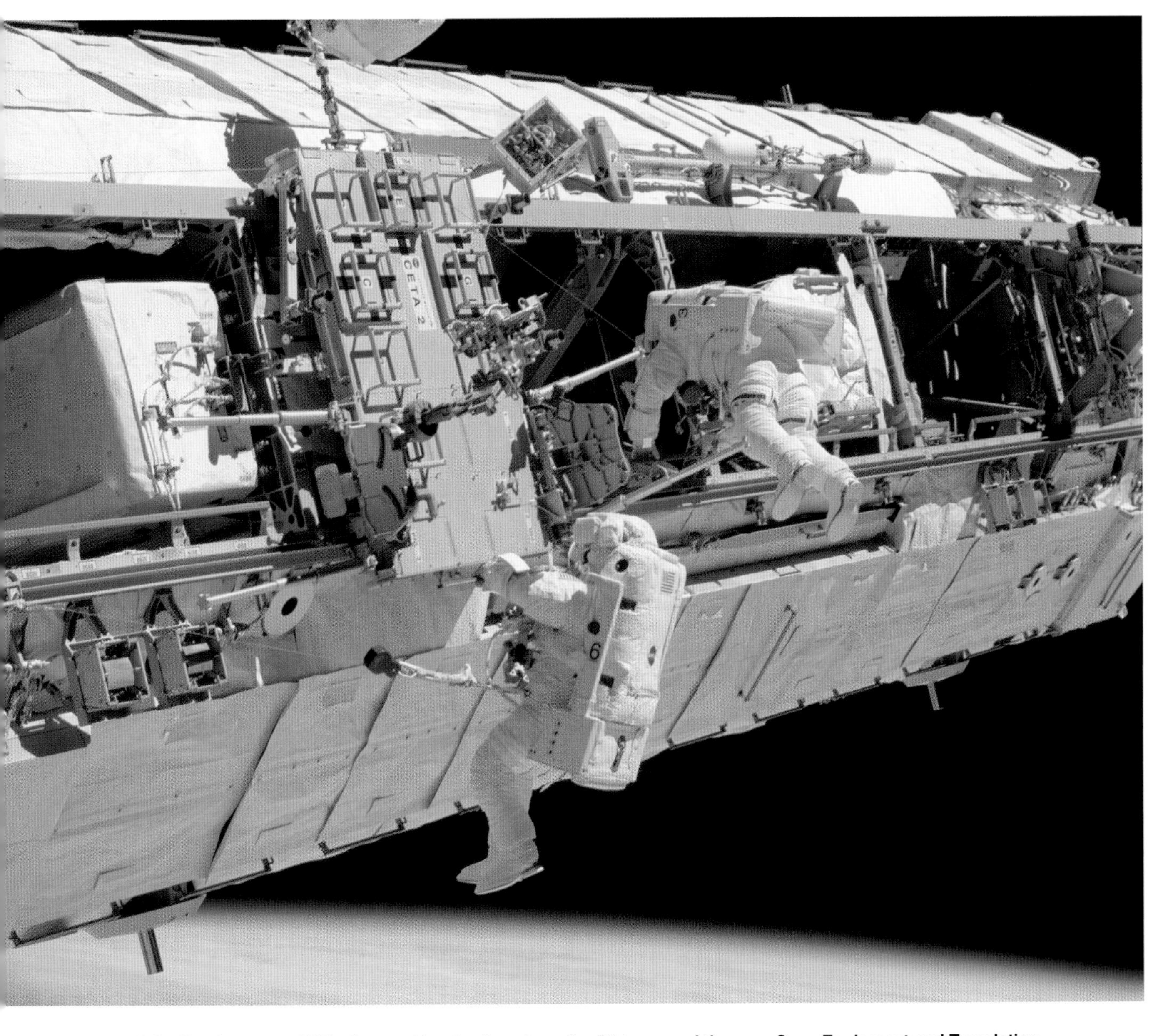

ABOVE: **John Herrington and Mike Lopez-Alegria at work on the P1 truss and the gray Crew Equipment and Translation Aid, a mobile cart.**

BELOW LEFT: **Herrington working on the P1 truss with *Endeavour*'s two star trackers visible just above the cockpit windows.**

BELOW RIGHT: **The newly installed P1 truss on the right in this view of the station, below the sunlight glint on the P6 solar array.**

Mission no. **113**

STS 107

Orbiter	Columbia
Launch	January 16, 2003
Landing	Lost during reentry, February 1, 2003
Duration	15 days, 22 hrs., 20 mins., 32 secs.
Crew	Rick D. Husband, William C. McCool, Michael P. Anderson, Kalpana Chawla, David M. Brown, Laurel B. S. Clark, Ilan Ramon

Mission: STS-107's crew undertook a microgravity and space research mission using the SPACEHAB module. *Columbia*'s left wing was damaged during ascent by a piece of external tank foam insulation. Reentry plasma penetrated the wing, causing structural failure and the loss of the orbiter and its crew.

Author's Note: Near the end of her *Columbia* mission, Laurel Clark wrote an email home to family and friends. Her thoughts represent the exuberant spirit of this close-knit crew.

Laurel Clark, Mission Specialist:

Hello from above our magnificent planet Earth. The perspective is truly awe-inspiring. This is a terrific mission, and we are very busy doing science 'round the clock. Just getting a moment to type email is precious, so this will be short, and distributed to many whom I know and love.

I have seen some incredible sights: lightning spreading over the Pacific; the aurora australis lighting up the visible horizon with the city glow of Australia below; the vast plains of Africa and the dunes on Cape Horn; rivers breaking through tall mountain passes; the scars of humanity; the continuous line of life extending from North America, through Central America and into South America; a crescent Moon setting over the limb of our blue planet. Mount Fuji looks like a small bump from up here, but it does stand out as a very distinct landmark.

Every orbit we go over a slightly different part of the Earth. Of course, much of the time I'm working back in SPACEHAB and don't see any of it. Whenever I do get to look out, it is glorious. Even the stars have a special brightness. I have seen my "friend" Orion several times.

I feel blessed to be here representing our country and carrying out the research of scientists around the world. All of the experiments have accomplished most of their goals despite the inevitable hiccups that occur in such a complicated undertaking. Some experiments have even done extra science. A few are finished and one is just getting started today.

The food is great, and I am feeling very comfortable in this new, totally different environment. It still takes a while to eat as gravity doesn't help pull food down your esophagus. It is also a constant challenge to stay adequately hydrated. Since our body fluids are shifted toward our heads, our sense of thirst is almost non-existent.

Thanks to many of you who have supported me and my adventures throughout the years. This was definitely one to beat all. I hope you could feel the positive energy that beamed to the whole planet as we glided over our shared Earth.

Love to all,

Laurel

The STS-107 astronauts exit crew quarters to board *Columbia*: (front) Kalpana Chawla, Willie McCool, Rick Husband, (back) Ilan Ramon, Mike Anderson, Laurel Clark, and Dave Brown.

Shuttle engineers lay out *Columbia's* wreckage on a hangar floor at Kennedy Space Center to assist the accident investigation.

Mission no. **114**

STS

114

Orbiter	Discovery
Launch	July 26, 2005
Landing	August 9, 2005
Duration	13 days, 21 hrs., 32 mins., 23 secs.

The STS-114 crew inside Destiny: (front row) Andy Thomas, Eileen Collins, and Soichi Noguchi; (back row) Jim Kelly, Charlie Camarda, Steve Robinson, and Wendy Lawrence

Crew	Eileen M. Collins, James M. "Vegas" Kelly, Charles J. Camarda, Wendy B. Lawrence, Soichi Noguchi, Stephen K. Robinson, Andrew S. W. Thomas

Mission: During ascent, *Discovery*'s external tank shed a one-pound (half-a-kilogram) chunk of foam; it missed the orbiter. After docking, the crew resupplied the ISS from the Raffaello logistics module. Spacewalkers tested heat shield-repair techniques and removed two small gap-fillers protruding from heat-shield tiles.

Steve Robinson, Mission Specialist:

When the *Columbia* accident occurred, we were thirty days from launch for a completely different mission, a Space Station construction flight. Post-*Columbia*, we became part of the restart of the space shuttle program, with many experimental techniques and tests of new equipment on the flight. NASA was relearning how to fly in space.

The core of the crew worked together for four-and-a-half years. That gave us a chance to try to recover from the loss of our friends, to recover our confidence in the NASA decision-making process, and to recover confidence in our own ability to recognize when something was about to go wrong.

Columbia was damaged because a big piece of foam came off the tank and hit the wing. All the foam folks worked very hard for a couple of years to make sure that wouldn't happen again; they were quite confident that it wouldn't—except that it did.

On Flight Day 2, Mission Control sent us a video showing another great big piece of foam coming off the tank and just missing *Discovery*. That video was shocking. With foam loss happening twice in a row, we thought for a while that we were the final shuttle mission. NASA conducted another year of mitigation work before the shuttle flew again.

We were the first to flip upside-down, pitching the nose up and around to expose the underside to the ISS cameras, a maneuver that became standard after that. Sure enough, the ISS photographs showed that there was something wrong with the heat shield. It wasn't a divot—it was something sticking out, something we'd never discussed before.

There were two thin gap fillers sticking out: ceramic cloth strips that prevent hot gas from flowing between the tiles. They had to be removed to give us a safe heat shield. So flight controllers came up with a Station robot arm trajectory that no one had flown before. Normally that would have taken months, but they developed it in about two days.

I don't think we would have burned up if we had not removed those protruding gap fillers, but we didn't have confidence that we wouldn't. One of the two protrusions had me worried that during entry it could generate a hot plasma vortex that might have swept up from the orbiter chine—the narrow, shelf-like extension of the fuselage, blending into the main wing—and right across the wing leading edge. The leading edge already had to endure higher temperatures than the tile-covered belly, and I was worried that its RCC panels would fail.

On the third EVA, Vegas and Wendy, both operating the Station arm, took me right where I needed to be. For quite a long time I could see nothing made by humans. I couldn't see any spacecraft, and couldn't see Earth; I just faced out into infinity. How many people get to do that?

Anchored by my feet with my hands free, seeing the slightly different colors of the stars, and not being distracted by the features of a very complicated spacecraft, I sensed I was an inhabitant of the universe, with nothing between it and me except my helmet's plexiglass. I'll never forget that.

I did retrieve the two gap fillers, clearing *Discovery* for reentry.

A last thought about STS-114: We believed that we flew our landing not only for us, but for the crew of *Columbia*. They didn't get to land. After wheels stop, we all felt a very strong connection with our friends.

ABOVE: ***Discovery*** **executes a nose-up "flip" so the ISS crew can inspect the orbiter's heat shield for any ascent damage.**

RIGHT: **Steve Robinson riding Canadarm2 to retrieve protruding gap fillers from *Discovery*'s heat shield.**

BELOW: **Eileen Collins, STS-114 commander, in the Zvezda service module at ISS.**

Mission no. **115**

STS

121

Orbiter	Discovery
Launch	July 4, 2006
Landing	July 17, 2006
Duration	12 days, 18 hrs., 36 mins., 47 secs.

The STS-121 crew (clockwise from bottom left) Mike Fossum, Lisa Nowak, Steve Lindsey, Mark Kelly, Stephanie Wilson, Piers Sellers, and Thomas Reiter

Crew	Steven W. Lindsey, Mark E. Kelly, Stephanie D. Wilson, Michael E. Fossum, Piers J. Sellers, Lisa M. Nowak Up: Thomas Reiter

Mission: *Discovery*'s crew used the orbiter boom sensor system with advanced imagers to examine the shuttle's heat shield, nose, and wings, and transferred 7,400 pounds (3,357 kilograms) of supplies to ISS. Spacewalkers performed Station repairs and tested application of a new tile repair putty.

Stephanie Wilson, Mission Specialist:

STS-121 was the second return-to-flight test mission after the 2003 *Columbia* accident. We tested a tile-repair material called NOAX and two kinds of RCC repair techniques to use if the orbiter's wing leading edge had been damaged, as it had been on *Columbia*.

We tested the shuttle arm grappled to the orbiter boom sensor system as a platform to support a spacewalker doing repair work. We put Piers, sometimes along with Mike, on the end of that platform to test its deflection and to see if it would be stable enough for repair work. Along with supplies for the Station, we brought up the ESA's Thomas Reiter to increase its crew size from two to three.

I was one of the robotic arm operators for both the shuttle's and the Station's arm. On Flight Day 2, we made a robotic inspection of the shuttle wing leading edges and the nose cap. To get the boom sensor system pointed at the nose cap, the arm elbow joint had to come very close to the crew cabin between the overhead and aft windows. The joint approached within two feet (half a meter) of the orbiter, just too close for my comfort, and I actually paused the auto sequence because it was not clear that it was moving away. We had quite an intense discussion with the ground, who assured us that the joint was about to move away from us. It did, but it put us on edge because a collision might have damaged both the orbiter tiles and the arm structure.

Mike Fossum, Mission Specialist:

We were out on the pad a long time for a hot, mid-afternoon launch. But I was so at peace with where we were that I took a nap, until they started rousing us about 20 or 30 minutes before launch. "Okay, time for everybody to get ready." We were lying there on our backs on the flight deck, looking up, when suddenly this big shadow swept across the cockpit. Either Steve Lindsey or Mark Kelly said, "What the heck was that?" It was the shadow of a vulture circling over the launch pad. Someone said, "That ain't a good sign." We all laughed—nervously.

The first spacewalk tested whether we could use the orbiter boom sensing system as a repair platform. Piers first made a solo run, then both he and I were lifted out of the payload bay; I was in the boom's foot restraint with Piers tethered alongside. When we were well clear of structure and facing out at deep space, my job was to excite structural modes with the biggest swaying and bouncing motions I could muster. It was horrifying to feel the resulting oscillations, but the engineers on the ground were happy. Later we were brought close to the Station exterior where I simulated heat shield repair actions with a force gauge.

At one point I had to egress the foot restraint to reposition it, then attempt to get my boots secured back into the footplate. Even though I was tethered, my heart was pounding as I slid my feet out, then pulled myself around, looking down a hundred feet of the boom's spindly stick at the orbiter and Earth below. Mission Control asked, "Mike, how's it going out there?" "Oh, it's going fine," I said. The doctors in the back row of Mission Control were laughing because they could see my heart rate. But that test gave us really good data that made us comfortable that if needed, with just a little learning curve, we could make repairs on the orbiter from the boom.

ABOVE: **Piers Sellers and Mike Fossum (end of arm) rode *Discovery*'s inspection boom to assess its stability as a work platform.**

BELOW: **An ISS view of high-altitude noctilucent clouds, seen just before dawn.**

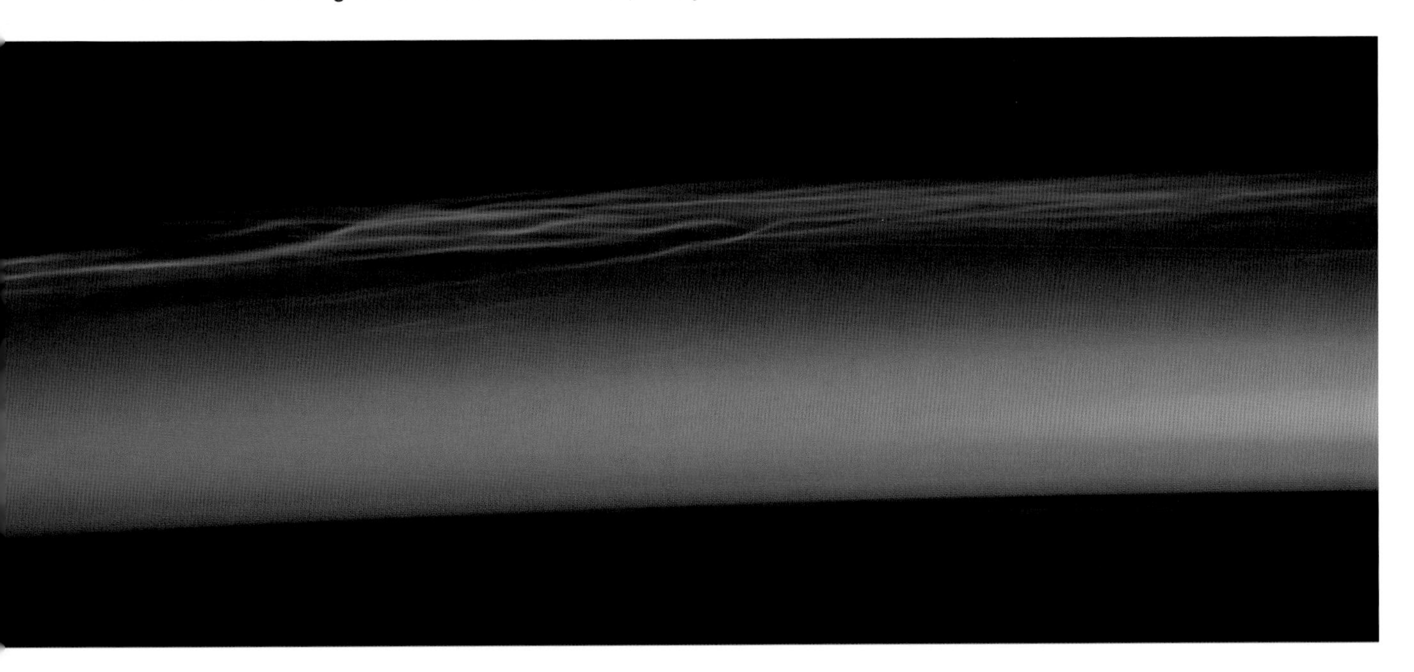

Mission no. **116**

STS

115

Orbiter	Atlantis
Launch	September 9, 2006
Landing	September 21, 2006
Duration	11 days, 19 hrs., 6 mins., 28 secs.

The STS-115 and Expedition 13 crews in the Destiny lab: (front row) Thomas Reiter, Pavel Vinogradov, and Jeff Williams; (middle row) Joe Tanner, Heide Stefanyshyn-Piper, Brent Jett; (back row) Chris Ferguson, Dan Burbank, and Steve MacLean

Crew	Brent W. Jett Jr., Christopher J. Ferguson, Steven G. MacLean, Daniel C. Burbank, Joseph R. Tanner, Heidemarie M. Stefanyshyn-Piper

Mission: STS-115 delivered the 17.5-ton (15.9-metric ton) P3/P4 truss, solar arrays and batteries to the ISS. Two teams completed three spacewalks to help extend the solar arrays and perform minor ISS repairs. An orbital debris strike on an *Atlantis* radiator nearly punctured a freon cooling line.

Heide Stefanyshyn-Piper, Mission Specialist:

We were the first flight after *Columbia* that was a true Station assembly mission. We brought up the P3/P4 truss that carried the next set of solar arrays to the Station. We conducted three EVAs: On the first two, we removed locks on the truss that protected the truss and solar arrays from vibration during launch. After releasing the locks, we were able to activate the Solar Array Rotary Joint (SARJ), the rotating bearing that enables solar arrays to continually track the Sun. We needed two spacewalking teams for that urgent pair of back-to-back EVAs.

My path to this mission was a long one, and in the beginning it didn't look as though I would ever fly in space. I remember Apollo 11's impact on me: My mother had bought my older brother a T-shirt printed with "The Eagle Has Landed," but I didn't get anything because in 1969, six-year-old girls wore dresses, not T-shirts. Six years later, during the 1975 Apollo-Soyuz Test Project, I saw a reporter interviewing a cosmonaut on TV. Because I'm Ukrainian-American I could understand bits and pieces of what the Russian was saying, and I thought, "Well, I could be an astronaut, and if we do things with the Russians, I can talk to them."

As a US Navy diver, I spoke to a fellow officer who had applied to the astronaut program, and I learned that you don't have to be a pilot. I thought, "Okay, I've got an engineering degree and I fix ships underwater." In 1990, I started looking into it more and more and thought, "I could do this."

I was in Astronaut Group 16, hired in 1996, with thirty-five US astronauts and nine international Our nickname was the Sardines. With the slowed Station assembl schedule and then *Columbia*'s loss, ten years passed between the time I showed up as a candidate to the time I flew in 2006. And I wasn't the last one in my class to fly.

At the start of the first EVA, I got outside, swapped my tether point from the airlock to the outside tether reel, and as I starte moving, Joe Tanner asked me, "So, how do your first two minute out in open space feel?" I looked around and said, "It does feel a lo like the Neutral Buoyancy Lab," that enormous water tank used fo astronaut training in Houston. "B here," I said, "there are no bubble and no divers."

In diving, it's very hard to get things moving, and you're always having to work against buoyancy and gravity. In space without thos effects, it's very easy to move around. But without all that water around to damp your motions, every little perturbation will be transmitted into your suit and body. It took a little time to realize how much effort I needed to put i when moving outside with that bi suit on, and it does take continuing muscle input.

I don't think I could have asked for a better crew; we had been together for four and a half years. After we undocked from the Station, we realized, "We did it. We got up here and in six crammed days at the Station, we got all of our work done." After landing, it was almost sad. We were togethe for so long, and we were not goin to be in the same crew office anymore.

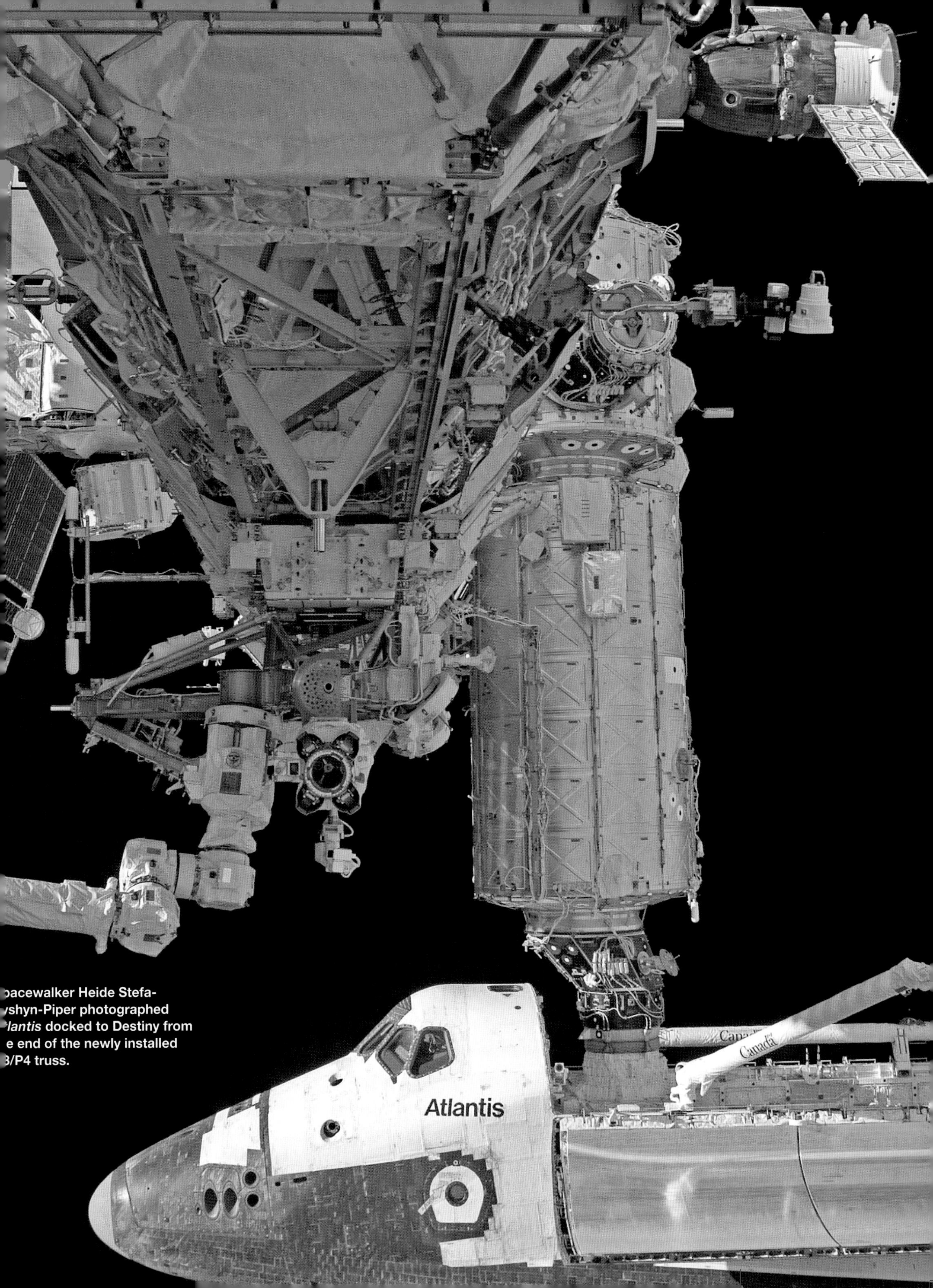

**ɔacewalker Heide Stefa-
ʋshyn-Piper photographed
ʼlantis docked to Destiny from
e end of the newly installed
3/P4 truss.**

Mission no. **117**

STS

116

Orbiter	Discovery
Launch	December 9, 2006
Landing	December 22, 2006
Duration	12 days, 20 hrs., 44 mins., 23 secs.

The STS-116 crew at Kennedy Space Center before launch: Joan Higginbotham, Bill Oefelein, Bob Curbeam, Christer Fuglesang, Nick Patrick, Suni Williams, and Mark Polansky

Crew	Mark L. Polansky, William A. Oefelein, Nicholas J. M. Patrick, Robert L. Curbeam, Jr., Christer Fuglesang, Joan E. Higginbotham Up: Sunita L. "Suni" Williams Down: Thomas Reiter

Mission: STS-116 exchanged Suni Williams for Thomas Reiter. *Discovery*'s crew installed the P5 truss and routed power from the new P3/P4 arrays into the ISS. Retraction of the P6 port solar wing initially failed due to kinking and billowing in the retracting panels. A manual assist from the EVA team finally tamed the balky segments.

Bob Curbeam, Mission Specialist: On STS-116, not only did we bring up the P5 truss, but we also worked outside to rewire the Station to its permanent electrical configuration. That was fun. We did it just as you'd move a lamp. Turn it off, unplug it, move it, plug it into another power source, then turn it on. We turned off half the Station, unplugged it from one solar array, and switched it to its new power source. Then we turned off the other half of the Station and did the same over there. We also had the pleasure of bringing Suni Williams up to the ISS on her first space flight.

We had a crew with five rookies and two experienced astronauts. We didn't know how the first-timers were going to feel that first day. I told them, "Just move slowly and do what you can. Concentrate on your task and get it done. Don't rush; inevitably, that's when you make mistakes. If we get behind, we get behind. But don't try to do things as fast as you would do them on Earth." I think the rookies did a very good job on day one. All of them adapted very well and just knocked out the work with little interference from Space Adaptation Syndrome.

During our work, we faced a solar array anomaly: The port-side P6 solar array didn't retract properly. There are little guidewires that run the length of the array and pass through grommets that are on each accordion-like segment of the array. The P6 guidewire was frayed with strands sticking out, so the wire wouldn't slide through the grommet. That's why the array was getting hung up.

On the third EVA, we tried to just shake the array as it came in, but that didn't work. Mission Control did a wonderful job over a couple of nights figuring out what tools we had and how we could configure them to allow me to manipulate the array, but not touch it. You don't want to touch 200 volts or so while you're in a pure oxygen spacesuit atmosphere. We took two different tools, Velcroed them together, then rolled Kapton tape over the whole length to keep me from electrocuting myself.

I rode the arm, and they took me out as far as they could so I could help guide the array in. Some of those array segments were just folding the wrong way, so I used one tool—we called it the hockey stick—to manipulate the panels and make sure they all folded the right way as they came in. Luckily, the guide wires were frayed in a way that the panels flowed in pretty well. But I thought "This isn't going to be too pretty coming out again."

Back inside after that last spacewalk, I had a tremendous feeling of accomplishment at knowing that we dealt with all that adversity and still met all the major mission objectives.

Near the end of the mission, I kept thinking, "Okay, this is the last time I'm going to see Earth from this perspective. This is the last crew I'm ever going to be on. This is the last deorbit burn, the last time I look back to see that really tight plasma trail behind us." I'd done some really wild things in space, and when those last moments came, I didn't realize I would feel that emotional.

LEFT: ***Discovery*'s brilliant exhaust lit the night sky after liftoff from Pad 39B.**

ABOVE: **The P6 solar array jammed and buckled as it retracted into the silver blanket box at right.**

BELOW: **Bob Curbeam shepherds the balky P6 array into the blanket box.**

Mission no. **118**

STS 117

Orbiter	Atlantis
Launch	June 8, 2007
Landing	June 22, 2007
Duration	13 days, 20 hrs., 11 mins., 33 secs.

The STS-117 crew in Destiny: (front) C. J. Sturckow, Lee Archambault, and Pat Forrester; (middle) Danny Olivas and Jim Reilly; (back) Clay Anderson and Steve Swanson

Crew	Frederick W. "C. J." Sturckow, Lee J. "Bru" Archambault, Patrick G. Forrester, Steven R. Swanson, John D. "Danny" Olivas, James F. "J. R." Reilly Up: Clayton C. Anderson Down: Sunita L. "Suni" Williams

Mission: *Atlantis*'s crew exchanged Clay Anderson for Suni Williams, and delivered the 35,677-pound (16,183-kilogram) S3/S4 truss section. Two spacewalking teams performed four EVAs to activate the truss, retract the starboard side of the balky P6 solar wing, and reattach a loose OMS pod thermal blanket.

Danny Olivas, Mission Specialist:

Challenges with our STS-117 mission started with the shuttle roll-out to the pad in March 2007. A freak hailstorm rolled over the pad and pelted the external tank with golf-ball-sized hail, damaging the tank's foam insulation. NASA, post-*Columbia*, agonized for weeks over whether or not they should repair the foam in place. The managers finally said, "Let's just roll back to the assembly building." Bru Archambault and C. J. Sturckow said, "Boy, if we just had 'gone ugly early,' we could have rolled this vehicle back to the pad weeks ago and been two weeks ahead."

The next STS-117 challenge was what we spotted after we opened the payload bay doors in orbit. There was damage to the left OMS pod thermal protection, and we didn't know what we were going to be able to do about it. We sent imagery to the ground and continued the mission. Once docked to the Station, we attached S3/S4, but the ISS lost three command and control computers on the Russian segment, affecting our ability to point efficiently at the Sun. The resulting electrical problems shut down half our power. That failure sent the flight plan right out the window.

Sizing up the damaged OMS blanket, Mission Control worried about reentry plasma intrusion affecting the pod's propellant tanks. After being selected to perform the EVA repair, I worried about causing more damage than the thermal blanket had already sustained. As soon as I laid hands on it, I noticed these little flecks of white material flaking off into space, and I knew I'd have to limit how much I touched that blanke I carefully pinned the leading edg down using wire brads, or pins, which had been developed for tile repair. Because the fasteners lacked stiffness, I had to swiftly jam the brads into the tile with or motion. I then used a surgical sta pler to anchor the blanket's side edges to the adjoining blankets. We hoped this fix would be good enough to get us home.

After we undocked from the ISS for the trip home, we gained an extra half day in orbit as we waited for better return weather. J. R. Reilly and I decided to spen it stargazing during a night pass. Over our heads, we saw more stars than I could ever imagine. Then, off the shuttle's right wing, saw two clouds in space. I pulle out the binoculars and peered into what I later learned were the Magellanic Clouds. With the Larg Magellanic Cloud in my field of view, I saw more stars clustered there than I could see out the overhead orbiter windows. I felt an overwhelming sense of my insignificance—not just my own but that of humanity itself. We are truly on this little-bitty speck out the middle of nowhere, and there so much more out there.

STS-117 was a testament to what makes NASA great. Despite the problems and failures, NASA did everything in such a way that—to the rest of the world—it looked like business as usual. We just worked the issues, from deployin the solar arrays and getting the P6 array retracted to stapling the blanket down and solving that Russian computer glitch. Nothing to me says "NASA" more than tha It doesn't matter how wide your eyes get when you see the failure, you just treat the problem as it is and work the issue.

ABOVE: **STS-117's exhaust plume sculpted by upper atmospheric winds near sunset.**

RIGHT: **Danny Olivas refastens the loosened thermal blanket on *Atlantis*'s port OMS pod.**

BELOW: **The forward edge of the OMS thermal blanket was lifted by aerodynamic loads during ascent.**

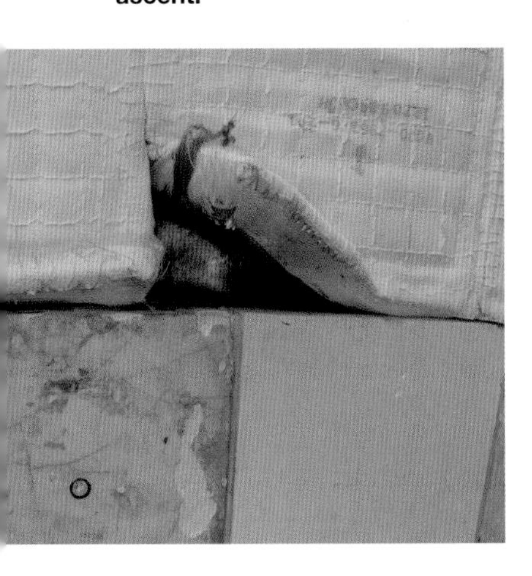

Mission no. **119**

STS 118

Orbiter	Endeavour
Launch	August 8, 2007
Landing	August 21, 2007
Duration	12 days, 17 hrs., 55 mins., 35 secs.

The STS-118 crew in Destiny: Alvin Drew, Charlie Hobaugh, Barbara Morgan, Tracy Caldwell Dyson, Rick Mastracchio, Scott Kelly, and Dave Williams

Crew	Scott J. Kelly, Charles O. "Scorch" Hobaugh, Tracy E. Caldwell Dyson, Richard A. Mastracchio, Dafydd R. Williams, Barbara R. Morgan, B. Alvin "Al" Drew

Mission: *Endeavour*'s crew installed the S5 segment at the ISS starboard truss, and transferred more than 5,000 pounds (2,268 kilograms) of cargo from SPACEHAB. Spacewalkers connected S5 to Station systems. A belly tile was damaged by foam impact but didn't warrant orbital repair.

Barbara Morgan, Mission Specialist:

NASA's theme for the mission was "Build the Station, Build the Future." Our construction mission to the ISS would install the S5 truss, to enable the next mission to install the final set of starboard solar arrays. We also brought up and installed an external stowage platform on the P3 truss, and replaced a broken gyro on the Z1 truss.

The "Build the Future" part of the mission was NASA, after many years, continuing its commitment to the earlier Teacher in Space program—a commitment to the education community, to teachers, and to students—to honor of the legacy of Christa McAuliffe and her *Challenger* crewmates.

Once, on orbit, I looked out the hatch window while we were on the nighttime side of the planet, and saw the port truss and its solar arrays. They were almost invisible, but I could see they were there because they were silhouetted against zillions of stars. Suddenly, in the distance, I watched a thin, bright blue curved line appear, and then another thin blue arc—a different hue—rose beneath it. Those two rose together and a third appeared beneath them. We were coming up on dawn and Earth's daytime side, and those thin blue lines were our planet's thin atmosphere. Then the solar arrays caught the sunlight and started to glow. Those blue layers multiplied—I think I saw about thirty, each with a different hue—and the solar arrays got brighter and brighter until they became a brilliant Inca gold against the black of space. They looked like golden sails.

I wasn't touching anything, just floating, trying to absorb this experience. There we were, 250 miles (402 kilometers) above the ocean, gliding along on this glowing, golden sailing ship. I looked again at the horizon, spotting jus above it a bright crescent Moon Although the laws of physics would make it impossible, I thought for a moment that all we had to do was push hard right o a tiller, and those sails would tak us straight to that Moon.

My favorite part of "Build the Future" was taking ten million basil seeds to orbit. We wanted something that would engage an challenge students, to literally pu a seed in their hands that would get them thinking and exploring and experimenting. Teachers wo receive these seeds along with packages of control seeds that h stayed on Earth. When the kids g the seeds, they would experimen with them in growth chambers th had designed and built. The seed served as both metaphor and reality for students, helping them understand that they don't have to wait until they are finished with graduate school to contribute to NASA's missions. There are thing that they can do right now.

I'm very proud of that whole program. We reached more than two million students through mo than 150,000 teachers.

We've all grown herbs over the years and know the wonderful aroma you get when you snap the stem of a basil plant. Al Drew pulled out from a locker a package that contained one million seeds, filming it so the kids woul know their seeds had really been in space, in microgravity. When he pulled out that package, the entire orbiter middeck immediate filled with an amazing basil arom Laughing, Al turned to me and said, "Barb, I could kill for some lasagna!"

Education is ongoing; every generation of students has its own future to work toward. There is nothing more important than space in keeping the future open for our kids.

ABOVE: **A spacewalker photographed *Endeavour* docked with the ISS at PMA-2 and Destiny.**

LEFT: **Barbara Morgan at work during robotic operations on *Endeavour*'s aft flight deck.**

BELOW: **Ascent debris gouged these two tiles, but they protected *Endeavour* through reentry.**

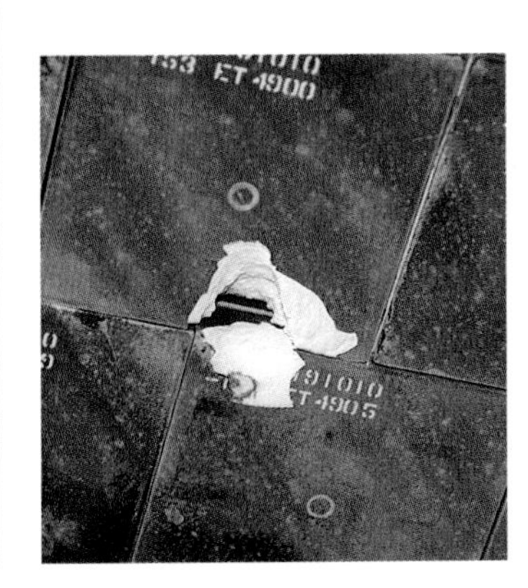

Mission no. **120**

STS

120

Orbiter	Discovery
Launch	October 23, 2007
Landing	November 7, 2007
Duration	15 days, 2 hrs., 22 mins., 58 secs.

The STS-120 and Expedition 16 crews in Harmony: (bottom row) Clay Anderson, Peggy Whitson, Yuri Malenchenko, and George Zamka; (middle row) Stephanie Wilson, Pam Melroy, and Paolo Nespoli; (top row) Dan Tani, Scott Parazynski, and Doug Wheelock

Crew	Pamela A. Melroy, George D. Zamka, Douglas H. Wheelock, Stephanie D. Wilson, Scott E. Parazynski, Paolo A. Nespoli Up: Daniel M. Tani Down: Clayton C. Anderson

Mission: STS-120 astronauts exchanged Dan Tani for Clay Anderson and berthed the Harmony module to the ISS. The joint crews relocated the P6 truss to the port side, but frayed guidewires damaged the extending array. Using scavenged ISS materials, spacewalkers stitched the torn array segments together and completed the extension.

Doug Wheelock, Mission Specialist:

Installing Harmony was the mission's number one objective: this Node 2 would eventually branch out into the JEM and the European Columbus module. Another goal was to relocate and extend the P6 solar array, the original ISS solar array first mounted atop the Z1 truss.

Our spacewalks would have us do the Harmony connection and the P6 relocation, then we would go out in the payload bay and practice a thermal tile repair with a caulking gun. On EVA 1, disconnecting that big P6 truss from Z1 was challenging but successful; Stephanie Wilson then moved it all the way over to the port wing.

On my first mission, I didn't realize that in space nothing goes the way you had planned. Nearing the end of our EVA 3, the P6 solar array had deployed and extended about thirty or forty feet (nine or twelve meters). Then the frayed cord caught—we heard the "Abort, abort!" call. We couldn't retract the array, and couldn't deploy it; the solar wing was just a floating mass out there. Following the EVA, everybody was dejected, asking, "What in the world are we going to do now?"

The ground, working in *Apollo 13* mode, came up with the "cufflink" solution: improvised wire links would bind the torn panel segments together. We did a videoconference with Mission Control, and they literally took the cufflinks they had built on the ground, put them in a box, and dumped them on the table, just like in the movie. Over the next several days, we turned Node 2 into a metal shop as we built these cufflinks out of sheet metal, tape, and cabling.

On EVA 4, Scott Parazynski put the foot restraint on the end of the combined arm and boom then just climbed up and into this thing from P6 and locked his feet in, with the boom flexin under his suit mass of about 40 pounds (181 kilograms). That array was like a sail: He'd push on it to put one of the cuff links through a grommet, and the arr would billow out a bit, then com back and hit his hockey stick-shaped probe, giving him a face full of array.

Using insulated pliers and snip Scott and I cut out the frayed guidewire. Then we retracted the excess cable. Scott got all five o the cufflinks in. When we were a finished, I grabbed my camera o of my bag and said, "Lean back and wave!"

The repositioned array had me mostly in shadow from the rest of the structure, and I got super, super cold; my hands and my teeth were chattering. I was havi a hard time gripping things, and could see a bit of my breath insic the helmet. I saw a little sliver of sunlight right up at the top of the mast canister where the array comes out, and I just shimmied out and laid my hands on this litt "rotisserie" to try to warm them up. Mission Control said, "Don't put your hands inside there!" I explained that I was just warmin my hands in that precious bit of sunlight.

On the EVAs, I was nervous about being able to manage my own fear—the fear of failing or losing something or breaking it, or of getting injured or dying. A big lesson was that if you just stick to the tenets of preparedness and working together, draw ing on the strengths of the team members, there's almost nothing that's impossible.

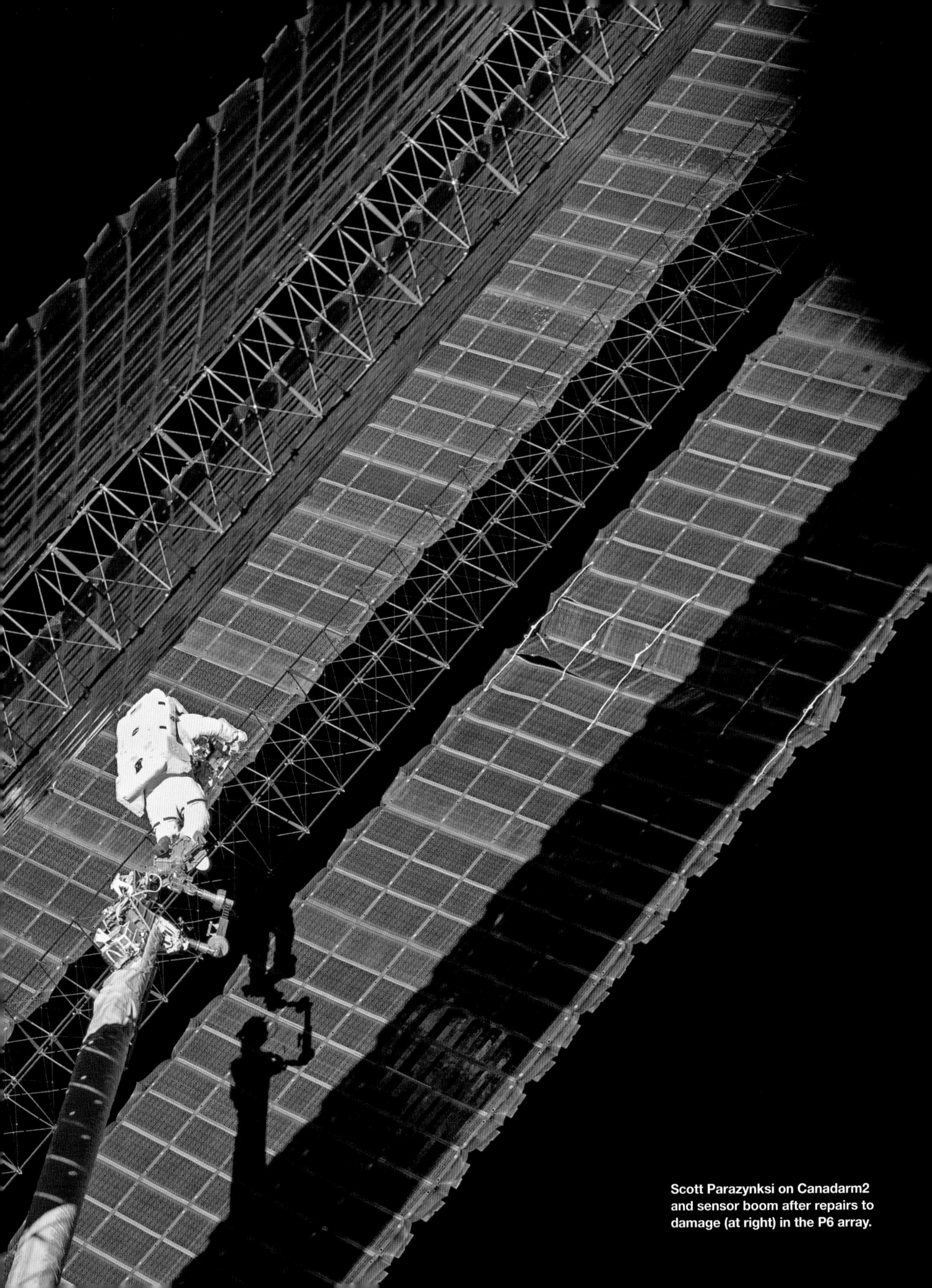

Scott Parazynksi on Canadarm2 and sensor boom after repairs to damage (at right) in the P6 array.

Mission no. **121**

STS

122

Orbiter	Atlantis
Launch	February 7, 2008
Landing	February 20, 2008
Duration	12 days, 18 hrs., 21 mins., 39 secs.

STS-122 and Expedition 16 crews in Columbus: (bottom row) Dan Tani, Steve Frick, Leo Eyharts, Peggy Whitson, and Yuri Malenchenko; (middle row) Stan Love, Hans Schlegel, Rex Walheim, and Alan Poindexter; (top) Leland Melvin

Crew	Stephen N. Frick, Alan G. Poindexter, Leland D. Melvin, Rex J. Walheim, Hans W. Schlegel, Stanley G. Love Up: Leopold Eyharts Down: Daniel M. Tani

Mission: STS-122 exchanged Leo Eyharts for Dan Tani. *Atlantis*'s crew, with spacewalkers' help, used the Station arm to berth the European Columbus laboratory to the Station's Harmony node starboard side. Two more spacewalks completed installation of Columbus's external experiments and upgraded Station systems.

Leland Melvin, Mission Specialist:

As we launched from the Cape and headed to the cosmos, I remembered what Dave Brown's father said after the *Columbia* tragedy: "We should honor the legacy of our friends who are no longer with us." So I thought, "I've got to be perfect, so I can honor the *Columbia* crew's legacy."

STS-122 was the first time—after a wait of ten years—that the ESA would have a laboratory on the Station where they could conduct ESA experiments. I was a first-time flier, coming off a major injury and not certain I would ever fly, now getting the chance to install that laboratory, a two-billion-dollar asset for doing human and materials research.

I was moving Columbus on the arm very slowly, and everything was just going perfectly. I got to the point where I was pushing the docking ring on Columbus against the ready-to-latch microswitches so slowly that the spring force from those ready-to-latch sensors on Node 2 just counterbalanced the arm motion. I thought, "What's going on?" Then just one more little push, and all four lights came on, and Columbus was docked. I gave a sigh of relief, thinking that would be my "ah-hah!" moment of the mission.

Instead, that moment came later, during the first meal together after we berthed Columbus safely. Peggy Whitson, the Station's first female commander, invited us over to the Russian segment to have dinner. We floated over with a bag of vegetables, and on the Russian segment, they had the beef and barley cooking. The moment when we all floated around the table, my perspective shifted. I was breaking bread with Russians and Germans—people we used to fight against. Joining them were an African American, an Asian American, and the first female commander—all traveling at 17,500 miles (28,164 kilometers) per hour, going around the world every 90 minutes, seeing a sunrise and sunset every 45, while listening to Sade's "Smooth Operator." The experience blew my mind.

We had an interview with Quincy Jones, the world-renowned musician and producer. I was talking directly to Quincy, and he said, "You know, Leland, math and music are the two true absolutes that use both the right and left parts of the brain. So, if you know music, you know math, and if you know math, you know music. This is a perfect moment when we're talking to kids and people around the world about using STEAM—science, technology, engineering, arts, and math."

Just before Quincy's remarks, Stan Love had an interview with his local TV station, and the reporter opened the interview with "So, you're in space?" Both of us thought, "That's it? That's the question?" We had been anticipating something a little more substantial. All Stan could answer was, "Y-e-a-a-ah?" Just think about the difference in the quality of those two conversations—that reporter and this profound, world-renowned musician who's riffing on math and music and exploration.

When we focus on all the technical stuff on a mission, we sometimes miss a critical part of the spaceflight experience: In the trenches, when we work together, all the geopolitical issues just melt away. We know that if Yuri Malenchenko, or Hans Schlegel, or myself, or anyone messes up, we could all be gone. And that's what brings us together technically, personally, and as a civilization—emissaries for humanity off-planet, working to bring people together on-planet.

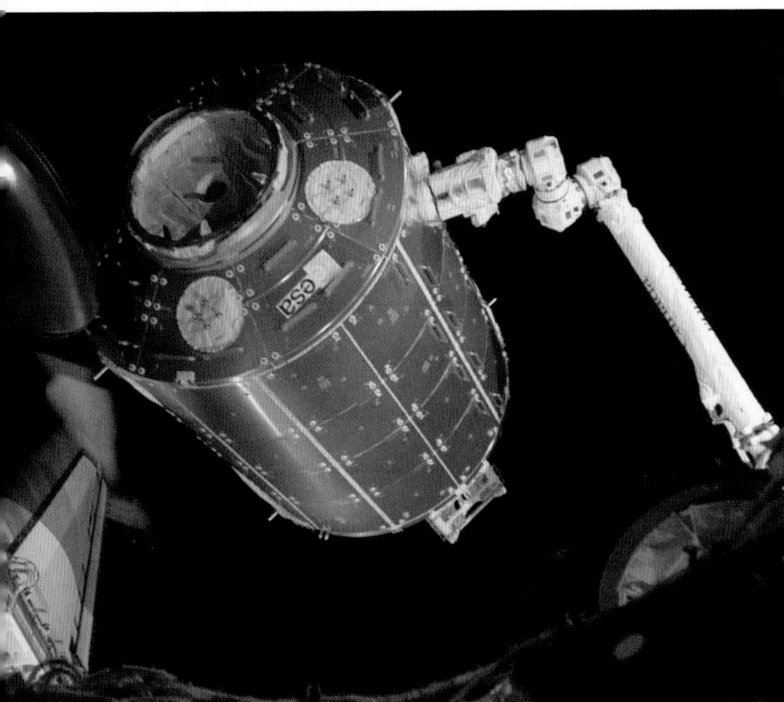

ABOVE: ***Atlantis*** **rises from Pad 39A, viewed by a high-flying F-22 fighter escort.**

LEFT: **Canadarm2 moves the ESA Columbus lab to its berth at Harmony.**

BELOW: **Leland Melvin operates the shuttle robotic arm on the aft flight deck.**

Mission no. **122**

STS 123

Orbiter	Endeavour
Launch	March 11, 2008
Landing	March 26, 2008
Duration	15 days, 18 hrs., 10 mins., 52 secs.

The STS-123 crew in Columbus (clockwise from bottom left) Rick Linnehan, Greg Johnson, Mike Foreman, Takao Doi, Dom Gorie, and Bob Behnken

Crew	Dominic L. Pudwill Gorie, Gregory H. "Box" Johnson, Robert L. Behnken, Michael J. Foreman, Richard M. Linnehan, Takao Doi Up: Garrett E. Reisman Down: Leopold Eyharts

Mission: STS-123 exchanged Garrett Reisman for Leo Eyharts. *Endeavour*'s astronauts delivered to the ISS the Japanese Experiment Logistics Module–Pressurized Section (ELM-PS), serving as a storage closet for the upcoming Japanese laboratory. Spacewalkers installed the Special Purpose Dexterous Manipulator, a multi-armed addition to the Canadarm2 robot arm.

Mike Foreman, Mission Specialist:

It took me eight tries to get into Navy test pilot school. Nine years later, I got into the NASA astronaut program—on my eighth try. I had a lot of rejection letters in my file cabinet, and after this first mission, I thought "Wow, I can forget about all those rejections." I was just loving the fact I'd finally made it.

On ascent, Dom Gorie had our astronaut wings in his suit pocket—mine, Bob Behnken's, and Greg "Box" Johnson's. Reaching orbit, we were all on the flight deck with Dom, and one of the first things he did was reach down, pull those wings out, and hand them over to us. I thought that was pretty cool.

We took up the first piece of the Japanese lab, Kibo. We also delivered the Special Purpose Dexterous Manipulator (Dextre). We took it up in pieces on a pallet that we first attached to the outside of the Station. STS-123 was one of the handful of shuttle missions that had five spacewalks, and it took three to put Dextre together.

On the mission's fourth spacewalk, Bob and I went out to do a few cleanup tasks around Dextre, then set up to do a shuttle tile-repair experiment. We took up broken tiles and a tool we called the "goo gun," and we worked about an hour and a half on applying a shuttle tile ablator compound to the damaged tiles. Then we put those repaired tiles back in the suitcase in the payload bay. *Endeavour* brought them home so that the engineers could cut them up and heat them to see if the repairs worked. The repair technique had to work if the STS-125 mission to service the Hubble Space Telescope was to go forward.

NASA was worried about what would happen if the mission's orbiter, *Atlantis*, sustained damage to tiles that would prevent a safe reentry. Because the Hubble telescope's orbit is much higher and at a different inclination than that of the Space Station, the *Atlantis* crew would not be able to reach the Station. So, we had to be able to fix tiles. After our repair demonstration, the team declared success. Our repair results opened the doors a bit more for managers to clear the last Hubble servicing mission.

After work was done one night we all went down to the Russian segment and had dinner together. Yuri Malenchenko offered some of his Russian food. We had gone to Russia a couple of times, and I like Russian food, but I had never had an opportunity to try their space food. It comes up in what looks like cat food cans. They're passing the cans around, and the contents looked like cat food, too. I thought, "Oh, my God, that does not look good." Of course, it was fine. Just zero presentation.

I retain two strong impressions from STS-123: First, getting to space for the first time and seeing Earth from that vantage point was truly a life-changing experience. If you're not an environmentalist before you go to space, you're at least partly one when you come back. We really do need to take care of Spaceship Earth.

Second, I'm a spiritual guy. My first spaceflight cemented my belief in a higher power. When I was up there, I thought, "Wow, this doesn't seem like an accident—that all of a sudden, Earth just appeared. Somebody is up there." One thing I will always cherish from that mission is my belief that because I've been to space and seen our Earth from out there, I am forever a different human being.

LEFT: ***Endeavour*** **pierces a cloud layer to bathe the Kennedy Space Center in light.**

ABOVE: **The Dextre manipulator robot installed on the ISS exterior.**

BELOW: **Bob Behnken and Rick Linnehan at work on the Dextre robot on pallet at upper left.**

Mission no. **123**

STS

124

Orbiter	Discovery
Launch	May 31, 2008
Landing	June 14, 2008
Duration	13 days, 18 hrs., 13 min. 6 secs.

The STS-124 crew in Kibo: (front row) Karen Nyberg, Garrett Reisman, Mark Kelly, Greg Chamitoff, and Ken Ham; (back row) Aki Hoshide, Ron Garan, and Mike Fossum

Crew	Mark E. Kelly, Kenneth T. Ham, Karen L. Nyberg, Ronald J. Garan Jr., Michael E. Fossum, Akihiko Hoshide Up: Gregory E. Chamitoff Down: Garrett E. Reisman

Mission: STS-124 exchanged Greg Chamitoff for Garrett Reisman. *Discovery* delivered the 32,000-pound (14,515-kilogram) Kibo lab to the ISS, berthed to Harmony's left side. Spacewalkers transferred the OBSS to *Discovery*, and replaced a nitrogen tank and worn parts of the starboard SARJ.

Karen Nyberg, Mission Specialist: We took up the pressurized portion of the Japanese laboratory. That was a huge deal for Japan. They had been working on the lab from the days we were planning Space Station *Freedom* in the 1980s. To be able to help them by installing Kibo on the Station was pretty neat.

The whole team—and not just the crew, but the flight directors and the Japanese and NASA engineering teams—was very competent, and Mark Kelly, the commander, believed in the competence of his crew. His leadership style was perfect for me. He assigned the tasks and trusted us to do them with little if any micromanaging. You could tell he cared about us personally too, without being "mushy." The Kelly brothers are anything but mushy.

I was assigned to take ET photos just after ascent. I struggled to get my straps, my gloves, and my helmet bag situated because I was encountering microgravity for the first time, and everything was floating. But I had that job to do, and I was very focused on it. Finally, when we were done with the ET photos, I took a deep breath and thought about how I had wanted to be an astronaut since I was a little kid.

Approaching the Space Station, as it got bigger and bigger, I was surprised. It looked different than I expected. Here was this manufactured object, built by all these countries, and it was so beautiful.

The Station's Russian toilet had broken down before our mission, and the Station crew was very much looking forward to our arrival. They had been using bags in the Soyuz toilet during that time, and it just didn't have the capacity for all the waste. We brought up spare parts for their toilet, and when we opened the hatch, Mark said, "Anybody looking for a plumber?"

Kibo is the biggest ISS module. When we brought it up, in fact, we could not carry the orbiter boom sensor system—the OBSS—that had launched on every mission since the *Columbia* accident. It's used to inspect the orbiter tiles, nose, and wings for reentry. The JEM would not fit in the payload bay with the OBSS. STS-123 had left their boom for us on the ISS, and we picked it up and brought back home.

I was the chief robotics operator, and Aki Hoshide and I shared duties for Kibo/JEM activation. Because of Kibo's size and mass, the robot arm operations were extremely slow. As we took Kibo out of the payload bay, the onset of arm motion caused oscillation—a little scary because the lab was nestled so close to the sides of the payload bay.

After we undocked from the Station, Mark told everybody to go up to the flight deck, saying, "Plant your face in a window." We turned the lights off, giving everyone a full orbit to look at Earth, take in a sunrise and sunset, view the night lights—all of that. During a jam-packed shuttle flight, you just don't get the opportunity to just look. I really appreciated the time Mark gave us to hang out.

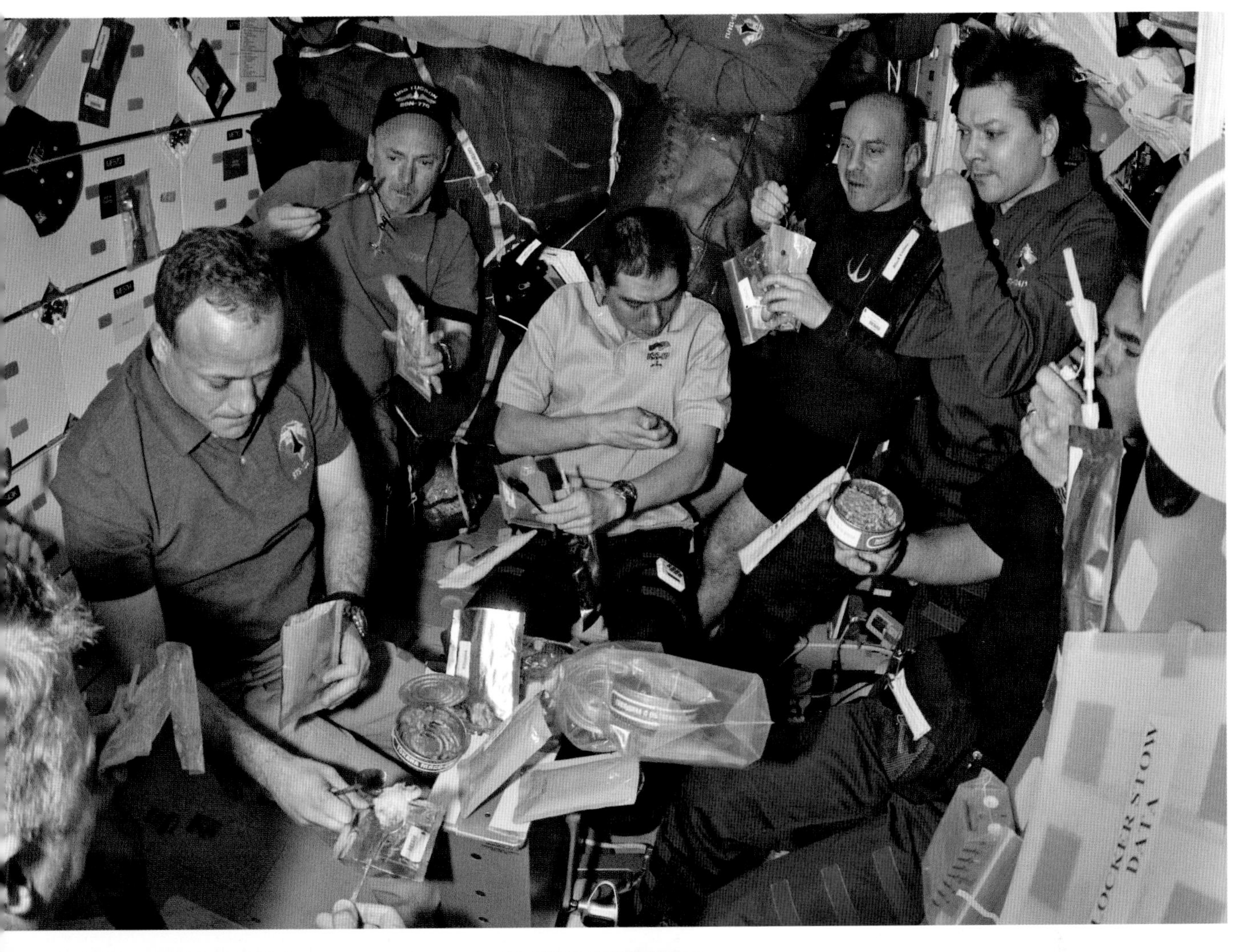

ABOVE: **Mike Fossum (far left), Ron Garan, Mark Kelly, Sergei Volkov, Garrett Reisman, Oleg Kononenko, and Greg Chamitoff gather in middeck for a meal.**

LEFT: **The Kibo lab (front) berthed at Harmony's port side, with *Discovery* at Harmony's forward docking tunnel.**

BELOW: **Karen Nyberg and Aki Hoshide fly Canadarm2 to berth Japan's Kibo lab at the Harmony module.**

Mission no. **124**

STS 126

Orbiter	Endeavour
Launch	November 14, 2008
Landing	November 30, 2008
Duration	15 days, 20 hrs., 29 mins., 30 secs.

STS-126 and Expedition 18 crews in Harmony: (front row) Sandy Magnus, Yury Lonchakov, Greg Chamitoff, and Mike Fincke; (middle row) Shane Kimbrough, Chris Ferguson, and Eric Boe; (back row) Steve Bowen, Heide Stefanyshyn-Piper, and Don Pettit

Crew	Christopher J. Ferguson, Eric A. Boe, Donald R. Pettit, Stephen G. Bowen, Heidemarie M. Stefanyshyn-Piper, Robert S. Kimbrough Up: Sandra H. Magnus Down: Gregory E. Chamitoff

Mission: STS-124 exchanged Sandy Magnus for Greg Chamitoff. *Endeavour*'s astronauts berthed Leonardo at the Harmony module and transferred 14,000 pounds (6,350 kilograms) of supplies and equipment to ISS, returning obsolete gear and trash. Spacewalkers conducted four maintenance EVAs, repairing the right truss's Solar Array Rotary Joint.

Don Pettit, Mission Specialist:

STS-126's mission at the ISS was to install the internal hardware, primarily regenerative life-support systems, to support six crew. The life-support equipment added the ability to reprocess urine into potable water. We also brought up crew amenities, such as sleep stations, to support expansion of the Station crew from three to six.

My most memorable task was working with Mike Fincke, the commander of Expedition 18, to get the regenerative life-support system up and running. From it, we produced our first liter of processed potable water from urine. I called it the coffee machine because the regenerative life-support system has the uncanny ability to take yesterday's coffee and, after it's consumed and excreted, turn it into today's coffee. Initially, it was running at 85 percent recovery, but they later backed that off to 80 percent. Now they've added a downstream process which has boosted recovery to 95 percent of the water in urine.

The system was like IKEA furniture—some assembly required. Two life-support racks were shipped up in Leonardo, and we installed them in the Lab. Maybe one-third of the rack volume was installed after the racks were at the Station because of the need to limit their weight for launch. So we had to install the catalytic converter, the distillation assembly, and the ion-exchange resin beds. Some were a little bit tricky to put in because of unplanned interferences, and it took a lot of my mechanical skills to get this system put together. We worked for about three days to assemble it and get it running, operated remotely by Mission Control. When we got our first liter of potable water from urine, we had a big press conference, with a crew photo of all of us holding clear bags of processed drinking water.

This was my second flight, and for Thanksgiving, I had Russian cuisine, which I first tasted on Expedition 6 in 2002. I had missed it ever since. Instead of the traditional American turkey, I had Tvorog, Takana meat, and garlic mashed potatoes. Tvorog is a Russian dairy product—my favorite for breakfast. It's slightly sweet and has some nuts in it; think of it as a blend between cottage cheese and yogurt. There's no western analog for Tvorog that I'm aware of—it's a Slavic, Russian kind of thing. Takana meat is mystery meat. It comes in a great big can about twice the size of a tuna can and it's really good. I've asked all my Russian counterparts, "What Takana meat?" And nobody could really tell me. They say, "Well, it's Takana." The Russians also make garlic mashed potatoes, which come in a great big pouch about the size of three shuttle servings—they're also very good.

What struck me on this flight was how adaptable human beings are to environments that they were never meant to live in. Think about this: In space, the acceleration forces on your body are a million times less than on Earth. Change any other environmental factor by a million: Change the temperature by a million, or change the pressure by a million. Most Earth creatures would just shrivel up and die. But we humans are amazingly able to adapt our mobility, our eating and sleeping habits, our toilet practices, all of these things so fundamental to life on Earth to this sudden, radical environmental change.

ABOVE: **Heide Stefanyshyn-Piper (right) and Shane Kimbrough service the starboard SARJ bearings during EVA 2.**

RIGHT: **Don Pettit flies the Canadarm2 from Destiny.**

BELOW: **Mike Fincke, Chris Ferguson (foreground), Eric Boe, and Pettit (background) demonstrate the "coffee machine."**

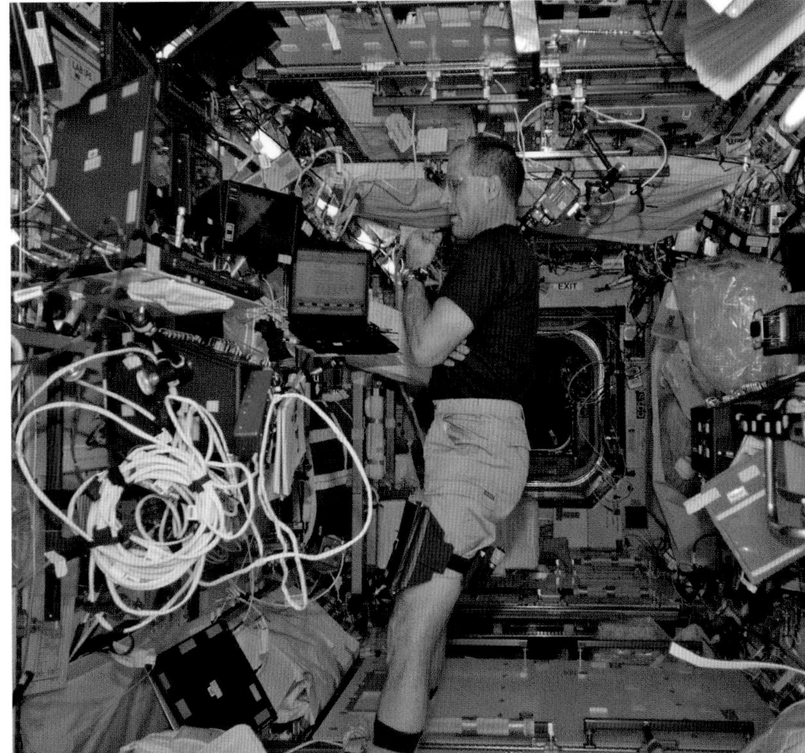

Mission no. **125**

STS

119

Orbiter	Discovery
Launch	March 15, 2009
Landing	March 28, 2009
Duration	12 days, 19 hrs., 29 mins., 42 secs.

The STS-119 crew in Harmony: (front row) Tony Antonelli, Lee Archambault, and Joe Acaba; (middle row) Koichi Wakata and Sandy Magnus; (back row) Steve Swanson, Ricky Arnold, and John Phillips

Crew	Lee J. "Bru" Archambault, Dominic A. Antonelli, Joseph M. Acaba, Steven R. Swanson, Richard R. Arnold, John L. Phillips Up: Koichi Wakata Down: Sandra H. Magnus

Mission: STS-119 exchanged Koichi Wakata for Sandy Magnus. *Discovery*'s crew delivered the 31,000-pound (14,061-kilogram) S6 truss, working with the ISS crew to berth the S6 far to starboard. After spacewalkers connected the S6 to Station systems, the crew unfurled the ISS's final set of large solar arrays.

Lee Archambault, Commander: We were nearing the end of the construction of the ISS. The ISS has a total of eight solar array wings, and in bringing up the S6 truss with the last two, we brought the Space Station to full power. After our mission, the ISS could support a full crew of six and their associated research capability.

Ours was a very junior crew. After one mission in the pilot seat on STS-117, I was now commander of STS-119, my first opportunity to lead a shuttle mission—my only one, as it turned out, because of the shuttle's looming retirement. Joining me on the flight deck, pilot Tony Antonelli and mission specialist Joe Acaba were both rookies. Steve Swanson, another mission specialist, flew his first mission with me earlier on STS-117. On the middeck, Ricky Arnold was also a rookie. Our most experienced flier was robotics operator John Phillips—he'd flown one shuttle and one long-duration Space Station mission. Veteran Japanese astronaut Koichi Wakata joined us on the way to his six-month expedition at the ISS. With such a junior team, our crew assignment spoke highly of the confidence that our astronaut office leadership had in our skills and training.

We had a very successful mission: With our very first spacewalk, using a combination of robotics and EVA coordination, we installed that S6 truss. The following day, we unfurled the solar array wings, bringing the Station to full power. By the end of EVA 3, we'd achieved all mission objectives.

As we were getting ready to come home on Flight Day 14, bad weather was rolling in at the Kennedy Space Center. We waved off the first landing opportunity. We all knew that the weather the following day was going to be even worse, so if we waved off our second try that first day, then the following day we were going to land at Edwards Air Force Base. Brent Jett was flying the STA, and he was critical in assisting flight director Richard Jones in making the decision to squeeze us in on our second de-orbit opportunity.

After I'd flown a thousand or more practice dives in the STA, it all came down to that one chance to actually fly the four-billion-dollar space shuttle to a full-stop landing at Kennedy. Our full team—Brent Jett flying in the STA with our instructors, the ground team, the Kennedy weather forecasters, and the Air Force guys—did a great job of threading the needle and getting us home on that second opportunity.

I think we broke through about an 8,000-foot (2,438-meter) ceiling—runway in sight—into very strong winds, about twenty-plus knots (ten meters per second) and gusty, but nearly straight down the runway. That space shuttle just cut right through that turbulence, smooth as silk with maybe a hint of a bump here or there—almost exactly the conditions Brent had predicted we'd find on our approach. We got the shuttle on the deck, and an hour or so later, rain showers started moving through, just as the forecasters predicted. No chance we would have gotten into Kennedy Space Center the next day, as the weather turned hellacious the following morning.

full Moon graces *Discovery*,
ady for launch on Pad 39A.

Mission no. **126**

STS

125

Orbiter	Atlantis
Launch	May 11, 2009
Landing	May 24, 2009
Duration	12 days, 21 hrs., 37 mins.,18 secs.

The STS-125 crew on *Atlantis*'s middeck: (front row) Greg C. Johnson, Scott Altman, and Megan McArthur; (back row) Mike Good, Mike Massimino, John Grunsfeld, and Drew Feustel

Crew	Scott Altman, Gregory C. "Ray Jay" Johnson, Michael T. Good, K. Megan McArthur, John M. Grunsfeld, Michael J. Massimino, Andrew J. Feustel

Mission: *Atlantis*'s crew upgraded the Hubble Space Telescope with the Cosmic Origins Spectrograph and Wide Field Camera 3. Spacewalkers repaired flight systems, and Mike Massimino manually tore free a handrail to reach and repair the STIS.

John Grunsfeld, Mission Specialist:

Hubble Servicing Mission 4 had been canceled in 2004 after *Columbia*'s loss. Administrator Sean O'Keefe believed that sending a lone shuttle on a mission without a rescue capability was too risky. He felt that if a crew were lost on a Hubble servicing mission, that would be the end of the shuttle program, and the end of US human spaceflight for a while.

We came up with the concept of a rescue shuttle. In this plan, *Endeavour*, the rescue shuttle, would rendezvous with *Atlantis,* and *Endeavour* would flip over and position itself at right angles above *Atlantis*. Then *Endeavour*'s robotic arm would grab a grapple fixture on *Atlantis*. The astronauts would then travel back and forth across the arm, getting the crew to *Endeavour* using just four space suits. Our team presented these ideas to administrator Mike Griffin, and after recognizing and assessing the plan's risks, he approved it—"Okay, let's go back to Hubble."

Our goals on the repair mission were to install the Cosmic Origins Spectrograph and the new camera, add the de-orbit fixture, install some new external insulation, fix all the things that were broken, and leave the telescope with all the instruments working, potentially giving it another five years of life. Here we are fourteen years later, and we're down to three gyros, but everything else is working great.

On our third EVA, we removed the COSTAR—that corrective optics instrument from 1993—and put in the Cosmic Origins Spectrograph. That went well, so Drew Feustel and I then tackled the Advanced Camera for Surveys repair. This was the first time that anybody had tried to remove and replace circuit cards, held in by these tiny screws. To get to them, we first had to cut through a metal plate. I became a brain surgeon, and Drew was my assistant, handing tools in and out, taking out parts I removed, and bringing me the new circuit cards. As an example of the precision required, for the tiny screws around the perimeter of a metal plate covering the cards, I took my tiny screwdriver and rotated each one to release the torque, then oriented the cruciform slots for each one up and down, so that when I came back later, I knew the correct orientation for inserting the mini power tool.

So many things went wrong in the details, but for each one, we had a solution that we'd already developed in training. On the other hand, some big things—like a stripped screw—didn't go wrong. A stripped screw would have ground the repair to a halt. But we finished in two hours and thirty-two minutes. We didn't know for sure that it would work, but we completed every task we had control over.

We hoped our work would have long-lasting value, and it surprised me that I felt emotional about it. I was just so happy that we'd repaired the Advanced Camera for Surveys (ACS). When I give talks, I say: "The greatest things in my life have been marrying Carol, having kids, and doing the ACS repair."

I believe we did something incredible together at great risk to our lives. Outside, leaving Hubble, I remarked, "As Arthur C. Clarke said, 'The only way to find the limits of the possible is by going beyond them into the impossible.' On this mission, we tried some things that many people said were impossible: fixing the Space Telescope Imaging Spectrograph (STIS), repairing ACS, achieving all the objectives we have on this mission. We wish Hubble the very best."

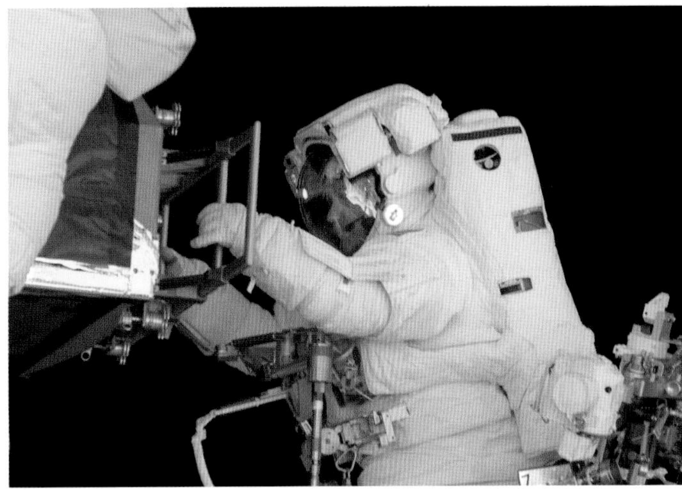

ABOVE LEFT: **John Grunsfeld and Drew Feustel ready for EVA in *Atlantis*'s airlock.**

ABOVE RIGHT: **Grunsfeld removes Fine Guidance Sensor 2 for replacement during EVA 5.**

BELOW: **Megan McArthur lifts Hubble on the shuttle arm for release after upgrades and servicing.**

Mission no. **127**

STS

127

Orbiter	Endeavour
Launch	July 15, 2009
Landing	July 31, 2009
Duration	15 days, 16 hrs., 44 mins., 58 secs.

The STS-127 crew at Pad 39A: Tom Marshburn, Chris Cassidy, Tim Kopra, Julie Payette, Doug Hurley, Mark Polansky, and Dave Wolf

Crew	Mark L. Polansky, Douglas G. Hurley, Christopher J. Cassidy, Julie Payette, Thomas H. Marshburn, David A. Wolf Up: Timothy L. Kopra Down: Koichi Wakata

Mission: Hydrogen leaks and bad weather postponed STS-127's launch five times. The mission exchanged Tim Kopra for Koichi Wakata. *Endeavour*'s astronauts attached the JEM Exposed Facility to Kibo's port side, and spacewalkers replaced P6 battery modules. The thirteen astronauts working aboard the ISS formed the largest single-spaceship crew to date.

Doug Hurley, Pilot:

With the multiple robotic operations and five EVAs, the sixteen-day STS-127 mission was among the most complex shuttle missions ever flown.

So many things have to go right on these intricate, choreographed missions—whether the Hubble-repair missions or that sequence of heavy, Station-assembly missions in the mid-2000s—that people may not realize just how many chances for failure there were. In delivering and assembling major pieces of the Space Station that were never put together on the ground, so many things had to go right to be successful. Most of these assembly missions had at least three or four highly choreographed spacewalks, and a couple had five EVAs. Going out the door five times over the course of ten days or less is unprecedented.

I won't say that the Apollo missions were easy, but if you compare the timelines for guys going to the Moon with those of these assembly or Hubble missions, you can see that they are comparable in complexity. Success in both types of missions required people to be in sync and equipment to work. There was little room for error.

After ascent on STS-127, when the engines cut off, we went from three g's to zero g's, and I was floating, hanging in my straps, looking out the window. I was struck by all those years that had gone into getting me to that point—not to mention the fact we'd scrubbed five times. I'm sure my facial expression as I looked at Mark Polansky was "What the hell just happened?" Mark looked back at me and said, "Are you with me, dude?"

Julie Payette, Mission Specialis

Being the flight engineer was phenomenal because I was righ in the middle of the action, seatec on the center Mission Specialist 2 seat behind the pilots, with the overhead windows behind me. I wondered if I'd see flame shoot out behind us at takeoff. Chris Cassidy was sitting to my right in the Mission Specialist 1 seat, holding a mirror so we could see will never forget the moment wh the engines lit: In a flash, I saw t boulevard of flame below us; the flame trench was just a mass of bright orange fire. Then we took off, and of course I had to conce trate on something else.

Once docked at the ISS, we us the Station's Canadarm2 to reach out to the batteries at the very enc of the P6 truss. Those batteries ar massive, and the EVA crew had tc go on the carrier platform to retrie a spent battery and replace it with a fresh one. Because the arm was so fully extended, every single tim they'd get on or off the platform, t arm would start flexing, swinging dangerously close to Station structure. And every time that arm started to swivel, I got a little scare we engaged the brakes on the arn joints and quickly floated down to the Kibo module porthole where w were able to see the actual distan between the tip of the arm carryin the battery platform and the very end of the Station structure. It was close. We continued, but asked th guys to be careful.

STS-127 was a very no-nonsense, "let's get the job done flight. Our joint crew at the Statio represented all five international partners. Because of this close cooperation, the ISS will go dow in history as a formidable examp of an effective foreign policy. It forced nations to work together, to think not from the microcosm of nationality, but as partners in a peaceful, collaborative spirit.

OVE: ***Endeavor*'s tail splits arth's glowing horizon at orbital usk.**

GHT: **Julie Payette suited for entry on the flight deck with lot Doug Hurley at left.**

Mission no. **128**

STS

128

Orbiter	Discovery
Launch	August 28, 2009
Landing	September 11, 2009
Duration	13 days, 20 hrs., 53 mins., 43 secs.

The STS-128 crew with *Discovery* at Pad 39A: Rick Sturckow, Danny Olivas, Christer Fuglesang, Kevin Ford, Nicole Stott, Pat Forrester, and José Hernández

Crew	Frederick W. "C. J." Sturckow, Kevin A. Ford, Patrick G. Forrester, José M. Hernández, John D. "Danny" Olivas, Christer Fuglesang Up: Nicole M. P. Stott Down: Timothy L. Kopra

Mission: STS-128 exchanged Nicole Stott for Tim Kopra. *Discovery*'s crew berthed the Leonardo MPLM to Harmony and transferred 15,200 pounds (6,894 kilograms) of equipment to the ISS. Spacewalkers installed a fresh ammonia tank, returned exposed experiments to the shuttle, and prepared connections for the future Tranquility module.

José Hernández, Mission Specialist:

We were nearing completion of the assembly of the ISS, and our mission brought seven tons (6.4 metric tons) of equipment to install both inside and outside. On a big construction project, you've got people who do the framing, others who do the roofing, still others for the drywalling—all those tasks. I felt that we were the movers, bringing up and installing experiment racks, cameras, antennas, and a spare ammonia tank assembly for the outside. It was a busy time at the ISS.

I was impressed that NASA was able to put together the ISS, LEGO-style, module by module. We'd bring in a module, berth it, hook up its electrical lines and cooling loops, and just keep building.

One maintenance task we practiced more times than I thought necessary was changing the shuttle cabin's lithium hydroxide—LiOH—canisters; they scrub CO_2 from the atmosphere. You undo four screws, take out the floor panel, remove the filter, put in the new one, and reinstall the panel, right? You practice it once, you know where the tools are, and you know how to unscrew the fasteners. It's righty-tighty, lefty-loosey. A monkey can do it, right?

I was the first one to replace a LiOH canister. "No problem. I'm just going to float over and get the tool. I'm going to float now, hovering over the panel. Yeah, there's the four screws, just like in training." I tried to loosen a screw. Well, Newton's third law of motion kicked in: "For every action, there's an equal and opposite reaction." So, as I turned the screw one way, my body turned the other way. I was kind of embarrassed—I looked around checking that no one was looking at me. I scratched my head there for a while, trying to not look so obviously stumped. Of course, we were in zero g, and I wasn't anchored to anything. I finally saw some gray tape on the floor, peeled it up, and sure enough, there were a couple of foot loops held there under the tape. I put my feet in the loop, anchored myself and away I went.

When I got home, the first thing I did was go to the mockup. Sure enough, the gray tape was there. I called the trainers together, and said, "Okay, I suppose you guys didn't tell us about this." They all started laughing: "We get the rookies all the time!" They didn't tell us because they wanted to let us figure it out in orbit, and then when we came back to complain they would enjoy a good laugh at my expense.

We were up fourteen days, and we each got to choose wake-up songs. My first one was a traditional Mexican song, "Son of the Town," by a famous Mexican artist who was popular during my father's time. It's about coming from humble beginnings and doing good things. I chose it as an homage to my father.

Florida's weather was bad on the day we were supposed to land. Our families were waiting for us there to watch the landing, but we had to divert to Edwards Air Force Base. Reentry was anticlimactic, just a lot of buffeting, equivalent to a bad airline flight over the Rockies. Gravity started taking hold, so my neck felt weak under that big helmet. But about an hour and a half after we stepped off the orbiter, we were on our way to dinner at Domingo's.

LEFT: **The nearly complete ISS emerges from orbital night, viewed after undocking.**

ABOVE: ***Discovery*** **carrying Leonardo pitches up for its pre-docking heat shield inspection.**

BELOW: **Christer Fuglesang on Canadarm2 transfers ammonia coolant tanks to the ISS.**

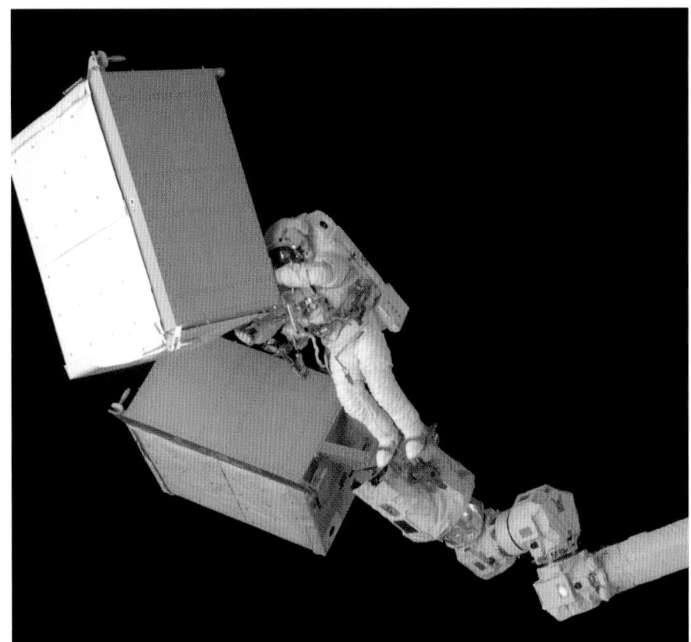

Mission no. **129**

STS

129

Orbiter	Atlantis
Launch	November 16, 2009
Landing	November 27, 2009
Duration	10 days, 19 hrs., 16 mins., 14 secs.

The STS-129 crew in Destiny: (front row) Leland Melvin, Charlie Hobaugh, and Bobby Satcher; (back row) Mike Foreman, Randy Bresnik, and Butch Wilmore

Crew	Charles O. "Scorch" Hobaugh, Barry E. Wilmore, Leland D. Melvin, Randolph J. Bresnik, Michael J. Foreman, Robert L. Satcher Jr. Down: Nicole M. P. Stott

Mission: *Atlantis*'s crew returned Nicole Stott and delivered critical spares to the ISS ahead of the shuttle's retirement. Spacewalkers installed a spare communications antenna and a new oxygen tank, and prepared a berthing port for the Tranquility module's arrival. STS-129 was the last shuttle mission to return a station crewmember.

Bobby Satcher, Mission Specialist:

We went up with six and came back with seven, returning Nicole Stott, who had been on the Space Station for three months. STS-129 was the only shuttle mission that flew two male, African American astronauts, Leland Melvin and me.

I got interested in going into space when I was young. I was inspired, as many were, by watching the Apollo missions on TV. And who doesn't want to do a spacewalk? I came in with the class of 2004, the first class selected after the *Columbia* tragedy, projected to work on the Station, Moon, Mars, and beyond. I realized that I would probably not get to the Moon, but at least I would be able to do some construction work on the Space Station, and with luck, do an EVA.

I went outside on EVAs 1 and 3. Coming out of the airlock in daylight and looking down towards Earth is an experience that always stays with me. I looked down between my feet, and we happened to be passing over Texas at the time, home of Mission Control and the Johnson Space Center. They timed that perfectly!

My EVAs were significantly easier than the training was because you're not working against the resistance of the water. I had to get used to the constant "micro-motion," the small, undamped disturbances induced by my movements in free fall that I had to correct for when not locked into a foot restraint. But I figured that out pretty quickly after getting over the magnificent view.

Because I was doing maintenance work on the Japanese robotic arm's end effector, they flew me over there on the Station robotic arm. That gave me another perspective because I could see the docked shuttle and an absolutely spectacular view of the whole Station with its radiator panels and solar arrays.

We were up over Thanksgiving Day, and enjoyed breaking bread with everybody there on ISS. Gathered in the Unity Node, we shared an international meal, not typical American menu. The ESA folks broke out some cow cheeks and the Russians brought some kind of canned fish. But we also had our turkey and stuffing.

Ours was one of the few all-male crews flying the shuttle. The interaction was good: We reverted to the relationships you might have on a basketball team. Each shuttle crew produces a mission highlights film, and in ours, we had this little pick-up football game with Leland Melvin, who was a former NFL receiver, and Butch Wilmore, who played linebacker. It was supposed to be touch football, but it wound up being close to tackle football in space. We threw the ball to Leland, and Butch didn't just touch him, he hit him. It was really fun, and nobody was injured.

Randy Bresnik's wife, Rebecca, was pregnant at the time we flew, and because our launch slipped a week or so, she delivered when we were on orbit. After the delivery on November 21, Mission Control set up a video conference with Randy's wife and his family, including baby Abigail. That was pretty special. Randy didn't see her in person until we got back down. We always tell Rebecca, "Thanks for letting him go."

LEFT: ***Atlantis*, backed by a Cape Canaveral sunset, waits on Pad 39A.**

BELOW: **Bobby Satcher works on ISS maintenance tasks on the third STS-129 EVA.**

Mission no. **130**

STS

130

Orbiter	Endeavour
Launch	February 8, 2010
Landing	February 21, 2010
Duration	13 days, 18 hrs., 6 mins., 22 secs.

The STS-130 crew afloat in the Cupola: (clockwise from top left) Steve Robinson, George Zamka, Terry Virts, Kay Hire, Nick Patrick, and Bob Behnken

Crew	George D. "Zambo" Zamka, Terry W. Virts Jr., Kathryn P. Hire, Stephen K. Robinson, Nicholas J. M. Patrick, Robert L. Behnken

Mission: *Endeavour*'s astronauts berthed the Tranquility module to the Unity node at ISS. Tranquility, the habitation module, mounts the Cupola observing station and supports crew hygiene and exercise. The joint crews then relocated the Cupola to its Earth-facing location on Tranquility.

Terry Virts, Pilot:

STS-130 was the final Space Station "core assembly" mission, the equivalent of putting the golden spike in the Transcontinental Railroad. We were also the only mission to bring up two modules—Node 3 and the ESA-built Cupola—at the same time.

In the shuttle front cockpit or in the pre-Cupola Station, we were always looking out a window surrounded by its frame. It was like driving through the Rockies while looking through a car window. But looking outside through the Cupola is like parking the car and getting out—the view is much more immersive. Your peripheral vision really adds to the emotional impact: "Wow, I'm in space."

I worked with Kay Hire on the robotic installation of the Cupola. On Flight Day 10, Nick Patrick performed a spacewalk to release the Cupola's shutter restraint bolts while Jeff Williams and I worked inside. Once Nick finished, I rotated a knob to swing open the seven window covers, one by one. It was cool to be the first human to look through the Cupola's windows. When Nick came in and got out of his suit, I grabbed him and we quickly floated back to the Cupola. His eyes were wide, his mouth open, and he said something like, "Wow, it's just like being out on a spacewalk."

We decided to leave all the covers open while we went back to work outfitting Tranquility, turning bolts and plugging in cables. Suddenly the module's walls turned red—very eerie, as if somebody had turned on red lighting. I floated down and looked o through the Cupola: We were ov the red Outback of Australia, anc its reflected glow completely lit u the inside of Tranquility.

Capcom Mike Massimino calle up and asked, "Hey, when you guys are in the Cupola on the bottom of the Station, do you fe like a bat hanging down, or do yc feel like a prairie dog popping its head up?" We took a poll, and it was fifty-fifty.

The Air Force requested that we perform a series of maneuver in the shuttle so that they could characterize the orbiter's optical signature from their Maui telescope complex. We got the opportunity on STS-130 to do it aft we had undocked. Zambo and I performed a tight, very scripted set of maneuvers—for example, Translational Hand Controller in for five seconds, then up for thre seconds, rotate. We were crazy busy with the hand controllers fo several minutes. The space shutt was spinning so fast it felt like a centrifuge. If we let go of a piece of gear, it quickly floated toward the orbiter's nose.

The flight was way better than I'd ever dreamed, the highlight of my professional life. I'd wanted t be an astronaut since I was five. As a boy, I grew up with posters the shuttle, an F-16, and the M31 galaxy on my bedroom wall. Afte landing, I thought that no matter what happens later in life, I was the pilot on STS-130, something to always be proud of. We worke so hard, got so much done, got everything installed. After we landed, I was so completely wipe out that the next night I slept for thirteen and a half hours. I hadn't done that since I was one.

ABOVE: **The first view through open shutters of the new Cupola Earth-viewing station, February 2010.**

BELOW: **With Cupola and Tranquility at left, *Endeavour* departs the ISS near sunset.**

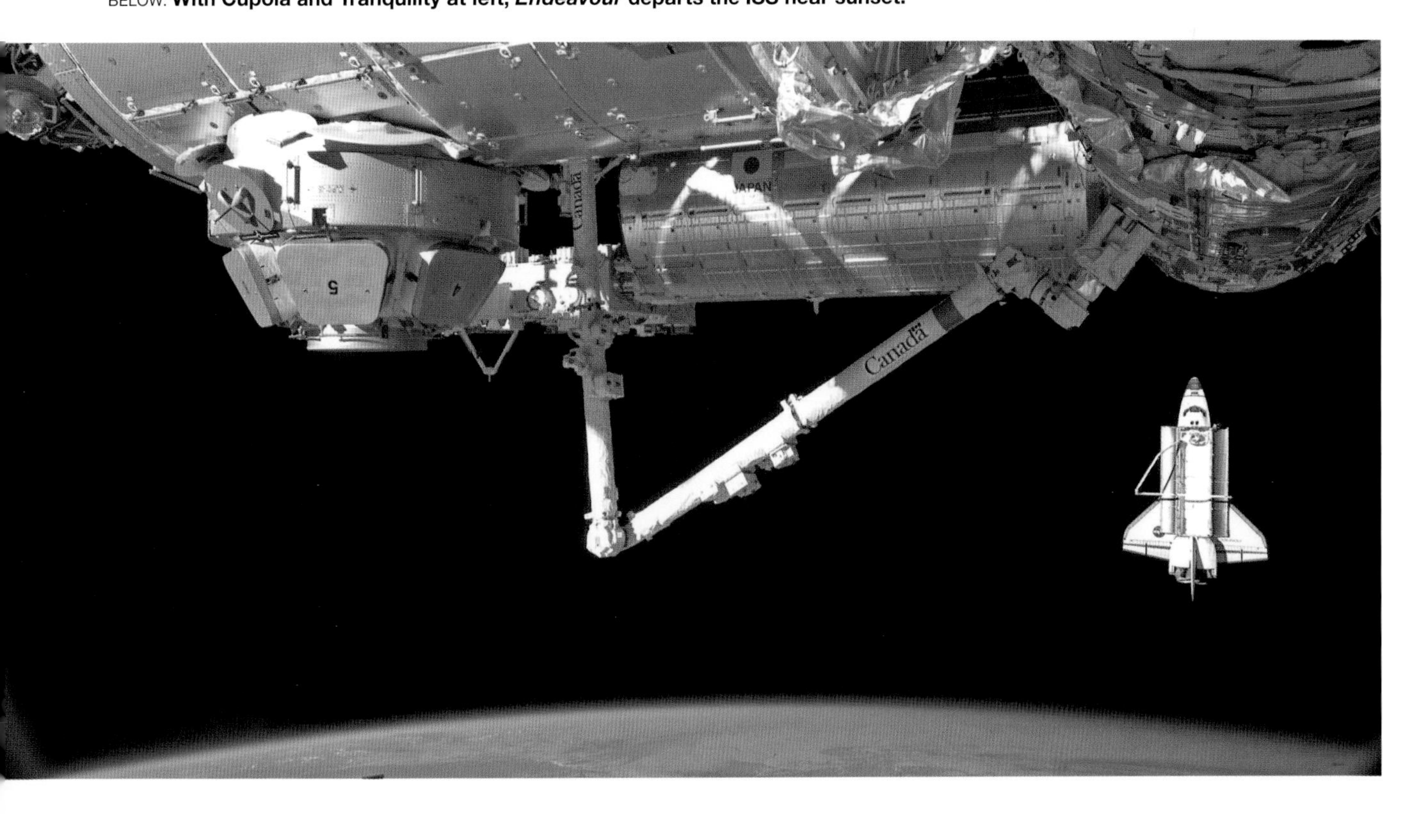

Mission no. **131**

STS

131

Orbiter	Discovery
Launch	April 5, 2010
Landing	April 20, 2010
Duration	15 days, 2 hrs., 47 mins., 9 secs.

The STS-131 crew on *Discovery*'s flight deck: (front row) Naoko Yamazaki, Alan Poindexter, and Stephanie Wilson; (back row) Jim Dutton, Clay Anderson, Dottie Metcalf-Lindenburger, and Rick Mastracchio

Crew	Alan G. Poindexter, James P. Dutton, Richard A. Mastracchio, Dorothy Metcalf-Lindenburger, Stephanie D. Wilson, Naoko Yamazaki, Clayton C. Anderson

Mission: *Discovery*'s astronauts berthed the Leonardo logistics module at the ISS, and transferred 17,000 pounds (7,711 kilograms) of supplies, a crew quarters, and four experiment racks. Spacewalkers swapped an empty ammonia tank for a fresh one, and replaced a failed gyro assembly. STS-131 was *Discovery*'s longest flight.

Dorothy Metcalf-Lindenburger, Mission Specialist:

STS-131 was the second-to-last flight of *Discovery* and also the next-to-last delivery of Leonardo, the roomy moving van filled with science experiments and one of the crew quarters, along with all the food and clothing planned for storage onboard as the shuttle neared retirement.

I grew up along Colorado's Front Range, so I have always liked rocks and I've always been able to look up at a big sky of stars. I was entering grade school when the first women joined the US astronaut corps. Sally Ride flew during my third-grade year, and Kathy Sullivan was right behind her with the first American female spacewalk. When I was a sixth grader, Santa brought me a telescope, and my parents decided to send me to space camp. As it happened, that same month—April 1990—*Discovery* carried the Hubble Space Telescope into orbit. Almost twenty years to the day after my space camp experience, I lived the dream and flew on *Discovery* myself.

Jim Dutton and Stephanie Wilson operated the robotic arms during the swap of the ammonia tank. On EVA 1, we had to take the new tank from the cargo bay and hand it off to the robotic arm, then stow it temporarily on the cart that travels along the truss. On EVA 2, we put the old tank on the cart, just strapping it down with adjustable equipment tethers. Then we had to get the new tank from the mobile base system and bolt it to the truss. Lots of these handoffs, and a lot of coordination and work.

During EVA 1, I was inside as the IVA crewmember, running checklists and coordinating the work of my friends outside. I set up on the flight deck, where I could see Rick Mastracchio and Clay Anderson working in the payload bay. I could also get ove to the front windows and keep eyes on them working on the fror of the truss during EVA 2.

On EVA 2, Rick was holding the new ammonia tank, and Clay was bolting it down, but the bolt wouldn't drive all the way in. We started troubleshooting, just like in the water. I referred to my cue cards and worked with the ground and the engineering tean Listening to Clay and Rick, I triec to make sure that we were thinking ahead and feeding Mission Control all the information. The extreme temperatures outside warped the back of the ammonia tank assembly, and the bolts wouldn't seat. Poor Rick held the tank for half an orbit—forty-five minutes—until the temperatures evened out, and then we got all four bolts in.

Tracy Caldwell Dyson was on orbit at the ISS, and then we thre women on STS-131—Stephanie, Naoko Yamazaki, and I—came up At the time, it didn't feel like a big deal to me, but now, when I spea about us, people really connect with that image of four women on orbit together.

Naoko Yamazaki joined Soichi Noguchi in orbit, the first time tha two Japanese were meeting in space. To mark the occasion, she and Soichi prepared a meal for us sent along by Japan's space agency. It was similar to having California rolls, but of course, we didn't have raw fish. But we did have the seaweed and tasty, sticky rice, with canned mackerel brought over by our Russian crew members. Alan Poindexter joined Oleg Kotov with their California rolls in a wasabi-tasting competition. It was one of those beautiful international moments representing what the Space Station is about—being our better selves.

LEFT: ***Discovery*'s ascent trajectory seen in a time-lapse photo from the Banana Creek Viewing Area at the Kennedy Space Center.**

ABOVE: **Dottie Metcalf-Lindenburger, Tracy Caldwell Dyson, Stephanie Wilson, and Naoko Yamazaki aboard Zvezda in 2010.**

BELOW: ***Discovery* displays its belly heat shield, speckled with newer, darker tiles, for inspection.**

Mission no. **132**

STS

132

Orbiter	Atlantis
Launch	May 14, 2010
Landing	May 26, 2010
Duration	11 days, 18 hrs., 27 mins., 59 secs.

The STS-132 crew in their preflight portrait: (front) Ken Ham; (middle) Garrett Reisman and Steve Bowen; (back) Mike Good, Tony Antonelli, and Piers Sellers

Crew	Kenneth T. Ham, Dominic A. Antonelli, Garrett E. Reisman, Michael T. "Bueno" Good, Stephen G. Bowen, Piers J. Sellers

Mission: *Atlantis*'s crew attached the Russian Mini-Research Module, Rassvet (Dawn) to the Russian Zarya module, using the robotic Canadarm2. Spacewalkers replaced aging P6 batteries and installed a spare antenna and stowage platform. *Atlantis* would fly one more mission, ending the flight campaign with STS-135.

Garrett Reisman, Mission Specialist:

STS-132 was part of the race to get the Space Station finished while we still had space shuttles to fly. In trying to finish the Station, we were putting in a new, redundant K_u-band antenna on top of the Z1 truss. Before our mission, the K_u system had only a single antenna and a single string of antenna electronics. We added a second string, important for the long-term relay of scientific data to the ground. We also replaced the power-storage batteries on the P6 truss, way out there on the far port side. The fresh batteries had much more capacity, whereas many of the original ones were reaching their life limits. We needed to get all the spares in place before we stopped flying the shuttle; anything we didn't get done on STS-132 was going to be very difficult to add to any of the remaining missions, with their own full workloads.

I don't mean to rank crews—it's like picking your favorite child—but compared to other crews I was part of, STS-132 was a very special group. We usually have one or two goofballs on a crew—guys who like to joke around—and the rest of the crew balances that out. But on STS-132 we had no counterweight. The guys on our crew—me, Piers Sellers, Tony Antonelli, and Ken Ham as our commander—we were all goofballs. The closest thing we had to parental supervision was probably Steve Bowen, but we pretty much corrupted him immediately, and Mike "Bueno" Good as well. The day after we were assigned, I ran into Steve Lindse the astronaut chief, up on the six floor. I said, "You put all of us on one crew! What the hell were you thinking?" And he said, sheepish "I don't know."

We had so much fun training, and we joked around a lot, right to the day of launch. We got to th launch pad, and we were standing there gazing up at the shuttle Tony came over and put his arm around me and said, "Garrett, you know all this fun we've been having during training?" I said, "Yeah, it's been awesome. " And he said, "Well, we're going to loo really stupid if we screw this up." Thankfully, we pulled it all off.

The most overwhelming senso experience I had was during EVA 3, when Bueno and I had no time line worries. By then we were abl to work so efficiently that we wer getting ahead, and we wanted to take a little time to appreciate the experience of being in space. We were coming out of eclipse, and the Sun was going to rise. I was out there along the truss, and I held a handrail, putting it behind me. I was facing into the velocity vector out there on the truss, and the only indication I had that ther was a Space Station or space shuttle or any spacecraft near me was the tactile sensation of this handrail—something was back there, but visually there was just nothing in front of me. I watched that Sun come up without any point of reference, intentionally taking the time to fully focus on it rather than just glimpsing something as I was working. I can still see it in my memory right now. That was the best sunrise I ever saw in orbit.

LEFT: ***Atlantis*** **soars toward orbit, shepherded by an Air Force F-15 fighter escort.**

ABOVE: **Garrett Reisman on Canadarm2 carries a K_u-band antenna dish for installation on the ISS.**

BELOW: **Piers Sellers and Steve Bowen assume their middeck launch positions while training in Houston.**

Mission no. **133**

STS

133

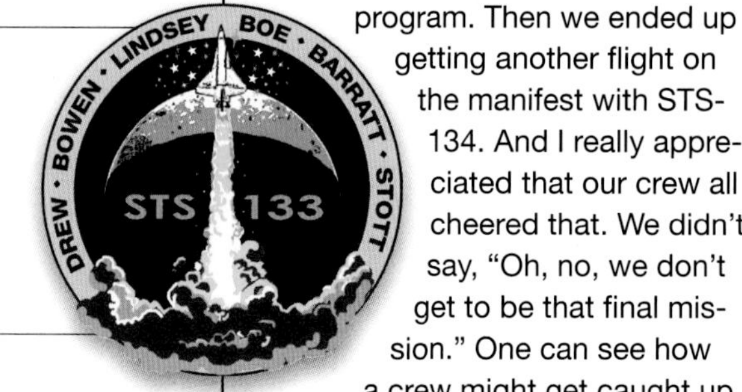

Orbiter	Discovery
Launch	February 24, 2011
Landing	March 9, 2011
Duration	12 days, 19 hrs., 3 mins., 53 secs.

The STS-133 crew on *Discovery*'s flight deck: (front row) Eric Boe, Steve Lindsey, and Mike Barratt (back row) Steve Bowen, Al Drew, and Nicole Stott

Crew	Steven W. Lindsey, Eric A. Boe, Nicole P. Stott, Benjamin Alvin Drew, Michael R. Barratt, Stephen G. Bowen

Mission: *Discovery*'s crew installed Leonardo and transferred its three tons (2.7 metric tons) of supplies to the ISS; Leonardo remained attached as a permanent storage closet. Spacewalkers installed cameras and power cables and retrieved a faulty ammonia pump for later return. STS-133 was *Discovery's* thirty-ninth and final mission.

Nicole Stott, Mission Specialist:

STS-133 was originally assigned as the final flight of the shuttle program. Then we ended up getting another flight on the manifest with STS-134. And I really appreciated that our crew all cheered that. We didn't say, "Oh, no, we don't get to be that final mission." One can see how a crew might get caught up in wanting to be on a historic final flight, but we were all genuinely excited that another shuttle flight was added to the program. So, STS-134 with the Alpha Magnetic Spectrometer (AMS) payload was added, and later STS-135, which became the last shuttle flight.

When we were still manifested as the final flight, we didn't want to adopt an "alpha-omega" theme. We wanted to look at *Discovery*'s last trip independent of that distinction. So, in designing our STS-133 patch, we went to Robert McCall and asked him to design it with the idea that it was to be the final flight of the program. But his idea was to express the beauty of our individual mission regardless of its position in the sequence, and we were thankful for that judgment.

Mike Barratt and I, who had both flown on the ISS before, ended up floating into the Station side-by-side. I was impressed at how much it felt like home. "I'm back," I thought. "I've returned to my second home." It felt so natural to move into that big, three-dimensional space again. Everything smelled and looked familiar.

Mike and I flew in, and as fast as we could, floated down to the Cupola to get that horizon-to-horizon Earth view we hadn't experienced on our first expedition. One of our first robotic tasks wa to pull a stowage pallet out of th shuttle and attach it to the Static but we got so caught up in just experiencing the Cupola view th the ground asked, "Are you guys going to start the robotics work get that thing out of the payload bay?" That snapped us out of it: "Oh, yeah...That's why we're here."

That view from the Cupola wit those seven windows was very different. I kept imagining myself one of those *Star Wars* TIE fighters. I could orient myself in a wa where everything was just openi up around me. Earth is stunning. It has this glow, this iridescence. Having those windows on every side of me just seemed to make Earth seem more alive, more thre dimensional.

After undocking, when we did the fly-around of the Station, I looked out a window trying to suck in every last detail: the colo on the solar arrays, a face in an ISS window as it shrank into tha tiny, miniature, fingernail-sized Space Station, before disappearing completely. Knowing that the might not be another time where would experience that...I could c about it right now.

When I was an engineer at the Kennedy Space Center, Jay Honeycutt was the shuttle manager there, and he came in to speak t all of us young engineers just ou of college, coming in to work on the shuttle program after *Challenger*, when we had just returne to flight. What he told us was, "Here's how we can, not why we can't." I've kept his words on my desk ever since. That should be the motto for all space program challenges: Go in with the positiv attitude that there is a solution—and go find it.

ABOVE: ***Discovery*** **approaches ISS, carrying the Leonardo cargo module for permanent installation as an ISS "closet.".**

BELOW LEFT: **During *Discovery*'s docking approach, the crew photographed the ISS through an overhead flight deck window.**

BELOW RIGHT: **Nicole Stott on the aft flight deck, running the EVA checklist for Steve Bowen and Al Drew outside.**

Mission no. **134**

STS

134

Orbiter	Endeavour
Launch	May 16, 2011
Landing	June 1, 2011
Duration	15 days, 17 hrs., 38 mins., 22 secs.

The STS-134 crew in Kibo (top) Greg H. Johnson and Mark Kelly; (middle) Mike Fincke and Greg Chamitof; (bottom) Roberto Vittori and Drew Feustel

Crew	Mark E. Kelly, Gregory H. Johnson, E. Michael Fincke, Roberto Vittori, Andrew J. Feustel, Gregory E. Chamitoff

Mission: *Endeavour*'s crew completed the construction of the US Orbital Segment by installing the cosmic-ray-detecting Alpha Magnetic Spectrometer at the ISS. Spacewalkers serviced the ammonia coolant system and added external scientific experiments. STS-134 marked the twenty-fifth and last flight for *Endeavour*.

Greg Chamitoff, Mission Specialist:

When I left the Space Station in 2008 after six months on Expeditions 17-18, I did not know if I would ever return to space. During STS-134, I had a chance to visit my old home-away-from-home, do some great work with my crew, and spend a little time outside to appreciate our view of Earth and the cosmos from the Station's back porch.

Just a few months before our flight, Commander Mark Kelly's wife suffered a horrible tragedy. Congresswoman Gabby Giffords was shot in the head while speaking at a public event in Arizona. Miraculously, she survived, but her condition was critical for a long time. Rumors of a replacement commander had us very concerned. After years of training together, this mission would not have been the same without Mark at the helm. In the end, he was able to command the mission, even as Gabby underwent major surgeries. It is a powerful testament to Mark that he was able to spend those months caring for Gabby, then command our mission before returning to her side.

After more than twelve years, thirty-five shuttle missions, and even more Soyuz and Progress crew and resupply flights, the Space Station was essentially complete. STS-134 performed the very last robotic operations and EVAs to install and assemble the final components of the US portion.

During STS-134, we had two Italian astronauts in space for the first time: our Roberto Vittori and Paolo Nespoli from the Station's Expedition 27. This called for an audience with Pope Benedict XV the first time a Pope made a real-time phone call to the heavens. All twelve of us crowded together in the JEM, Kibo, for th very special call. It was easiest for the Pope, Roberto, and Paol to converse in Italian, so for mos of the fairly long call, the rest of us just smiled and used up all the oxygen around us. Kibo, however, was not really designe to support twelve people at a time, so gradually we all started feeling like we were going to pas out. I don't know what the Italia conversation was about, but the English conversation was about ending the call before we all star ed floating away.

During the first EVA, I found myself holding onto a truss hand rail during a pass across Earth's dark side. I let go with one hand turn and look at the stars behind me. I was staring into this infinite void. It was like the feeling you get standing at the edge of a cliff it beckons you, even as you pull back in fear. The entire universe, everything unknown, and the answers to so many questions were all in that direction. But everythin I knew—my life, my family, all of humanity and our world—was represented by that single handra in my grasp. I felt I was on the ultimate edge of our existence, al most as if I had a choice of which way to go. I chose to come home of course, but I will never forget that moment.

It was the twenty-fifth flight for *Endeavour*, and the orbiter looke spectacular attached to the ISS. Space was where *Endeavour* belonged, and it was hard on all o us to know that after landing, this amazing machine would never fly in space again.

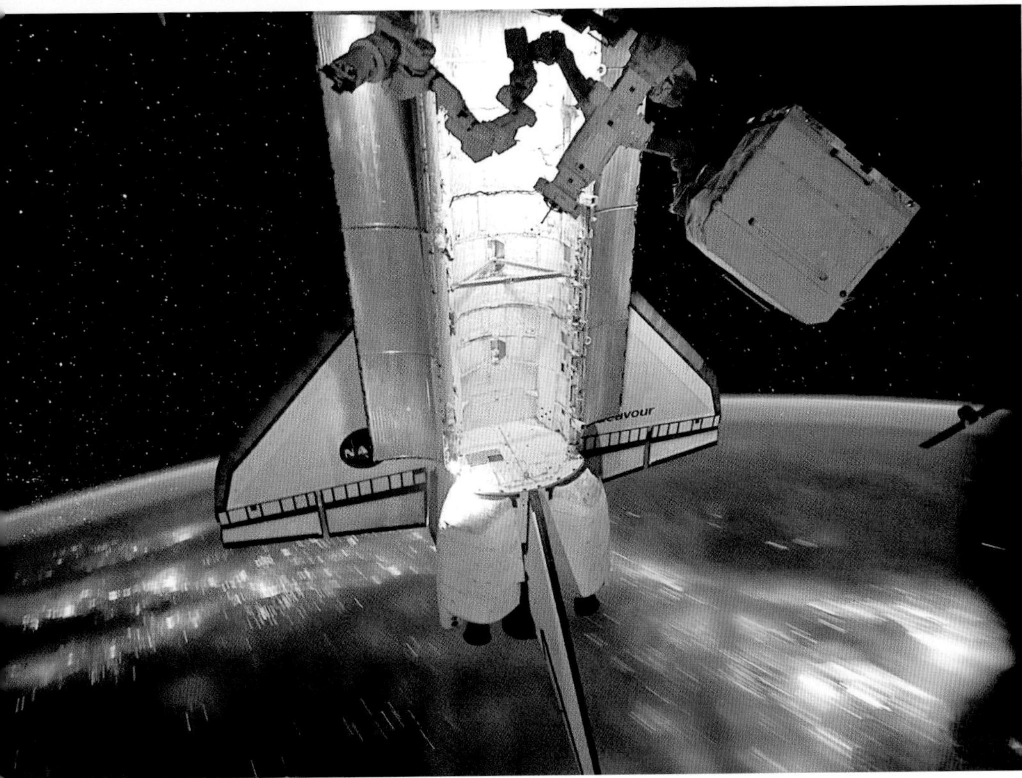

ABOVE: ***Endeavour* docked to the completed ISS, as photographed from the departing Soyuz TMA-20 crew transport.**

LEFT: **A docked *Endeavour* set against Earth's glowing horizon and a starry sky.**

Mission no. **135**

STS

135

Orbiter	Atlantis
Launch	July 8, 2011
Landing	July 21, 2011
Duration	12 days, 18 hrs., 27 mins., 56 secs.

STS-135, the final shuttle crew, on *Atlantis*'s flight deck in 2011: (front) Doug Hurley and Chris Ferguson; (back) Sandy Magnus and Rex Walheim

Crew	Christopher J. Ferguson, Douglas G. "Chunky" Hurley, Sandra H. Magnus, Rex J. Walheim

Mission: The STS-135 crew berthed Raffaello at the ISS and transferred 9,400 pounds (4,264 kilograms) of supplies for future Station expeditions. After touchdown, Commander Chris Ferguson called, "Mission complete, Houston. After serving the world for over thirty years, the shuttle has earned its place in history, and it has come to a final stop."

Chris Ferguson, Commander:

As we reached the tail end of the shuttle program, there were some stock and resupply tasks at ISS that needed to be done to prepare for "the long ocean voyage ahead." So they added two missions that were on paper but weren't congressionally approved. We went through the better part our training flow as what I'll call a "pink flight": Everybody believed it was going to happen, but it hadn't officially been sanctioned. We didn't learn that we were official until within three months of launch.

Our deliveries were mostly food and long-term supplies—unglamorous but necessary. With just a crew of four, our crewmembers were all multitasked. The nice thing was that everyone had flight experience. There were no rookies onboard: It was a crew who knew what to do.

Saying goodbye to both the orbiter and the ground control teams was hard. If we think back to Mercury, Gemini, and Apollo, there was always something coming right after each mission, and you knew what came next. Ending a program in the 1960s and early 1970s felt like a temporary goodbye. But on STS-135, we had no idea what was happening next. The Constellation program had been canceled. Maybe it was to be resurrected as something else, but that was still undefined. There were budding commercial enterprises beginning to take shape, and there was a new commercial concept for getting NASA astronauts back and forth to the ISS, but nothing had gelled.

Before we came home, there was a little part of me that was sad. I thought "This is an amazir vehicle. It seems to have so muc more life in it, and it's just a shar to put it to bed."

I wanted to make a good, successful landing, so I told Do Hurley, "Look, we've got to mak this one look okay." We were flying an ascending entry north-east across the Gulf of Mexico, coming up over the Yucatan anc then across the southern part o Florida. I don't remember reentering on so sharp a northbounc track before. I think we were about Mach 7 or so, and while looking out the window, I radioed something like, "Is that the Yucatan?" Big Butch Wilmore in Mission Control came back and deadpanned, "No, that's Miami. My answer was something like, "Then I guess I'd better get reac to land."

Our mission extension pushed our landing time from somewhere in the daylight to sometime at nighttime. Landing is a little more challenging at night, and this was my first. The brain works just a little bit slower after a space mission. But Chunky made a real good call in the final flare, and the end product was just fine.

After landing, I radioed what I thought our whole crew was feeling: "There's a lot of emotion toda but one thing is indisputable: Ame ica's not going to stop exploring. Thank you, *Columbia*, *Challenger*, *Discovery*, *Endeavour*, and our shi *Atlantis*. Thank you for protecting us and bringing this program to such a fitting end. God bless the United States of America."

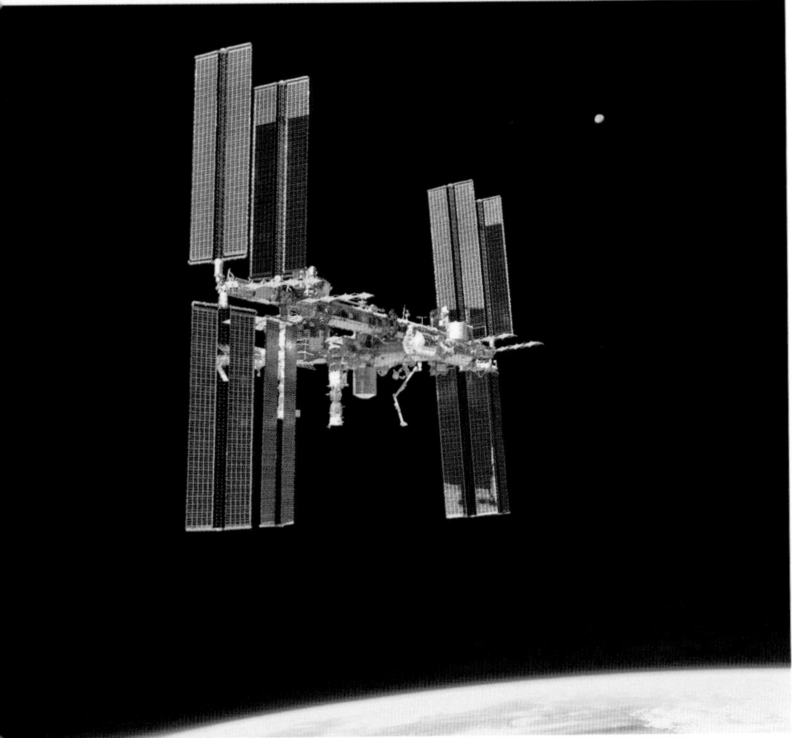

ABOVE: **On the aft flight deck, Sandy Magnus and Chris Ferguson pause to smile for a crewmate photographer.**

LEFT: **A view of the ISS and the Moon from *Atlantis.***

BELOW: **A shuttle recovery technician marks the stopping point of *Atlantis*'s main landing gear on Kennedy Space Center's shuttle landing facility Runway 15.**

Atlantis touches down on Kennedy Space Center Runway 15, a precise and graceful end to the shuttle's flight career.

Discovery
Space Shuttle
Discovery

The Space Shuttle in Perspective

When STS-135 *Atlantis* returned to Earth on July 21, 2011, its touchdown closed out thirty years of the most important program in US space history. Three decades of regular trips to orbit fixed the space shuttle in the public mind as the essence of "America in space." The vehicle was so far ahead of its time that no other nation has flown anything that even approaches its capabilities.

Few appreciate just what the shuttle accomplished during its career. The shuttles spent 1,323 days in space; deployed 180 satellites, payloads, and space station components; and returned another 52 payloads from orbit. Thirty-seven shuttle flights docked at the ISS, and another nine at the Russian space station Mir. With the Spacelab module aboard, the shuttles served as small but capable research stations in orbit. The orbiters launched 833 crewmembers on 135 shuttle missions and flew 355 individual astronauts and cosmonauts—306 men and 49 women. A few senior shuttle crewmembers will, because of spaceflight experience gained on the orbiters, likely walk the surface of the Moon.

The shuttle was retired long before the end of the orbiters' useful or structural lifetimes because the compromises built into its original design finally rendered it unsustainable. It fell well short of its advertised goal of providing airliner-like efficiency and safety to and from low Earth orbit. The shuttle cost three billion dollars annually to maintain, whether NASA launched one mission or eight, a budgetary burden that continually prevented development of a successor. Astronauts recognized that the shuttle's design lacked a capable crew-escape system, and the spaceship's vulnerability would likely have resulted in another lost orbiter and crew had operations continued. The complex and compromised shuttle finally became too unaffordable and too risky to fly.

But as shuttle commander Jim Halsell wrote, "When the shuttle worked as designed—which was most of the time—it was something to behold. All of us who were blessed with the privilege to fly the shuttle know deep down we were among the luckiest people in all of human history."

The surviving orbiters are housed at museums across the United States: *Enterprise* at the Intrepid Sea, Air, and Space Museum in New York; *Discovery* at the Smithsonian National Air and Space Museum in Virginia; *Atlantis* at the Kennedy Space Center Visitor Complex in Florida; and *Endeavour* at the California Science Center in Los Angeles. When you visit any of the space shuttles, you will look upon the craft, that, more than any other, launched the dreams of millions of Earthlings into space.

rbiter *Discovery* at the nithsonian National Air and pace Museum's Udvar-Hazy enter in 2012.

Acknowledgments

The *Space Shuttle Stories* project originated in a conversation with Dennis Jenkins, author of several authoritative histories on the development, testing, and operations of the Space Transportation System. My goal was to complement the many extant volumes covering the shuttle's history with lived astronaut experiences. Dennis is the world's most widely published expert on shuttle engineering and deserves an external tank-full of thanks for patiently answering my many technical questions and offering valuable advice. When shuttle *Endeavour* goes on impressive vertical display at the California Science Center, it will have been Dennis's engineering leadership that put *Endeavour* back on the pad for visitors to view.

This collection of space shuttle stories depends, of course, on the generous and extensive contributions of my astronaut colleagues, who agreed to face-to-face or online interviews to discuss their memorable shuttle experiences or who provided written narratives. These 133 interviews comprise the raw material from which these shuttle stories were drawn. I hope my editing of their interview transcripts captures my colleagues' excitement and pride in their achievements and their awe at the experience of working and living in Earth orbit. Any errors made in recounting their stories are entirely mine.

It was impossible for me to interview each of the more than 350 individuals who flew on the space shuttle. To those I missed, I express my sincere regrets and unreserved admiration. I encourage those I didn't interview to go on to preserve and publish their stories, and I encourage historians to pursue those unrecorded experiences as the shuttle era inexorably recedes.

I'm grateful to the Association of Space Explorers, with its Registry of Space Travelers, for putting me in touch with many far-flung colleagues. I also thank Sheva Moore at NASA HQ, who helped me locate audio and video with the voices of the *Challenger* and *Columbia* crews. I thank transcriber Joel Hines, who assisted in putting several of these interviews into written form. In researching the book's images, I had the help of many dedicated professionals. At NASA's Johnson Space Center, thanks go to Jane Buckley, Will Close, Cadan Cummings, Cindy Evans, Kenton Fisher, Warren Harold, Diane Laymon, Maura White, Mary Wilkerson, and Megan Dean. At Kennedy Space Center, I'm grateful to Amber Notvest, Daniel Casper, Josiel Torres, and Isaac Watson. Several spectacular IMAX images of the orbiters and their crews were made available through IMAX Corporation with the instrumental help of Gayle Bonish. Kate Igoe of the Smithsonian Institution generously provided an iconic *Discovery* image from the National Air and Space Museum.

Wayne Hale, in his worthy foreword for this book, cites the many tens of thousands of Americans who created the successful space shuttle team. These included the NASA leaders who conceived and gained approval for the shuttle, the industry engineers and managers who designed and built it, flight controllers and flight directors who shepherded the orbiters in space for three decades, the instructor and simulator teams who prepared the astronauts for flight, the shuttle technicians, ground support specialists, and launch controllers who prepared and launched the shuttle, the industry team who maintained and upgraded the fleet, and the thousands of NASA and international partner managers and staff who managed the shuttle and its payloads.

I also thank those military and law enforcement staff who provided security, safety, and rescue services to the shuttle and its crews. I thank these and other shuttle team members not mentioned; they succeeded in building and operating a national space transportation system which led the world in space for more than three decades.

My agent, Deborah Grosvenor, guided this project to a successful home at Smithsonian Books, and I'm grateful for her advice and criticism.

Thanks to Smithsonian Books Director Carolyn Gleason for giving us the "Go!" Editors Jaime Schwender and Linda Shiner helped me forge a much more readable and interesting collection of shuttle adventures. These able writers were a pleasure to work with throughout our production process. David Griffin created the bold design for the book and chose, from my selections, just the right mix of dramatic shuttle images. Smithsonian Books marketers Matt Litts and Sarah Fannon are experts at making *Space Shuttle Stories* ever more widely known on Earth and beyond.

Thanks to colleagues and friends Dennis Jenkins, Bill Barry, and Tony Reichhardt, who graciously reviewed portions of the manuscript and made many constructive suggestions. I am grateful to my shuttle colleagues Carl Walz and Linda Godwin, who gave the book a critical read for technical accuracy. Any errors that remain in this work are mine alone.

Finally, *Space Shuttle Stories* would not have taken flight without the help of my wife, Liz. After supporting me unconditionally during my astronaut career and shepherding our family through four stressful missions, she encouraged me through the three years needed to communicate the experiences of my friends and heroes. Through their eyes, Liz, you've helped me leave Earth again.

he astronauts interviewed for this project are, in alpabetical order:

oren Acton
m Adamson
ndy Allen
y Apt
e Archambault
an Barry
ike Bloomfield
uy Bluford
o Bobko
harlie Bolden
ance Brand
an Brandenstein
oy Bridges
urt Brown
an Burbank
an Bursch
ob Cabana
ohn Casper
ob Cenker
reg Chamitoff
ranklin
Chang-Diaz
eroy Chiao
evin Chilton
ary Cleave
ike Coats
en Cockrell
ady Coleman
ileen Collins
ob Crippen
ob Curbeam
ancy Currie
an Davis
rian Duffy
Bonnie Dunbar
Mary Ellen Weber
John Fabian
Chris Ferguson
Anna Fisher
Bill Fisher
Mike Foale
Mike Foreman
Pat Forrester
Guy Gardner
Mike Gernhardt
Linda Godwin
Fred Gregory
John Grunsfeld
Sid Gutierrez
Fred Haise
Greg Harbaugh
Bernard Harris
Rick Hauck
Steve Hawley
Susan Helms
Tom Henricks
José Hernández
John Herrington
Rick Hieb
Dave Hilmers
Jeff Hoffman
Scott Horowitz
Doug Hurley
Marsha Ivins
Tammy Jernigan
Janet Kavandi
Wendy Lawrence
Mark Lee
Dave Leestma
Steve Lindsey
Rick Linnehan
Paul Lockhart
Jack Lousma
Shannon Lucid
Steve MacLean
Sandy Magnus
Mike Massimino
Bill McArthur
Mike McCulley
Bruce Melnick
Leland Melvin
Dottie Metcalf-Lindenburger
Barbara Morgan
Mike Mullane
Story Musgrave
Jim Newman
Claude Nicollier
Carlos Noriega
Karen Nyberg
Ellen Ochoa
Danny Olivas
Steve Oswald
Scott Parazynski
Bob Parker
Julie Payette
Don Pettit
Charlie Precourt
Ken Reightler
Jim Reilly
Garrett Reisman
Dick Richards
Steve Robinson
Kent Rominger
Jerry Ross
Mario Runco
Bobby Satcher
Winston Scott
Rhea Seddon
Brewster Shaw
Loren Shriver
Steve Smith
Woody Spring
Bob Springer
Heide Stefanyshyn-Piper
Bob Stewart
Susan Still Kilrain
Nicole Stott
Kathy Sullivan
Dan Tani
Joe Tanner
Norm Thagard
Gerhard Thiele
Don Thomas
Andy Thomas
Kathy Thornton
Pierre Thuot
Dick Truly
Ox van Hoften
Terry Virts
Jim Voss
Carl Walz
Doug Wheelock
Stephanie Wilson
Jeff Wisoff

lso contributing via their oral histories or published interviews were:

aurel Clark
. O. Creighton
oe Engle
Hoot Gibson
Greg Jarvis
Ken Mattingly
Christa McAuliffe
Ron McNair
Ellison Onizuka
Judy Resnik
Dick Scobee
Mike Smith

Recommended Reading

In addition to the recommended books listed below, many fine astronaut memoirs and biographies centered on the shuttle era expand on the flyer experiences recorded in *Space Shuttle Stories*. Many can be found at these links:

www.nasa.gov/centers/hq/library/find/bibliographies/astronaut_biographies

www.space-explorers.org/bookshelf

www.collectspace.com/resources/books_astronauts.html

DK Publishing. *Space Shuttle: The First 20 Years.* London: DK, 2002.

Columbia Accident Investigation Board. *Columbia Accident Investigation Board Report.* Washington, DC: National Aeronautics and Space Administration and the Government Printing Office, 2003.

Cooper Jr., Henry S. F. *Before Lift-Off: The Making of a Space Shuttle Crew.* Baltimore: Johns Hopkins University Press, 1987.

Craig, John, John Krige, and Roger Launius, eds. *Space Shuttle Legacy: How We Did It and What We Learned.* Reston, VA: American Institute of Aeronautics and Astronautics, 2013.

Hale, Wayne, Gail Chapline, Helen Lane, and Kamlesh Lulla, eds. *Wings in Orbit: Scientific and Engineering Legacies of the Space Shuttle, 1971–2010.* NASA, 2011.

Heppenheimer, T. A. *History of the Space Shuttle, Vol. 2: Development of the Shuttle, 1972–1981.* Washington, DC: Smithsonian Books, 2002.

Houston, Rick. *Wheels Stop: The Tragedies and Triumphs of the Space Shuttle Program, 1986—2011.* Lincoln, NE: University of Nebraska, 2014.

Jenkins, Dennis R. *Space Shuttle: The History of the National Space Transportation System—The First 100 Missions.* Dennis R. Jenkins, 2002.

Jenkins, Dennis R. *Space Shuttle: Developing an Icon—1972–2013.* 3 vols. Dennis R. Jenkins, 2016.

Jenkins, Dennis R. *The History of the American Space Shuttle.* Atglen, PA: Schiffer, 2019.

Jones, Tom. *Ask the Astronaut: A Galaxy of Astonishing Answers to Your Questions on Spaceflight.* Washington, DC: Smithsonian Books, 2016.

Jones, Tom. *Sky Walking: An Astronaut's Memoir.* Washington, DC: Smithsonian Books, 2016.

Legler, Robert D., and Floyd V. Bennett. *Space Shuttle Missions Summary.* NASA, September 2011, https://historycollection.jsc.nasa.gov/JSCHistoryPortal/history/reference/TM-2011-216142.pdf

Leinbach, Michael D., and Jonathan H. Ward. *Bringing Columbia Home: The Untold Story of a Lost Space Shuttle and Her Crew.* New York: Arcade, 2018.

NASA. *Columbia Crew Survival Investigation Report.* NASA, 2008, https://ntrs.nasa.gov/api/citations/20090002404/downloads/20090002404.pdf

Neal, Valerie. *Spaceflight in the Shuttle Era and Beyond: Redefining Humanity's Purpose in Space.* New Haven: Yale University Press, 2017.

Rogers, William P. ed. *Report of the Presidential Commission on the Space Shuttle Challenger Accident.* Washington, DC: United States Government Printing Office, 1986.

Vaughn, Diane. *The Challenger Launch Decision: Risky Technology, Culture, and Deviance at NASA.* Chicago: University of Chicago Press, 1996.

White, Rowland. *Into the Black: The Extraordinary Untold Story of the First Flight of the Space Shuttle Columbia and the Astronauts Who Flew Her.* New York: Atria, 2017.

eferences

addition to the more than 130 astro-
ut interviews conducted to provide the
ories in this book, some quoted content
me from earlier oral histories, published
orks, or press interviews.

TS-4

: "last mission of the flight-test pro-
am": Thomas K. Mattingly III, interview
Kevin Rusnak, *NASA Johnson Space
enter Oral History Project,* April 22, 2002,
tps://historycollection.jsc.nasa.gov/
SCHistoryPortal/history/oral_histories/
attinglyTK/MattinglyTK_4-22-02.htm.

TS-7

: "laughing for quite a bit of the time":
ally Ride, interview by Melanie Wallace,
ova, PBS, November 27, 1984, www.
os.org/video/nova-interview-sally-ride/

TS-51D

: "as the Syncom rotated slowly above
": Rhea Seddon, *Go for Orbit: One of
merica's First Women Astronauts Finds
er Space* (Murfreesboro, TN: Your Space
ress, 2015), 262-264.

TS-51G

: "The engines start to come alive":
ohn O. Creighton, interview by Jenni-
r Ross-Nazzal, *NASA Johnson Space
enter Oral History Project,* May 3, 2004,
tps://historycollection.jsc.nasa.gov/
SCHistoryPortal/history/oral_histories/
reightonJO/CreightonJO_5-3-04.htm _

TS-51I

2: "we had three satellites": Joe H. Engle,
terview by Rebecca Wright, *NASA John-
on Space Center Oral History Project,*
une 3, 2004, https://historycollection.jsc.
asa.gov/JSCHistoryPortal/history/oral_
istories/EngleJH/EngleJH_6-3-04.htm

TS-51L

2: "you realize the joy was in the journey
ong the way": June Scobee Rodgers,
ilver Linings: *My Life Before and After
hallenger 7* (Macon, GA: Smyth and Hel-
ys Publishing, 2011), 116–17, 144.

72: "whole planets to explore": Ben Bova, *Vision of the Future: The Art of Robert McCall.* (NY: Abrams, 1982), 62.

72: "getting the secret handshake": Michael Smith, *NASA Challenger Anniversary Resource Tape,* 1996, www.youtube.com/watch?v=IviOm71Iml0

72: "the world will be a better place because you tried": Ellison S. Onizuka, "A Message to the Future Generations," 1980, Astronaut Ellison Onizuka Space Center at the Japanese Cultural Center of Hawaii, www.jcch.com

72: "It's always a challenge": Judy Resnik, interview by Tom Brokaw, *Today,* NBC, April 9, 1981, www.youtube.com/watch?v=EaafRyuwA8w.

72: "believing in oneself": Ron McNair, Commencement Address at the University of South Carolina, August 1984, www.youtube.com/watch?v=if8SHZ-K4lY.

72: "their payload specialist": Greg Jarvis, *NASA Challenger Anniversary Resource Tape,* 1996, https://www.youtube.com/watch?v=IviOm71Iml0

72: "it's a wonderful thought": Christa McAuliffe, STS-51L *Crew Pre-flight Press Conference,* December 13, 1985, www.youtube.com/watch?v=YvAnUPx3lrw

STS-26R

80: "Today, up here where the blue sky": Rick Hauck, In-flight Tribute to the *Challenger* STS-51L Crew, October 2, 1988, https://forum.nasaspaceflight.com/index.php?topic=21959.0.

STS-27R

82: "We didn't know it": Robert Gibson, *Astronaut Tales: Tell Me a Story,* Kennedy Space Center, April 25, 2015, www.youtube.com/watch?v=BswkvaAaqSM

STS-36

96: "as if we were in a ghost ship": Mike Mullane, *Riding Rockets: The Outrageous Tales* of a Space Shuttle Astronaut, (New York: Scribner, 2006), 335.

96: "I was part of it": Ben Evans, "Now You Don't: 25 Years Since the Mysterious Mission of STS-36 (Part 2)," AmericaSpace, 2015, www.americaspace.com/2015/03/01/now-you-dont-25-years-since-the-mysterious-mission-of-sts-36-part-2/

STS-77

184: "to take a break": John Howard Casper, *The Sky Above: An Astronaut's Memoir of Adventure, Persistence, and Faith,* (West Lafayette, IN: Purdue University Press, 2022), 223.

STS-93

224: "I wonder what that was": Wayne Hale, "STS-93: We Don't Need Any More of Those," Wayne Hale's Blog, October 26, 2014, https://waynehale.wordpress.com/2014/10/26/sts-93-we-dont-need-any-more-of-those/

STS-107

260: "Hello from above": Laurel B. S. Clark, "From One Laurel to Another: A Letter from Columbia," Universe Today, February 1, 2013, www.universetoday.com/99712/from-one-laurel-to-another-a-letter-from-columbia/.

Space Shuttle in Perspective

311: "When the shuttle worked as designed": James Halsell, message to author, November 26, 2022.

Index

Published by Smithsonian Books
Director: Carolyn Gleason
Senior Editor: Jaime Schwender
Assistant Editor: Julie Huggins
Edited by Linda Shiner
Designed by David Griffin

This book may be purchased for educational, business, or sales promotional use. For information, please write: Special Markets Department, Smithsonian Books, P.O. Box 37012, MRC 513, Washington, DC 20013

Library of Congress Cataloging-in-Publication Data

Names: Jones, Tom, 1955 January 22- editor.
Title: Space shuttle stories : firsthand astronaut accounts from all 135 missions / Tom Jones.
Description: Washington, DC : Smithsonian Books, 2023. | Includes bibliographical references and index.
Identifiers: LCCN 2023013350 | ISBN 9781588347541 (hardcover)
Subjects: LCSH: Space Shuttle Program (U.S.)—Anecdotes. | Space Shuttles—United States—Anecdotes. | Astronauts—United States—Anecdotes.
Classification: LCC TL795.5 .S6125 2023 | DDC 629.44/10973—dc23/eng/20230510
LC record available at https://lccn.loc.gov/2023013350

Paperback ISBN: 978-1-58834-803-6

Printed in China, not at government expense
29 28 27 26 25 1 2 3 4 5

Interviews within text have been edited for space and clarity.

For permission to reproduce illustrations appearing in this book, please correspond directly with the owners of the works. Unless otherwise noted below, all images are courtesy of NASA. Smithsonian Books does not retain reproduction rights for these images individually, or maintain a file of addresses for sources.

Courtesy of IMAX: 177, 189; **IMAX/Lockheed Martin/NASA:** 235t, 235bl; **IMAX/Lockheed Martin/Smithsonian Institution:** 49t, 53t, 91tl, 99, 144–145; **Smithsonian National Air and Space Museum:** 310, Photo by Dane Penland (NASM 2013-01344); **USAF:** 279t, 301tl.

FRONT AND SPINE: **Orbiter *Atlantis* carrying the Destiny science lab to ISS on STS-98, February 7, 2001.**

BACK COVER: ***Columbia*, STS-61C, lifts off from Pad 39A at Kennedy Space Center on the morning of January 12, 1986.**